WORKS ISSUED BY
THE HAKLUYT SOCIETY

THE LETTERS OF
F. W. LUDWIG LEICHHARDT

VOLUME II

SECOND SERIES
NO. CXXXIV

ISSUED FOR 1968

HAKLUYT SOCIETY

PATRON
H.R.H. THE DUKE OF GLOUCESTER, K.G., P.C., K.T., K.P.

COUNCIL AND OFFICERS, 1967–68

PRESIDENT
SIR GILBERT LAITHWAITE, G.C.M.G., K.C.B., K.C.I.E., C.S.I.

VICE-PRESIDENTS

J. N. L. BAKER, M.A., B.Litt.   Sir ALAN BURNS, G.C.M.G.
Professor C. R. BOXER, F.B.A.   Dr E. S. DE BEER, F.B.A.
Professor D. B. QUINN

COUNCIL (*with date of election*)

W. E. D. ALLEN, O.B.E. (1967)
Dr K. R. ANDREWS (1966)
Professor C. F. BECKINGHAM (1964)
Professor J. S. BROMLEY (1964)
Dr J. S. CUMMINS (1967)
Dr C. I. JACKSON (1967)
Sir HARRY LUKE, K.C.M.G. (1967)
B. F. MACDONA, C.B.E. (1967)
J. F. MAGGS (1966)
Mrs DOROTHY MIDDLETON (1966)

Rear-Admiral G. S. RITCHIE, D.S.C. (1965)
Royal Commonwealth Society (1966, D. H. SIMPSON)
Royal Geographical Society (G. R. CRONE)
R. A. SKELTON (1966)
Lieut-Cdr. D. W. WATERS, R.N. (1965)
Dr G. WILLIAMS (1964)
Dr D. P. J. WOOD (1964)

TRUSTEES
J. N. L. BAKER
Sir GILBERT LAITHWAITE, G.C.M.G., K.C.B., K.C.I.E., C.S.I.
The Right Hon. LORD RENNELL OF RODD, K.B.E., C.B.

HONORARY TREASURER
J. N. L. BAKER

HONORARY SECRETARIES
Miss E. M. J. CAMPBELL, Birkbeck College, London, W.C.1
Dr T. E. ARMSTRONG, Scott Polar Research Institute, Cambridge

HON. SECRETARIES FOR OVERSEAS

Australia: G. D. RICHARDSON, The Public Library of New South Wales, Macquarie Street, Sydney, N.S.W. 2000
Canada: Professor J. B. BIRD, McGill University, Montreal
India: Dr S. GOPAL, Ministry of External Affairs, 3 Man Singh Road, New Delhi
New Zealand: C. R. H. TAYLOR, Box 5102, Tawa
South Africa: Professor ERIC AXELSON, University of Cape Town, Rondebosch, South Africa.
U.S.A.: Dr W. M. WHITEHILL, Boston Athenaeum, 10½ Beacon Street, Boston, Massachusetts, 02108

CLERK OF PUBLICATIONS AND ASSISTANT TREASURER
Mrs ALEXA BARROW, Hakluyt Society, c/o British Museum, London, W.C.1

PUBLISHER AND AGENT FOR SALE AND DISTRIBUTION OF VOLUMES
Cambridge University Press, Bentley House, 200 Euston Road, London, N.W.1

THE LETTERS OF
F. W. LUDWIG LEICHHARDT

COLLECTED AND
NEWLY TRANSLATED BY

M. AUROUSSEAU

VOLUME II

CAMBRIDGE
Published for the Hakluyt Society
AT THE UNIVERSITY PRESS
1968

Published by the Syndics of the Cambridge University Press
Bentley House, P.O. Box 92, 200 Euston Road, London, N.W. 1
American Branch: 32 East 57th Street, New York, N.Y. 10022

© The Hakluyt Society 1968

Library of Congress Catalogue Card Number: 68–13538
Standard Book Number: 521 01026 8

Printed in Great Britain
at the University Printing House, Cambridge
(Brook Crutchley, University Printer)

CONTENTS

VOLUME II

THE LETTERS

III Scientific Reconnaissance in New South Wales *page* 425
    A. Around Sydney   427
    B. Around Newcastle on foot   525
    C. In the Hunter-Goulburn valley on horseback   611
IV To the Moreton Bay District   651

MAPS

5 New South Wales in Leichhardt's time   426
6 Surroundings of Sydney in 1842   473
7 Journeys from Newcastle, 1842   524
8 Journeys in the Hunter-Goulburn valley   610
9 Journey to the Wide Bay District and back to Port Stephens, 1843–4   *facing p.* 651

APPENDIXES

Contractions and abbreviations used   783
  I Table of events and movements March 1842–July 1844   787
  II Annotated calendar of letters March 1842–August 1844   798

III

SCIENTIFIC RECONNAISSANCE IN NEW SOUTH WALES

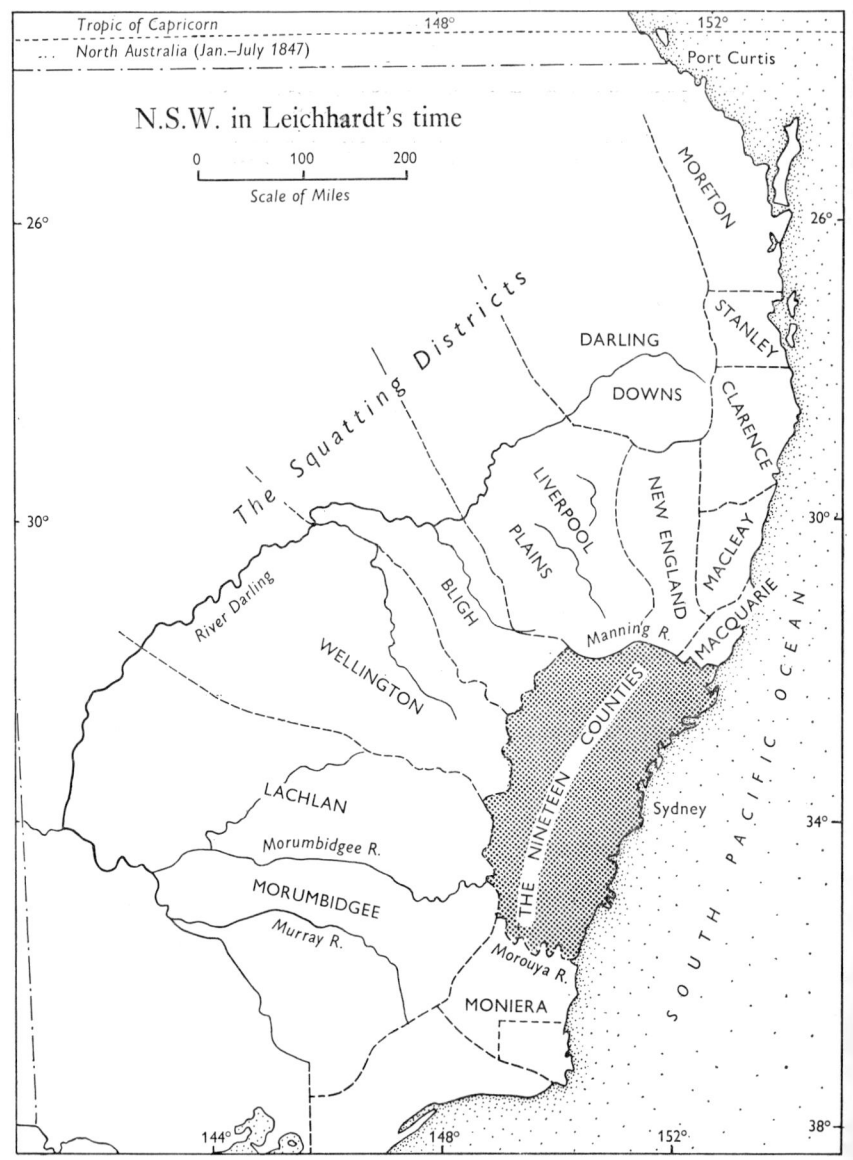

Fig. 5. New South Wales in Leichhardt's time.
From *The Church in Australia*, S.P.G. London, 1845.

A. AROUND SYDNEY

50 [TO CARL SCHMALFUSS, Cottbus.]

Sidney den 23ten März 1842

Mein theuerster Schwager

Weite Meere habe ich durchkreutzt, heftigen Stürmen habe ich getrozt. Die Sonne zog über meinen Scheitel von Süden nach Norden und nun liegt die ganze Eerde zwischen mir und Euch, nun ist Tageszeit, Jahreszeit, ja der Karakter der Himmelsgegenden verschieden, der Süden ist kalt, der Norden heiß, nd der Schatten meines Körpers fällt am Mittage nach Süden. Ich verließ London am ersten October 1841; wir hatten ein ausserordentl. stürmisches Wetter von London nach Cork; doch das Verlangen, die Ferne zu sehen, war so groß, daß Gefahr nd Mühsal vergessen wurden, und daß wir unter dem Sausen des Sturmes und dem Brausen des Meeres von Manteln und Segeltüchern geschützt oft munter Wanderlieder in die Mondhellen Nächte hinaussangen. O, daß ich Euch in Worten ausdrücken könnte, wie mich das großartige Schauspiel der Seenatur auf das Tiefste bewegte! Am klaren Himmel tauchten die herrlichen Himmelsbilder im Osten über die scharfe Linie des Horizontes hervor, schritten in ihrer erhabenen Ruhe über das bewegte Meer nd sanken so im Westen allmählig nieder. Wie wir uns dem Aequator näherten, verschwanden die alten befreundeten Sternbilder nd ungesehene, glänzendere begrüßten uns. — Oft war unser Schiff, ein so leichtes Menschenwerk, Gegenstand meines Nachdenkens nd meiner Bewunderung. An 300 Menschen schwammen auf unermeßlicher Öde mit dem Gefühle fast vollkommen Sicherheit, von vielen Lebensbequemlichkeiten umgeben einer ferner Heimat zu und dieser kleine bevölkerte von Wind un Wetter bewegte Punkt wurde selbst wieder nach dem Compaß geleitet, einem kaum 3″ langen Stückchen Eisen, doch der wunderbarsten Eisenbrücke, welche sich über die weitesten Meere von einem Continent zum andern spannt. — Wir sahen einige Inseln, wie z. B. St. Antonio eine der Cap verdischen Inseln, Trinidad an der Küste von Brasilien, Sant Paul zwischen dem Cap der guten Hoffnung un Neuholland; doch landeten wir nirgends. Das Leben auf dem Schiffe war eigenthümlich. Ausser den 250 Auswanderern befanden sich 20 unabhängige Passagiere am Bord. Wir lebten fast wie eine Familie mit einander, da wir zusammen frühstückten, Mittag aßen nd Thee tranken; dennoch blieben wir getrennt genug, um uns in der Zwischenzeit nach unsern eigenen Neigungen zu beschäftigen. Ich studirte vorzüglich Seekunde, lernte die Länge un Breite bestimmen, beobachtete die Veränderungen des Barometers nd Thermometers. Dieses letztere war, während wir die Linie kreutzten nie höher als $23\frac{1}{2}$ Grad. Ueberhaupt litten wir mehr von Kälte im Norden un später im Süden des Aequators, als von der Hitze. Da die Matrosen sich gewöhnlich ziemlich handgreifliche Scherze erlauben, wenn sie das Aequator kreutzen, so machte der Kapitain es zum Geheimniß nd man erfuhr es nicht eher, als da das Schiff schon 5° Südl. war. Die Matrosen pflegen die Passagiere zu taufen, welche zum ersten Male die Linie kreutzen, d. h. sie begiessen sie mit Wasser, machen ihnen Bärte mit Theer pp. — Es konnte nicht fehlen daß zwischen 20 Menschen, die beständig mit einander waren, vielfache Reibungen statt finden müßten. Wir hatten überdieß einige Elemente, die eine dauernde Gährung unterhielten. Auf der andern Seite brachte diese Gährung die wahren Karaktere zu Tage und wurde also für das Menschenstudium von ausserordentlicher Wichtigkeit. Von Seethieren wurde wenig gesehen: obwohl das Meer vielleicht am reichsten bevölkert ist, indem Polypen nd Krustenthiere un Würmer auf seinem Grunde sich befestigen oder dort hausen nd indem Fische nd Weichthiere in unzähligen Schaare seine weiten Räume durchziehn. Oft blickte ich vom Schiff in das Meer, bedauernd, daß es nur nicht vergönnt war, alle die Wunder näher kennen zu lernen welche eine Tiefe von vielleicht 2–3000 Fuß bedeckte. — Dennoch ergötzten uns oft Heerde von Delphinen, welche mit halbem Körper aus dem Wasser hervortauchten nd besonders bei bevorstehenden stürmischen Wetter sich sehen ließen — oder der bedächtige Haifisch, welcher in langsamen Bewegungen unser Schiff umschwamm, um sich jedes Abwurfes sogleich gierigst zu bemächtigen. Seevögel wurden in Menge gesehen nd gefangen, besonders das Albatross, welches mit 10 bis 12 Fuß Flügelbreite fast immer auf seinen Flügeln zu leben scheint, hunderte von Meilen vom Festlande entfernt, in der Nacht zwischen Wellenbergen Ruhe haltend. Vier nd ein halber Monat (Oct. Nov. Dec. Jan. u $\frac{1}{2}$ Febr.) verstrichen, ehe wir zwischen den beiden Felsenpforten von Port Jackson einfuhren. Als der Pilot auf einem von Neuseeländern[a] gezogenen Boote zum Schiffe kam, hätte ich das alte olivenfarbige mürrische Seekind als ersten Herold der neuen Erde umarmen mögen. Port Jackson ist eine Buchtenreiche weite Flußmündung, rings von Hugeln nd Felsen eingeschlossen, die infolge eines langersehnten Regens von frischgrünenden Bäumen bedeckt waren, aus denen überall freundliche Gehöfte oder Lusthäuser hervorblickten. Man hatte fast 18 Monate hindurch an ununterbrochenen Dürre gelitten. Schaaf nd Rinderheerden waren, an den vertrockneten Tränkplatzen vor Durst niedergesunken. Zwei Tage vor unserer Ankunft begann es endlich zu regnen, und so regnete es 14 Tage hindurch unaufhörlich. Dieß machte die erste Zeit meines Aufenthaltes etwas unbehaglich, doch da es den Anblick der Natur so wesentlich veränderte nd verbesserte, trug ich die unanne[hm]lichkeit willig. Die Ufer dieses herrlichen Hafens waren vor 52 Jahren von wilden Menschen bewohnt, die nie vorher einen weißen Mann gesehen hatten. Jetzt erhebt sich hier eine große Stadt von 42000 Einwohnern, nach allen Seiten von den Lusthäusern ihrer reichen Einwohner umgeben. Sie ist zum Theil in einem Thale, zum Theil an zwei Bergen hinaufgebaut, regelmäßig, in der Regel mit großen Häusern nd breiten Straßen. Der Sandsteinfels, welche die ganzen Gegend zusammensetzt, kommt in den Straßen oft zu Tage nd oft sind sie selbst durch ihn hindurch gehauen. Ich wurde in Sydney wohl empfangen nd obwohl große Mittel nöthig sind, um etwas Tüchtiges zu wirken, so hoffe ich doch mit der Zeit gleichfalls mein Ziel zu erreichen. Das Klima ist ausserordentlich mild nd angenehm, besonders zu jetziger Jahreszeit, welche eigentlich nur über plötzliche[n] Wechsel klagen ließt. Es herrscht in dieser Stadt eine erstaunliche Thätigkeit nd eine übermäßige Speculations eifer. Eine große Menge von Schiffen füllen den Hafen; täglich kommen nd gehen sie .ch England, .ch China, nach Neusee, land, Van Diemens land, Port Philipp nd nach verschiedenen Küstenplatzen. Dampfboote laufen nach Hunters river, nach Moreton bay, .ch Pt Philipp. Jeder Luxus, jede Bequemlichkeit läßt sich in Sydney kaufen. Gute Straßen führen nach den verschiedenen Städten, welche ein wenig mehr im Innern des Continents liegen: So nach Bathurst an der andern Seite der blauen Berge, ungefähr 50 Meilen von Sidney entfernt, nach Liverpool, Windsor am Hawkesbury Fluß u.s.w. Und wer hat diese wunderbaren Veränderungen in so kurzer Zeit bewirkt? — 1788 führte Arthur Philipp 850 Verbrecher hierher, gründete die Kolonie und begann durch Verbrecherkraft dieses Land urbar zu machen. Von jener Zeit wurden fortdauernd Verbrecher hierher gesendet, zu öffentlichen Bauten verwendet, oder aber freien Ansiedlern zugestanden um sich ihrer als Dienstboten zu bedienen. Verbrecher die ihre Strafe erduldet hatten, wurden frei nd siedelten selbst sich an, ja selbst während ihrer bedingten Freiheit konnten sie schon Güter erwerben und sich unabhängig bereichern. Die reichsten Männer der Kolonie waren Verbrecher oder stammen von Verbrechern. Allmählig wandern mehr n mehr freie Ansiedler ein nd jetzt finden sich an 100000 auf Australiens Boden. Diese Auswanderer werden fast alle von der Begierde sich Schätze zu erwerben angetrieben. Sie wollen einige Jahre ihres Lebens diesem Geschäfte widmen, um sodann wieder in ihrer Heimath zurück zu kehren, und ruhig geniessend ihrem Tode entgegen zu harren. Wenige kommen hier, um hier zu bleiben: viele verändern indessen ihre Absicht wie sie die Schönheiten dieses reichen Landes besser kennen und die Unan, nehmlichkeiten leichter ertragen lernen. Solche Familien freier Ansiedler, die Interesse an die Colonie nehmen und sie als ihr Vaterland betrachten sind denn auch allein der wahre Schatz des Landes und aus ihnen wird sich allmählig ein mächtiges Volk entwickeln, welches uns das alte Europa vergessen machen möchte. — Es ist indessen natürlich, daß der gegenwärtige

Reconnaissance in New South Wales

Gesellschaftliche Zustand von Sidney nicht befriedigend sein kann, da eben so unharmonische Elemente diese Gesellschaft bilden. So liberal man auch sein mag, stets wird sich in der Gesellschaft früherer Verbrecher der Gedanke aufdrängen, daß man es eben mit Menschen zu thun habe, die einst fähig waren, schwere Verbrechen zu begehen. Es ist wahr sie haben ihre Strafe erduldet, die sind in die Gesellschaft weiß gewaschen zurück gekehrt, — doch haben sie deßhalb auch sich wahrhaft gebessert, verdienen sie unser Vertrauen? Solche Betrachtungen, die sicher- lich nicht ohne Grund sind, haben denn oft die freien Ein- wanderer bestimmt, sich als eine getrennte bessere Klasse an zu sehen — und dieß hat auf der andern Seite die emancipirten Verbrecher mit einander vereinigt, so daß jetzt zwei Parteien, zwei Gesellschaften einander gegenüber stehen, die sich vielfach anfeinden nd den Fortschritt der Kolonie hemmen. — Wenn früher ein *Verbrecherschiff* nach Sidney kam, meldeten sich die- jenigen Personen, welche Dienstboten bedürften, und man gab ihnen nach der Große ihres Geschäftes *Verbrecher*. Sie hatten polizeiliche Aufsicht nd große Gewalt über diese, welche sie denn auch häufig mißbrauchten. Der große Vortheil war daß sie für diese Verbrecher nichts zu bezahlen hatten und nur für Nahrung nd Kleidung sorgten. — Da man indessen später wohl einsah, daß in der ⌜sich erhebende⌝ Kolonie nichts mehr der Auswurf Europas eingeführt werden dürft, gab man das System Verbrecher an die Kolonisten ab zu treten, auf, errichtete eine andere Strafcolonie zu Moreton bai (mehr gegen Norden) nd auf Norfolk Island. Die Erstere ist jetzt gleichfalls freien Ansiedlern überlassen. — Hatte der Verbrecher nun seine Strafe d. h. 2–3–20 nd mehrere Jahre Dienst erlitten, nd hatte er sich gut aufgeführt, so gab man ihm eine Erlaubniß karte nd er konnte nun unter der Aufsicht der Polizei sein eigenes Geschäft beginnen — unter der Bedingung, daß ein neuer Fehler ihn in seinen frühern Zustand von Dienstbarkeit oder Gefangenschaft zurück führen würde. Unbedingte Freiheit erhielten nur wenige nd reiche Männer nd Familien väter in Sidney sind Verbrecher, welche bedingungs weise ihre Freiheit genießen. Seitdem ich mein eigenes Vaterland,

seitdem ich Euch verlassen habe, habe ich mich nie so heimath‚
lich gefühlt, als hier. Einer meiner Reisegefährten war ein
Musiklehrer,[b] ein junger verheiratheter Mann ohne Kinder, der
nach Sidney seinen Schwager folgte. Als er hier ankam, miethete
er ein Haus für den enormen Prieß von 100 Thalern jährlicher
Rente nd da er ein Stübchen übrig hatte, bot er mir an mit ihm
zu leben nd so einen Theil seiner Kosten zu tragen. Dieß wurde
angenommen, meine Stube einfach doch behaglich eingerichtet
nd so lebe ich denn verhältnißmäßig ausserordentlich zufrieden,
ganz mit meinem Studium beschäftigt in diesen mir neuen
Verhältnissen. Dann hat es unstreitig einen großen Reiz für mich,
dieses sich entwickelnde Volk zu beobachten, welches sich
vielleicht in weniger als einem Jahrhunderte gleich den Vereinig‚
ten Staaten Nordamerik[as], von England loß reißen wird, um
einen unabhängigen Staat od. Staaten bund zu bilden. In einem
milden Klima, in einer reichen Natur, in für den Handel sehr
günstigen Verhältnissen muß ein thatkräftiges Volk [g]leich dem
Englischen schnell ausserordentliches leisten. In Italien habe ich
eine solche Race oft mit Schmerzen vermißt nd ich glaube daß
Italiens Zustand [sic]h nicht eher besser wird, als bis eine
friedliche Einwanderung von arbeitsamen Männern des Nordens
statt findet.

Früchte sind in Fülle zu haben. Von unserem Apfel, der
Birne, den Pfirsischen, den Feigen, Trauben bis zu den tropischen
Früchten: der Ananas, der Banane, der Cocos nuß gedeihen sie
alle, sei es in Sidney, welches indessen ausserordentlich sandig
⌜nd unfruchtbar⌝ ist, sei es in den fernen Ansiedlungen nd
Städten gegen das Innen oder an der Seeküste. Sie sind indessen
sehr theuer, wie auch die Gemüse, während das Fleisch besonders
jetzt ausserordentlich billig ist. Der Lohn der Dienstboten ist
ausserordentlich hoch ud selbst Knaben erhalten 56–60 Rtl.
jährlich mit Kleidung nd Nahrung. — Die Dienstboten sind
deßhalb auch sehr unabhängig und man muß sich vorsehen, von
ihnen nicht in Stich gelassen zu werden. — Von Zeit zu Zeit
sind Verbrecher entsprungen, um in der Wildniß frei zu sein und
nicht arbeiten zu müssen. Diese haben theils allein, theils zu

Reconnaissance in New South Wales

mehrere vereinigt Angriffe auf die Reisenden nd auf die einsamen Ansiedlungen gemacht. Sie bemächtigen sich nur des Geldes der Uhr nd dergl. Kostbarkeiten; doch niemals oder sehr selten waren ihre Raubzüge blutig. Diese Menschen nennt man bushrangers (Buschläufer) und die berittene Polizei ist ihnen so dicht auf den Hacken daß ihre Unternehmungen immer seltner werden nd mit der sich vermehrenden Bevölkerung bald gänzlich verschwinden werden. Die Verbrecher werden jetzt hier, wie in irgend einem großen Gefängniße bei uns behandelt. Im Gefängniße unter strenger Aufsicht und unter dem Verbot zu sprechen, werden sie jeden Morgen zur Arbeit hinaus geführt, sei es um öffentliche Gebäude auf zu führen, Straßen zu bauen oder Wälder aus zu roden. Sie haben mit den Bewohnern jetzt keine Gemeinschaft mehr. Konnt Ihr nun aus dem früher Erwähnten abnehmen, daß der gesellschaftliche Zustand in Sidney wegen des beständigen Gegensatzes Emancipirter Verbrecher nd freier Einwanderer nicht eben der angenehmste ist, so sind dennoch eine Menge gebildeter Familien hier, welche es mich völlig vergessen lassen, daß ich in einer Verbrechercolonie oder so weit von Europa entfernt bin. Ich machte während meiner Reise die Bekanntschaft der Familie eines Captain Marlow, der von Schottland nach Sidney versetzt wurde. Man behandelte mich ausserordentlich freundschaftlich nd es scheint, als hätten sie die Absicht, mir die eigenen fernen Verwandten einigermassen ersetzen zu wollen. Ich wurde bei einer Menge von Personen eingeführt, was mich nicht wenig verwirrte, da ich stets ein zurückgezogenes Leben geführt hatte nd nur sehr wenige Bekanntschaften suchte. Ich kann noch nicht gehörig unterscheiden, wen ich zu lassen oder zu suchen habe nd die daraus entspringenden conventionellen Pflichten sind mir sehr unangenehm. — Ich erfuhr hier, daß an eine Expedition ins Innere unter einem Jahre schwerlich zu denken wäre und da ich während dieser Zeit unabhängig zu leben wünschte, gedachte ich zuerst, einige Zöglinge zu nehmen, für die man gut bezahlt. Doch das ungebundene Freiheits gefühl ist so stark in mir, und der Ärger mit trägen Zöglingen ist meiner Vorstellung so gegenwärtig nd zu wider, daß ich diese Idee auf gegeben habe. Es befindet sich hier eine Schule der Künste nd ich werde während des Winters wahrschl. einige Vorlesungen über Botanik nd Zoologie halten. Ich kann indessen wenig über meine gegenwärtige Stellung sagen. Jeder Tag kann wesentliche Veränderungen hervorrufen. Mag es Euch genügen, mich bis auf die Entfernung von Euch zufrieden und deßhalb auch glücklich zu wissen, obwohl ich nicht leugne, daß ich oft ungeduldig werde, wenn ich nicht so rasch nd großartig ausführen kann, wie ich möchte. Dieß ist denn auch die Ursache, daß ich oft mit Schmerzen an Williams veränderten Pläne zurück denke, da wir mit seinen Mitteln, wenn er sich nur ein wenig für die Wissen‚ schaft erwärmt hätte, um so viel schneller nd sicherer unser Ziel hätten erreichen können. Selbst sein Bruder scheint Neuholland wieder verlassen zu haben, da er mir auf zwei Briefe den Antwort schuldig geblieben ist. William wurde besonders durch den Gedanken zurück geschreckt hier eine Wildniß, Kannibalen und rohe Ansiedler zu finden. Hätte er die blühende Stadt gesehen, versehen mit allem, was sich Europaeischer Luxus wünschen kann, ja selbst für den wissenschaftlichen Mann nicht ohne Mittel — er würde wahrscheinlich mit mir gekommen sein nd mein Schicksal getheilt haben. Ich leugne nicht, daß mir das allgemeine Bestreben, sich Geld zu erwerben, diese durchaus materielle Richtung, welche höhere geistige Bedürfniße nicht zum Bewußtseinkommen läßt, keines weges gefällt; doch bin ich ja vielleicht selbst im Stande, mich durch Erweckung jener Bedürfnisse nützlich zu machen nd dieser Gedanke reizt mich. Es sind hier mehrer Männer, welche sich durch besondere Umstände eine schöne Unabhängigkeit erworben haben und nur wenigstens von Zeit zu Zeit sich mit den Wissenschaften beschäftigen. Ich hatte einen Empfehlungsbrief an Sir Thomas Mitchell,[c] dem Surveyor General von Neu Süd Wallis (dem Chef der Land vermessungs commission). Er führte mich bei Doct. Nicholson[d] ein, der mit meinem William indessen nichts zu thun hat. — Sie sind beide einflußreiche Männer nd ich hoffe allmählig Grund zu gewinnen.

Ich habe mehrere Deutsche hier kennen gelernt. Hr. Kirchner,[e] ein junger Mann aus Frankfurth am Main ⌈kam vor zwei Jahren hierher, arbeitete als Commiz im Hause eines reichen Kaufmanns, heirathete seine Tochter nd macht jetzt recht schöne selbstständige Speculatione.⌉ Hr. Schmidt aus Stargardt ⌈kam als Missionär⌉ nach Neusudwallis und befindet sich jetzt in Moreton bai, einige 80 Meilen gegen Norden. Er brachte seine Frau mit: sie scheinen beide recht zufrieden. So sind noch eine Menge von Landsleuten hier.⌉ Der Deutsche ist seiner Bescheidenheit nd Mäßigkeit wegen gewöhnlich sehr geachtet und mein Vaterland allein dient mir oft zum Empfehlungs׳ briefe. — Was ich indessen auch Gutes nd Löbliches über dieses Land schreiben möchte, das indessen wie jedes andere seine fühlbaren Mangel hat, Ihr würdet sicherlich dem unruhigen Wanderer folgen, um in diesem reizenden Klima Eueres Daseins Euch mehr zu erfreuen oder in einer für Euch neuem Schöpfung einen neuen Gottesdienst zu beginnen. — Denn Ihr steht am andern Ufer des weiten, sturmbewegten Meeres nd die Gefahren [all]ein, die geträumten Gefahren, würden Euch zurück׳ schrecken. Lebt denn wohl meine Theuern, grüßt meine liebe Mutter, ⌈der ja dieser Brief gleichfalls gilt,⌉ und alle die Uebrigen.

Euer herzlich Euch liebender Ludwig

Sydney, 23 March 1842

My dear brother׳in׳law,

I've crossed the vast waters of the deep and defied its storms, the Sun has passed over my head from the South into the North, and now the whole mass of the Earth lies between you all over there, and me. The time of day, the seasons, the very quarters of the globe, are different. It's cold in the South, warm in the North; and my own shadow falls to the South at noon. I sailed from London on the 1st of October 1841, and we had an unusually stormy passage as far as Cork, but I was so eager for a glimpse of distant places that I hardly noticed the danger and the trial. Cloaks and canvas shielded us from blustering wind and drenching spray, and when the Moon was shining clear, we used to join in singing the songs that rovers sing. If only I could find words to tell you how deeply I was stirred by the majestic course of things in mid-ocean! When the sky was clear, the constellations rose in splendour in the East, above the sharp line of the horizon, strode in sublime silence over the restless sea, and declined so gently in the West. As we got nearer to the Equator old, familiar clusters disappeared behind us, and brighter stars, that we had never seen before, came up to greet us.—I used often to think about the ship—so frail a product of man's handiwork. It was carrying about 300 people across boundless wastes to a remote, new homeland, yet they were surrounded by many of the comforts of life and they felt quite safe; and this lonely little world, propelled by wind and weather, depended in turn upon the compass, a needle of iron barely 3" long, but in its way a kind of magical iron bridge linking continent to continent across the openest seas. We saw a few islands, like St Antonio [Santo Antão] in the Cape Verde group, Trinidad off the coast of Brazil, and St Paul between the Cape of Good Hope and New Holland; but we landed nowhere. We led a peculiar life in the ship. Besides the 250 emigrants there were 20 independent passengers on board. At breakfast, dinner and tea we lived together like one family, but the distance between us was enough to allow us to follow our own inclinations between times. I made a special study of navigation, learned how to determine longitude and latitude, and observed the changes in atmospheric pressure and temperature. Whilst we were near the Equator the temperature never rose above $23\frac{1}{2}°$ [R?]. On the whole, we suffered more from cold when North of the Equator and again when South of it, than from heat. As the seamen usually indulge in rather crude practical jokes when crossing the line, the captain did not disclose the crossing until we were $5°$ farther South. They baptise passengers who have never crossed the line before, by hosing them, tarring their chins, and so on. Between 20 people who are constantly cooped up together there was bound to be friction. Besides that, there were a few who were bent on making trouble. This at least brought people out in their true colours, and had real importance for the student of human character. Of life in the sea we saw little. The sea, nevertheless, is probably teeming with life, considering the polyps and crustaceans and worms that live either freely on the bottom or attached to it, and the countless hosts of fishes and molluscs that course through its immense volume. Often, when looking over the rail down into the sea, I've regretted that it's simply not granted to us to learn much about strange

Reconnaissance in New South Wales

things that are hidden under perhaps 2–3000 feet of water.—Still, we were often delighted to see schools of dolphins, each rising and plunging in a way that exposed half of its body above the surface. They're very likely to appear when there's stormy weather ahead. And there were the cautious sharks, which circled slowly around the vessel, ready to snap up refuse with ravenous haste. We saw, and captured, plenty of sea birds, particularly albatrosses. These birds have a wing-span of 10 to 12 feet, seem to live for ever on the wing hundreds of miles from land, soaring between the huge rollers all night without a sound. Four and a half months went by (Oct. Nov. Dec. Jan. and ½ Feb.) before we sailed between the rocky headlands at the entrance to Port Jackson. When the pilot came on board, from a boat manned by Maoris,[a] I could have hugged the dour, weather-beaten old salt as if he were the herald of a new world. Port Jackson is a spacious, indented estuary, enclosed between cliffs and hills covered with trees. Thanks to long-awaited rains, the trees were springing green, and the homes and mansions you could glimpse between them looked friendly. For nearly 18 months there had been unbroken drought, and herds of cattle and flocks of sheep had been dying of thirst around the parched watering places. The drought broke at last, two days before our arrival, and it rained for 14 days on end, without ceasing. This rendered my first days here somewhat uncomfortable, but it worked such a radical change for the better in the look of the country that I gladly put up with the inconvenience. 52 years ago the shores of this magnificent harbour were inhabited by savages who had never before seen a white man. They are now the site of a big town of 42,000 inhabitants. It is surrounded by the mansions of its wealthy citizens. It extends partly up into a gully and partly over two ridges, and is, on the whole, regularly laid out with big buildings and wide streets. The massive sandstone of which this part of the country is composed, outcrops in many of the streets. In fact, some streets have been hewn through the rock. I was well received in Sydney, and, although one needs considerable means to do anything ambitious, I hope, given time, that I shall succeed like others in doing what I want to do. The climate is extraordinarily mild and pleasant, particularly at this time of the year, sudden change being all one could really complain about. The activity of the town is astonishing, and a spirit of feverish speculation prevails. The harbour is crowded with ships and there are daily arrivals and departures from and to England, China, New

Zealand, Van Diemen's Land, Port Phillip and different places on the coast. There are steamboats going to Hunter's River, Moreton Bay and Port Phillip. You can buy any article of luxury or convenience in Sydney. Good roads lead to different towns a little farther inland—to Bathurst, for instance, about 50 [German] miles away over the Blue Mountains, to Liverpool, to Windsor on the Hawkesbury River, etc. And to whom do we owe such astonishing change in so short a time? In 1788 Arthur Phillip brought 850 convicts here, founded the colony, and began, with convict labour, to make the country productive. From then on, they continued to send convicts here, where they were employed either on public works or were assigned as servants to the free settlers. Convicts whose sentences had expired were set free and themselves became settlers. Even those granted conditional pardon could acquire property and become financially independent. The richest men in the colony were convicts, or are the descendants of convicts. But more and more free settlers have gradually been coming here, and there are now about 100,000 of them on the land in Australia. Nearly all these immigrants have come here to make their fortunes and for nothing else. They are willing to devote years of their lives to the purpose, in the hope of being able to return Home to enjoy life quietly until they die. Few come here to stay. But many change their minds when they have come to discern the attractions of this rich country, and to find its disadvantages less irksome. Families of this kind, who are taking an interest in the colony, and have come to look upon it as their own country, constitute, in fact, its only real wealth. And it is through them that a powerful state will gradually arise, a state which may possibly consign old Europe to oblivion. Naturally, however, you can't expect existing social conditions in Sydney to be satisfactory, considering the incompatible elements of which society here is composed. No matter how liberal you are, if you're in the company of ex-convicts you'll never banish the thought that you are in fact dealing with people who were once capable of committing serious crimes. True, they have paid the penalty and have returned, white-washed, to society—but has it redeemed them? Do they really merit our confidence? Such considerations, which are by no means groundless, explain why many free emigrants have come to regard themselves as [people of] a distinct and superior class. The effect of their opinion of themselves has been to unite the emancipated convicts. In consequence, we now find two parties, two kinds of society confronting each other. There is hostility between them which often impedes the progress of the colony. In earlier days, when a convict ship arrived at Sydney, people who needed servants used to notify the authorities, who would assign them a number of convicts according with the size of their business. They were responsible for their surveillance and they had a degree of authority over them that was often abused. The great advantage was that the convicts cost them nothing except for food and clothing. Later on, when the colony began to develop, it became clear that it could not continue to receive the scum of Europe. The system of assigning servants was discontinued, and other convict stations were established, at Moreton Bay (farther North) and Norfolk Island. The former has also been given over to free settlers.—If a convict had served his sentence of 2 or 3 up to 20 or more years, and had behaved well, he was given a ticket-of-leave. He could then, under police supervision, set about making his own living, but on condition that he revert to his former state of servitude or imprisonment should he commit a new offence. Very few were granted unconditional freedom, and there are wealthy men and heads of families in Sydney who are really convicts on conditional pardon. Since I left my native land and left you all behind, I have never felt so much at home as I do here. One of my fellow passengers was a music teacher,[b] a young married man with no children, who had followed a brother-in-law to Sydney. When we arrived he took a house at the exorbitant rent of 1000 [rix] dollars [£143] a year. As he had a small room to spare, he asked me to lodge with him to help him to meet his expenses. I agreed, and now, my room simply but comfortably furnished, I find myself in many ways very contented. I'm absorbed in study under conditions quite new to me. I have to admit, moreover, that I'm fascinated when I watch what is happening here. A state is coming into being which may, perhaps in less than a century, break loose from England as did the United States of North America, and so establish an independent nation or federation. In a mild climate, where Nature is generous, [in a place] very favourably situated for trade, an energetic people like the English ought to make rapid and remarkable advances. In Italy I used often to deplore the lack of such energy, and I believe that conditions will never improve there until a peaceful invasion of industrious people from the North takes place.

There's fruit here in plenty. From our own apples, pears, peaches, figs *and grapes to the pineapples, bananas and coconuts of the tropics, they all do well, either in Sydney, though the soil is very sandy ⌜and sterile⌝, or near the towns and settlements farther inland, or on the sea coast. Fruit's very dear all the same, and so are vegetables, but meat, particularly now, is extraordinarily cheap. Servants' wages are surprisingly high, and even boys get 56–60 rix dollars [£8 to £8. 10s.] a year with food and clothing. Even so, servants are very independent and you have to see that they don't leave you in the lurch.—Now and then convicts have escaped into the bush, seeking freedom from work and restraint. Sometimes alone, sometimes in small bands, they've made attacks on travellers or the lonelier settlements. They take possession of nothing but money, watches and similar valuables, and only on rare occasions have they committed robbery with violence. They call these fellows bushrangers. The Mounted Police are so hot on their heels that their activities are declining, and, as population is increasing, they'll soon come to an end. Convicts here are now treated in the same way as prisoners in any of our big prisons. They're under strict discipline and are forbidden to speak. Every morning they're led out to work either on public buildings, on road-construction or at clearing the land. With the public they have no intercourse whatever. Lest you assume, from what I've been telling you, that social conditions in Sydney are none too pleasant on account of antagonism between convicts, emancipists and free settlers, let me assure you that there are families enough of cultivated people to make me forget not only that I'm living in a convict colony, but even that I'm so far from Europe. On the passage out, I got to know the family of a Captain Marlow, who was transferred from Scotland to Sydney. They were so extraordinarily kind to me that I think they were to some extent trying to take the place of my own kin whom I had left so far behind me. Quite a number of people took me up, which was rather embarrassing, because I've led a withdrawn kind of existence, and I did not want to make many acquaintances. I can't yet tell for certain whom I can correctly approach, whom not; and I find the appropriate matters of form very distasteful.—I've heard that it's hardly likely that any expedition into the interior will be despatched for at least a year, and, as I want to remain independent in the meantime, I thought at first of taking a few pupils, as it pays well. But the sense of untrammelled liberty means so much to me, and I'm so keenly aware of, and repelled by the vexation of*

Reconnaissance in New South Wales

dull pupils, that I've given up the idea. They have a School of Arts here, where I shall probably deliver a few lectures on Botany and Zoology during the Winter. I can't say much about how my affairs stand just now. They could change radically over night. May it suffice you to know that I am not worried, even about being parted from you all, which means that I'm happy. I admit that I often get very impatient at my inability to act as quickly and ambitiously as I would like. For the same reason, it often hurts to think of the change that William made in his plans. With his means, had he only been a little more interested in science, we would have been able to attain our end so much the quicker, with so much the less risk. Even his brother seems to have left New Holland, for I have written to him twice but he has not replied. What most deterred William was the thought of living in a wilderness amongst cannibals and rough settlers. Could he have seen this thriving town, which has all that European luxury could desire, even for the student of science who can afford it, he would most likely have come out here to share my future with me. I don't deny that the materialism, the devotion to making money, which preclude any interest in the needs of the mind, don't please me in the least. But I may perhaps myself be of some service in arousing that interest, and the idea is encouraging. There are several people here who have been able, through special circumstances, to achieve an admirable degree of independence, and who show some interest, even it be only occasional, in scientific matters. I had a letter of introduction to Sir Thomas Mitchell,[c] the Surveyor General (head of the land-survey department) of New South Wales. He introduced me to Dr Nicholson[d] (who is not related to my William, however).—They're both influential men, and I hope that I'll gradually gain some footing.

I've made the acquaintance of several Germans here. Herr Kirchner,[e] a young man from Frankfurt am Main, ⌜came here two years ago, worked as a clerk for a wealthy business man, married his daughter, and is now a successful speculator and quite independent.⌝ Herr Schmidt from Stargardt ⌜came out to New South Wales as a missionary[f] and is now stationed at Moreton Bay, about 80 [German] miles farther North. He brought his wife out with him and they both seem to be very happy about it. And there are quite a lot of others from Home.⌝ Germans on the whole are highly respected here on account of their sobriety and lack of pretension, and my nationality has often served as my letter of introduction. If I said all I would like to say in praise of the excellence of this country, which of course has its palpable shortcomings like any other, you'd certainly be following the restless wanderer out here, so that you might either rejoice in living in this delightful climate, or set up a new form of worship in what would seem like a new creation. There you are, though, on the far side of vast and stormy seas; and the dangers, the imaginary dangers, would hold you back. Well, good-bye, my dear ones. Give my regards to mother ⌜for whom this letter is also meant,⌝ and remember me to all the others.

Your affectionate Ludwig

51 DOCTOR LITTLE, Finsbury Square, London

Sidney den 25tn März 1842

Mein theuerster Freund

Lassen Sie mich wenigstens einen Brief an Sie jetzt schon anfangen, damit Sie sehen, daß ich mich Ihrer fortdauernd erinnerte und daß ich mich mit einem so verehrten Manne gern und häufig unterhalten möchte. Ich kann Ihnen über bestimmten Aussichten, über begründeten Hoffnungen zu erfüllender Pläne leider nichts mittheilen: doch was ich während meines Abschieds von Ihnen erlebt, war im Durchschnitt wohlgeeignet, meinen Muth aufrecht zu erhalten und an endliches Gelingen vertrauungsvoll zu glauben. Aus einem Briefchen, welches ich Ihnen von Cork sandte, da ich mir schmeichelte, Sie nähmen einigen Antheil an den nun verwitterten Wanderer, haben Sie ersehen, eine wie harte Lehrzeit ich von London nach Cork durch zu machen hatte. ⟨Meine Reise von Cork nach Sydney wurde nur einigemale durch begegnende Schiffe und durch sichtbar werdende Felsen inseln unterbrochen. Die Begegnung des 'Planters'[a] von Neuseeland war in so fern die interessanteste, als auf jenem Schiffe Doct. Dieffenbach nach Europa zurück kehrte. Er zeigte mir eine Menge von Scizzen, welche er von Bergen und Landschaften im Innern von Neu zeeland gemacht und sprach mir viel über die gute Erndte, welche meiner harrte. Doct. Dieffenbach hatte indessen den großen Vortheil als Naturforscher von der New Zealand Company unterstützt zu werden, was mir

Reconnaissance in New South Wales

abgeht, der ich mich auf eigenen Füßen mühsam zu halten strebe.⟩ Meine Schiffsgesellschaft war ein wissenschaftliche durchaus todter Körper, welcher einer Insecten larve gleich nur frühstückte Mittag aß und Thee trank und dessen Mitglieder sich durch Gezänk und Verläumdung best möglichst die Zeit vertrieben. ⟨Da an Beobachten von Seethieren wegen der Indolenz der Schiffsführer nicht zu denken war, hielt ich mich an Navigation, an Bestimmung der Länge und Breite, an Beobachtung des Barometers und Thermometers, wobei endlich mein ausdauerndes Eifer auch unsern Capitain ein wenig zu meiner Hülfe anregte. Am Cap der guten Hoffnung wurden Albatrosse secirt, in der Basses Straße ein Haifisch und Barracuda fische (welche zur Familie der Makarelen gehören) und bei unserer Ankunft in Port Jackson einige *Ascidien*. Während der ersten drei Monate nahm ein junger Phthysicus einen großen Theil meiner Zeit in Anspruch, da der arme Knabe in letzten Stadium der Krankheit ohne Wärter, ohne Freund sich fast heimlich und unter falschem Namen auf das Schiff begeben, um durch die lange Seereise oder durch das milde Klima Neuhollands seine Gesundheit wieder zu gewinnen. Ich strebte ihm Bruder und Mutter zu ersetzen, doch das Üble war zu tief und am letzten Tage des Jahres 1841 senkten wir seinen fast fleischlosen Leichnam ins feuchte Grab. Wir hatten das Scharlack fieber epidemisch am Bord; doch hatte es mit wenigen Ausnahmen einen milden Karakter und kostete von vielleicht 70 Kranken nur einem jungen Mädchen von 22 Jahren das Leben; indem ein Fieber mit typhösem Karakter sich der ursprünglichen Krankheit zugesellte fast allgemein fingen die Kinden nach der Krankheit ausserordentlich zu schwellen an und dieser Zustand einer allgemeinen Wassersucht dauerte oft 4 Wochen. Anwendung von Senfpflastern auf allen Theilen des Körpers zeigte sich sehr wohlthätig.⟩ Unter dem Aequator hatten viele der unverheiratheten Mädchen hysterische Zufälle und zwei verloren gänzlich den Verstand, was bis zu unserer Ankunft in Sydney anhielt. — Sehr häufig waren diese Zufälle folge des Sonnenstichs, denn die Mädchen, die vermehrte Kraft der fast senkrechten Sonne vergessend, kamen ohne Bedeckung auf das Verdeck, soviel Vorsicht man auch nahm, und litten nun zuerst an Ohnmachten und später an Lach und Weinkrampfen pp. Hierzu kam der unbeschäftigte Zustand dieser Mädchen, welche eine Menge unverheiratheter Männer, obwohl immer beobachtet, um sich sahen und plötzlich in ein heißeres Klima versetzt, die animalischen Begierden lebhafter finden mußten. ⟨Um die Köpfe der unbeschäftigten Männer ein wenig an zu regen gab ich ihnen, solange wir den sanften Passatwinden durch die südliche Tropenzone geführt wurden, eine Art Vorlesung über das was sie täglich sahen, das Meer, die Licht, Wolkenbildung Zusammensetzung der Erdrinde und einen allgemeinen Ueberblick über die Thierklassen. Doch wie wir zu höhern südlichen Breiten gelangten, bewegten starke westliche Winde unser Schiff zu heftig und ein kaltes Regenwetter machte den Aufenthalt auf dem Verdeck so unangenehm daß ich von meinen Zuhörern verlassen wurde. So war ich entweder für mich selbst oder für andere immer beschäftigt und die Zeit flog mir fast unbemerkt unter den Händen weg. Man erkannte meinen guten Willen und so sehr man sich auch untereinander zankte und anfeindete, so ließ man mich doch in Ruhe. In der That war diese dauernde Gährung eine gute Schule für Menschenerkenntniß, da ich das erstemal in meinem Leben 20 fremde Menschen so in der Nähe sah und dieses dauernde fast chemische Spiel von Anziehung und Abstoßung zwischen ihnen so recht in Muße beobachten konnte. Man lebt en famille und dennoch gesondert genug, um seinen eigenen Neigungen zu folgen, die Kinder und die Alten werden alle so *familiar*, daß man wirklich am Ende glaubt, man gehöre zu einander. — So fuhren wir denn endlich durch die enge Felsenpforte von Port Jackson in diesen schönsten sichersten geräumigsten Hafen, der auf seiner Nord und Sudseite eine Menge von Buchten und nebenbuchten bildet und rings von anstrebenden baumbedeckten Hügeln und Sandsteinfelsen eingeschlossen wird.⟩ Ich kann Ihnen nicht beschreiben, mit welchem Enthusiasmus ich zuerst über diesen Boden hinsprang, mit welcher Freude ich jedes fremde Gewächs begrüßte und wie

Reconnaissance in New South Wales

endlich die Menge rings andringender neuer Anschauungen mir fast den Kopf verrückten. Denn aber erinnerten mich die einge՛ wanderten Pflanzen wieder an das heitere Italien, wo unter demselben milden Himmel die Traube die Feige, der Citronen und Apfelsinen baum, die Olive, die Aloë wuchsen, zu welchem hier sich noch die Banane, der Ricinus baum, das Bambus rohr nd manche andere tropische Gewächse gesellten. Allen waren Einwanderer, wie ich selbst — sie standen den eingeborenen Kindern dieser Erde, den weiß rindigen Eucalyptus, den Banksias gegenüber und den mannichfachen Gliedern der Myrthen familie und der Epacrideen, welche alle Gebüsche um Port Jackson zusammen setzen. — Ueber den Boden hin sprangen und krochen ungesehene Formen von Orthopteren; Truxalis, Mantis, Phasma wurden schnell bemerkt und ge՛ fangen. — Die Ausdehnung von Sidney mit seinen 42000 speculierenden thäthigen Menschen setzte mich in Erstaunen. Solchen ungeheuern Wechsel haben 52 Jahre bewirkt. Eine vom weißen Manne nie vorher betretene Wildniß, in welcher nur das Coo՛eè (der Zuruf) der wilden erschallte, hat sich in eine weite, wohlgebaute Stadt verwandelt, umgeben von Lusthäusern, erfüllt mit dem Luxus Europaeischer Civilisation. — Ich ging mit meinem Empfehlungsbrief zu Sir Thomas Mitchell. ⟨Er empfing mich ziemlich freundlich und sagte mir daß eine Expedition vor 12 Monaten nicht zu denken sei. Ich bot dem Museum, meine Mineralogische sammlung als Geschenk an. Sie wurde dankbar angenommen; doch da das Museum, wie fast alle Museen die ich in meinem Leben gesehen, in Unordnung ist und man beabsichtigt das Local zu verändern, so hat man mein Geschenk noch nicht in Empfang genommen. Ich erhielt von Sir Thomas einen Empfehlungsbrief an Doct. Nicholson, dessen ominöser Name mich mit nicht geringen Hoffnungen erfüllte; doch beide haben außerdem noch keine weitere Annäherung versucht.⟩ Ich habe indessen die Bekanntschaft einiger Familien gemacht, die mir das materielle Leben so angenehm, als möglich zu machen suchen. Dann finden sich eine Menge von Lieb՛ habern der Wissenschaft, die anfingen hier und da zu beobachten und zu sammeln und die am Ende wohl zu größer Thätigkeit und zu größerem Nutzen angetrieben werden können. — Der botanische Garten ist wenigstens recht schön ausgelegt, wenn auch für die Wissenschaft wenig geschickt — es könnte wenigstens etwas geschehen, denn die Gelegenheit ist da. Das Museum ist wenigstens begonnen, die Bibliothek ist in mancher beziehung schon reich — leider nicht in den Naturwissenschaften. Eifer findet sich genug im jungen Australischen Volke — sie mögen wenigstens einigemale hören, wenn ihnen auch zum anhaltenden Studium die Geduld fehlt. In einer School of Arts werden öffentliche Vorlesungen gehalten und ich habe die Absicht, mich dort nützlich zu machen. Als Decandolle[b] seine Vorlesungen über Botanik in Genf begann, fand er das Volk ausserordentlich un botanisch. Doch sein eigener Eifer wirkte so ausserordentlich auf die sich schnell vergrößernde Zahl seiner Zuhörer, daß ganz Genf in kurzer Zeit sich schien in ein botanisches Collegium verwandelt zu haben. So haben wir in den Wissenschaften häufige Beispiele von eifrigen thätigen Männern, die dem Magnete gleich ihr eigenes Streben auf alles, was ihnen nahe war übertrugen. Kann ich ihnen nicht gleichkommen, so sollen sie mir wenigstens als Muster stets vor Augen schweben. Meine allgemeine Richtung und die Bekanntschaft mit fast allen Zweigen der Naturwissenschaften, läßt mich die zerstreuten Bestrebungen in dieser Colonie leicht auffinden und giebt mir die Hoffnung sie einst zur gemeinsamen Thätigkeit vereinigen zu können. Ich werde mich bald meinem Lieblingszweige zu wenden und über die andern nur streben, eine anregende Aufsicht aus zu üben. — In Hobart town sind sie uns schon in so fern voraus, als sie ein recht reichhaltiges Journal[c] begonnen haben, von welchem gegenwärtig 2 Nummern erschienen sind. — Für Geologie und Mineralogie ist in der Nähe von Sidney sehr wenig zu thun. Der Sandstein fels bildet in undurchdrungene Mächtigkeit mit sehr wenigen Ausnahmen den Boden in ausserordentlicher Erstreckung. Hunters river, New Zeeland und Port Philipp sind bei weitem interessantere Puncte. Illawarra ist reich an guterhaltenen fossilen Muscheln, welche in einem bläulichen fast Lias ähnlichen Thone gefunden werden. Wenn sie die Straße von Botany Bay mit mir folgten, würden sie ungefähr 2 Meilen von Sidney, sich von wellenförmigen Sandhügeln umgeben sehen, die ohne eine vorwaltende Richtung, innerlich aus hartem Sandstein zusammengesetzt, an ihre Oberfläche durch ein über ihnen ruhendes Meer in einen blendend weißen losen Sand verwandelt scheinen. Unbekannt, aber vielleicht noch fortdauernde Ursachen haben endlich diesen Meeresboden zu Tage hervorgehoben[d] und eine eigenthümliche Vegetation hat sich seiner bemächtigt. Sie finden hier keinen Rasen, sehr wenige krautartige Pflanzen — fast alle Gewächse sind holzig und strauchartig, ungefähr 4′–8′ hoch. häufig mit Schlingpflanzen bedeckt. Die Anordnung der Blüthen ist im allgemeinen sehr auffallend und die Blüthen selbst sind oft von ausserordentlicher Schönheit. Es ist indessen jetzt Herbst und Flora ist nicht in eben reichem Schmucke. Wie die Gewächse den Boden, so haben zahlreiche Insecten arten jeden Theil der Gewächse in Beschlag genommen. Niemals trat mir die Pflanzenwelt als die größere Basis der Insecten welt so lebhaft vor Augen. Indessen, selbst dieß mag theilweise von der Jahreszeit abhängen. Wenn die Dipteren durch ihre Menge beschwerlich fallen, so beruht dieß mehr auf der ungeheueren Anzahl der Individuen einiger Arten, wie z. B. der Muskitos, der Fliegen, als auf den Arten reichthum der Diptern im allgemein. So will es mir wenigstens scheinen. Sie haben nur Ihre Freunde zu fragen, welche in Sidney gewesen sind und sie werden nicht ermangeln, Ihnen über die Lästigkeit der Muskitos wunderbare Geschichten zu erzählen. Da vielleicht die höhere Temperatur die Insecten eier rascher und schon im Leibe der Mutter entwickelt, haben sie oft das unangenehme Schauspiel, völlig entwickelte Maden aus der Legeröhre hervortreten und auf das vor Ihnen stehende Fleisch fallen zu sehen. — Die Schmetterlinge sind sehr zahlreich, doch stehen sie in der Pracht der Farben den Südamerikanischen bei weitem nach. Die Hemipteren sind ausserordentlich mannichfaltig; am zahlreichsten *Orthopteren* und Käfer. Für die Mollusken sind die Ost Küsten von Australien besonders günstig. In der Torres Straße und bis zum 23° südlicher Breite haben Polypen eine Menge gefährlicher Riffe gebildet, welche den Indien und China fahrer zwingen südlich um Neuholland herum zu steuern, wenn er von Sidney nach Batavia Calcutta oder Canton geht. Zwischen jenen Riffen wimmelt es, wie in künstlichen Teichen, von Weichthieren, von Anneliden, von Crustaceen. Der sich dort ansiedelnde Artenjäger würde einer reichen Erndte gewiß sein. Die Riffe brechen die Gewalt der Wellen, ohne doch den regelmäßigen Wechsel der Ebbe und Fluth zu hemmen und die Tropen sonne erzeugt die der Thierentwicklung so ausserordentlich günstige Temperatur von 22°–23° Reaumur. Herr Stewart,[e] Chirurg am Militair hospital hat seine Aufmerksamkeit den Fischen von Port Jackson zugewendet, und da er zu gleicher Zeit ein tüchtiger Zeichner ist und tuscht, hat er eine Menge von Zeichnungen von unbekannten Seefischen angefertigt. Er möchte indessen wohl sehr wenige ganz neuen Arten haben. Da der Einzelne mit beschränkten Mitteln nur unvöllständig oder langsam seinem Ziel näher kommt und die reichen Leute in Sidney noch so ganz von materiellen Bestrebungen beherrscht werden, kann es nicht auffallen, daß die reiche Meeresschopfung noch so wenig bekannt ist. Das Meer ist unserer unmittelbaren Beobachtung nicht zugänglich oder nur theilweise so. Die Fische, die Mollusken müssen ihrer Freiheit beraubt werden, um uns vor Augen zu kommen und indem sie so zu gleicher Zeit ihr Lebenselement verlassen, wird es uns unmöglich, ihre Lebenserscheinungen zu beobachten. Wäre ich reich, so würde ich Fischer beschäftigen, künstliche Meeresteiche gleich denen anlagen, welche ich beim Ritter Rocca Romana in Neapel gesehen habe und so durch Beobachtung der Meeresthiere in ihrem Elemente ihrer wenig bekannten Natur nachspüren. — Sidney ist auf zwei parallelen gegen das Meer in Landzungen auslaufenden Hügeln und in die den zwischen liegende Thale gelegen. Gegen Sud Westen scheinen die Hügel durch einen Quer hügel verbunden — gegen N. Osten laufen sie in zwei Spitzen aus, zwischen denen Sydney Cove liegt. Sydney Cove ist also der Meer gefüllte Theil des ebenerwähnten Thales, in welchem ein unbedeutender Bach niederkommt.ᶠ Auf dem östlichen Zunge hat die Regierung einen Park angelegt und einen Theil dieses Parks für den Botanischen Garten bestimmt. — Der Park besteht zum Theil aus weiten Rasenplätzen, von Eucalyptus bäumen und Casuarinas umgeben, aus mit Myrthen gebüsch bedeckten Sandstein felsen und aus dichtern Wald parthien. Wohlgebaute Straßen erlauben zu Wagen, zu Pferde und zu Fuße sich der reizenden Anlage zu erfreuen. Gehen Sie in den Botanischen Garten und zum Meeresufer, so können sie sich dort, wie überall am Ufer dieses gesegneten Hafens, eigenhändig Austern von den Felsen brechen, welche Sie durch ihren delicaten Geschmack an die *natifs* ihrer Heimath erinnern; sie können sich in einer Bananen pflanzung an tropischen Pflanzenformen ergötzen oder dem zischelnden Geräusche biegsamen Bambusrohres lauschen, neben welchem Trauerweiden ihre langen Zweige über einen stillen Weiher niederhängen. Lustwandeln Sie dagegen in den wilden Partien des Parks, unter dem Schatten der weißrindigen Eucalyptus, so wird ihre Aufmerksamkeit bald durch das Geraschel hürtig entfliehenden Eidechsen (wenigstens zwei Arten) — bald durch auffallend gefärbte Vögel in Anspruch genommen, welche eben im Allgemeinen durch den Reichthum ihrer Farben den Wohllaut der Stimmen ersetzen. Die Reptilien sind zahlreich, die Vögel in der Umgegend von Sidney selten. Der Ornothologe muß zu den blauen Bergen und zum Hunters river, wo seiner eine reiche Erndte wartet. Eingeborene Saugthiere sind bis auf die Phalangirte in der Umgegend von Sidney jetzt eine Selten‑ heit. Die Kinder der weißen Männer haben ihre Stelle genom‑ men.

Könnten Sie mich nur einen vom Mondlichte erhellten Abend in diesen Park begleiten! Vielleicht würde die zum Genuß geneigte Menschen natur auch in Ihnen für den Augenblick das eifrige Streben für die Wohlfahrt ihrer Londoner Mitmenschen zurückdrängen. Sie würden dem Klaren Monde ins Auge schauen; Sie würden dem Gesange der Cicaden und Heimchen zu horchen; Sie würden über jenem blauen Wasserspiegel, von dunkeln Baum massen umgeben, hinschweifen; Ihr ganzer

Körper würde in dieser milden Temperatur vom innigsten Behagen ergriffen werden. Sie würden ausrufen 'Hier lasset uns Hütten bauen!'ᵍ — Und fehlt der Mond, so spannt sich ein glänzender Himmel voll leuchtender Sternbilder über uns. Der Centaurus, das südliche Kreuz, das Schiff Argo mit Maja Placida und Canopus, der große und kleine Hund und der Orion, der Stier und die Zwillingen schreiten nächtlich durch den Himmel, die Magellanischen Wolken erregen nächtlich meine Aufmerksamkeit. — Die Hitze des Tages ist groß, doch die Trockenheit der Luft bewirkt eine schnelle Absorbtion der Ausdünstung der Haut, welche hierdurch beständig sich abkühlt und die hohe Temperatur bei weitem erträglicher macht als im feuchten China West oder Ostindiens. Auf der andern Seite hat diese schnelle Abkühlung den Nachtheil, besonders bei eben angekommenen Europaeern Erkältungen und rheumatische Zufälle zu bewirken. Sie sind sehr geneigt, sich wie in unserem kalten Norden rasch zu bewegen, ein dichter Schweiß drängt aus allen Poren hervor und durchnässt ihre Kleider; sie setzen sich nieder ohne die Kleider zu wechseln und die schnelle Verdunstung thut nun das Ihrige. Ebenso ist der Wechsel der Temperatur oft ausserordentlich. Während die Sonne um ungefähr 2 Uhr Nach mittags die Quecksilbersäule zu ihrem Maximum ausdehnt, bewirkt der heitere Nacht himmel eine so freie Eradiation der erhitzten Erdoberfläche, daß die Abkuhlung gewöhnlich einen sehr starken Thau niederschlägt. Der Unter׳ schied des Maximums und Minimums nach den auf dem South head angestellten beobachtungen, wo indessen das Seeklima bei weitem fühlbar sein muß, als in Sidney, ist indessen nur 6°–8° Fahrenh. (2½ Uhr Nach mittag 9 Uhr Abends): während des Sommers ist es wahrscheinl. bedeutender. Sie werden unter den Erscheinungen, die in Neuholland das Gegentheil der Euro׳ paeischen sind, auch die Beobachtungen finden, daß Regen fast ohne Ausnahme mit höhern Barometerständen verbunden ist. Ich zweifelte sehr an der Richtigkeit dieser Behauptung; doch habe ich es selbst mehrere Male und zwar auf sehr auffallende Weise beobachtet.

Reconnaissance in New South Wales

Ich sollte Sie um Entschuldigung bitten, Ihnen in einem langen Briefe bekannte Sachen geschrieben zu haben, da vielleicht was sie hier beisammen finden schon längst in Wort und Schrift Ihnen vor Augen kam. Doch wünsche ich, daß Sie sehen, wie ich mich allmählig in dieser neuen Welt einbürgere, da ja eben jeder Mensch nach seiner eigenthümlichen Weise anschaut und diese Anschauungen verarbeitet. Nach zwei Monaten werde ich Ihnen vielleicht mehr mittheilen können. Ich würde mich nicht wenig freuen nach 6 bis 7 Monaten eine Antwort von Ihnen zu erhalten. Fragen von Ihrer Seite würden meine Aufmerksamkeit auf unbeachtete Puncte lenken und ihre Lösung[en] würden Ihnen nützlich werden. Horchen Sie überall herum und lassen Sie kein Wörtchen über Neuholland verloren gehen. Ich habe noch eine Bitte: Briefe von und nach Deutschland müssen eine besondere Addresse nach England haben. Erlauben Sie mir, Briefe an meine Angehörigen an Sie zu schicken, und auch Briefe an mich an Sie addressiren zu lassen. Meine Deutsche Correspondenz ist leider so selten, da ich gezwungen bin die Kasse meiner Angehörigen zu sparen, indem keine Post convention zwischen Preussen und England besteht — daß Sie sicherlich nicht häufiger als alle zwei Monate von Briefen beunruhigt werden sollen.

Leben Sie wohl, mein verehrter Freund, grüßen Sie Ihre liebe Frau und Kinder und erinnern Sie sich Ihres ergebensten Freundes

Ludwig Leichhardt

[P.S.] Meine Addresse: Mr Ludwig Leichhardt Bligh Street at Mr Marsh's Sidney Wenn sie Owen[h] sehen machen Sie ihm mein Compliment damit er mich nicht wieder vergißt, wie das erstemal.

Sydney, 25 March 1842

My dear friend,

Let me at least begin a letter to you this soon, to show that I have never ceased to think of you, and that I would be a willing and ready correspondent with your esteemed self. Unfortunately I've nothing yet to tell you about definite prospects or about reasonable hopes of being able to carry out my plans; still, what I was able to learn when I was leaving was decidedly encouraging, and enabled me to look forward with confidence to eventual success. From a note I sent you from Cork—for I flattered myself that you were still interested in the now seasoned traveller—you will have learnt of the hard apprenticeship I had to serve between London and Cork. ⟨The passage from Cork to Sydney was seldom relieved by the sight of another vessel or the glimpse of a rocky island far away. Bespeaking the Planter from New Zealand was the most interesting incident, as she had Dr Dieffenbach[a] on board. He was on his way back to Europe. He showed me a number of sketches he had made, of mountains and landscapes in the interior of New Zealand, and told me what a harvest was in store for me. Dr Dieffenbach, however, enjoys the great advantage of being employed as a naturalist by the New Zealand Company, whereas I have no such support and am painfully trying to stand on my own feet.⟩ My shipmates were quite dead to any interest in scientific matters. They ate their way like caterpillars through breakfast, lunch and tea, and did their best to kill time by bickering and back-biting. ⟨Lack of interest on the part of those in charge of the vessel put marine biological observation out of the question, so I took up navigation, determined longitude and latitude, and read the barometer and thermometer. My persistent application to the work was enough, in the end, to stir even the captain into giving me a little help. Off the Cape of Good Hope we dissected some albatrosses, in Bass strait a shark and some barracouta (which belong to the mackerel family), and, on arriving at Port Jackson, some Ascidians. During the first three months a consumptive young man took up a good deal of my time. The poor fellow was in the last stage of the disease and had joined the vessel secretly, under an assumed name, without an attendant or even a friend, hoping that the long sea passage or the mild climate of New Holland would restore him to health. I tried hard to do what a mother or brother could have done for him, but his case was hopeless and on the last day of 1841 we cast his almost fleshless body into the deep. We had an epidemic of scarlet fever on board, but, with few exceptions, they were mild cases, and, out of 70 patients, the only one to die was a young unmarried woman of 22. As a fever of the typhus type almost invariably accompanied the primary illness, children, when recovering, became extraordinarily puffy, and this condition of general dropsy often lasted for four weeks. Application of mustard plasters to all parts of the body proved to be very beneficial.⟩ Between the tropics many of the single young women had attacks of hysteria. Two of them completely lost their reason and had not recovered it when we reached Sydney. These attacks were mostly caused by sunstroke, for the girls, forgetting the greater power of the Sun when right overhead, used to come out on deck bare-headed in spite of all our precautions. At first they had fainting fits, which gave way later to paroxysms of laughter and tears, etc. Part of the trouble was that the girls had nothing to do. They were passing rapidly into a hotter climate and must have felt the quickening of their natural instincts. Although they were in full view all the time, they took good notice of the numerous unmarried men. ⟨To give the idle men a little to think about, I used to deliver a kind of lecture on what was daily before their eyes—the sea, the sunlight, the formation of clouds, the constitution of the Earth's crust, a review of the animal kingdom—whilst the light trade winds were carrying us through the southern part of the torrid zone. But as we passed into higher southern latitude stiff westerly winds made the ship too lively, and raw, wet weather made it so unpleasant to be on deck that my audience forsook me. Somehow, I was always occupied, either on my own behalf or that of others, and time slipped through my hands almost unnoticed. They gave me credit for my good intentions, and, no matter how much they wrangled and contended amongst themselves, they left me in peace. As a matter of fact, this continual state of turbulence was a good course in psychology, for it was the first time I had ever seen 20 different people in such close association, and I was able to watch the almost chemical play of attraction and repulsion between them at leisure. It was life en famille, yet we were independent enough to follow our own inclinations. Adults and children got to know each other so well that they really came to think that they belonged to one another.—Well, at last we sailed through the narrow, rocky entrance to Port Jackson, into this most beautiful, secure and commodious of harbours, which branches out into numerous bays and coves on its northern and southern sides, and is surrounded by sandstone cliffs and abrupt, tree-covered hills.⟩ I can't tell you with what excitement I leapt ashore, with what delight I greeted every botanical novelty, or how new impressions came crowding in from all sides until I felt positively dizzy. Fairly soon, however, the plants that have been introduced began to remind me of sunny Italy, where grapes, figs, bitter and sweet oranges, the olive and the aloe flourish under skies as serene as here; but here, the banana, the castor-oil tree, the bamboo cane, and many other tropical plants, were growing side by side with them. And they were all immigrants, like myself, confronting what was native to this soil: the white-barked Eucalyptus, the Banksias, the diverse members of the Myrtle family and of the Epacridae, of which the bush is composed all around Port Jackson.—And there were undescribed species of Orthoptera creeping around or jumping about on the ground. It did not take me long to identify and capture Truxalis, Mantis, and Phasma.—The amount of ground covered by Sydney and its 42,000 busy speculators astonished me. 52 years have wrought incredible changes here. A remote place, where the white man had never before set foot, and which resounded to nothing but the coo-ee (the hailing call) of the blacks, has been transformed into an extensive, well-built town surrounded by villas and replete with the products of European luxury.—I took my letter of introduction to Sir Thomas Mitchell. ⟨He received me quite pleasantly, but informed me that an expedition was out of the question for 12 months. I offered my collection of minerals to the Museum as a gift, where it was accepted with thanks; but, as the Museum, like every other museum that I've seen, was in disorder, and the question of moving it to another place is under consideration, they've not yet taken custody of my gift. Sir Thomas gave me a letter of introduction to Dr Nicholson, whose portentous name seemed to hold out some promise, but neither of them has taken any further notice of me.⟩ However, I've made the acquaintance of several families who are doing their best to help me to enjoy myself. Besides these, there are a few amateurs of science, people here and there who began to record observations and to make collections, and whom it may ultimately be possible to encourage into greater and more useful efforts.—The Botanic Gardens, if nothing else, are decidedly well laid out though they're little adapted to scientific study—this could lead to something, as the opportunity is there. The Museum has at least been established; but the Library, in some respects very well stocked, is unfortunately not so for natural science. Zeal is by no means wanting in this young country. There are times when people are at least willing to listen though they have not the patience to apply themselves to study. There's a School of Arts where public lectures are delivered, and

Reconnaissance in New South Wales

I have it in mind to make myself useful there. When Decandolle[b] began to give his lectures on botany in Geneva, he found that there was extraordinarily little public interest in the subject, yet his own enthusiasm worked such a change over his rapidly increasing audiences that very soon the whole of the city seemed to have been transformed into a college of botany. In science there have been many such instances of zealous, energetic men drawing those around them into the field of their own efforts, like magnets. If I can't equal them, I should at least keep their example constantly in mind. My general orientation and my acquaintance with nearly all branches of natural science [should] make it easy for me to find out about the dispersed efforts [that are being made] in this colony, which encourages me to think that I might succeed in coördinating them. I shall soon be turning to my own pet subject, but will do as much as I can to encourage interest in the rest.—They're so far ahead of us in Hobarttown as to be publishing a really copious journal,[c] of which 2 numbers have already appeared. Little work can be done in geology or mineralogy around Sydney because an impenetrably thick sandstone forms the surface of the country nearly everywhere and is of remarkable extent. Hunter's River, New Zealand and Port Phillip are more interesting by far. Illawarra is rich in well-preserved fossil mussels, which occur in a bluish, almost Lias-like clay. Were you to follow the road to Botany Bay with me, you'd find yourself among wave-like sandhills 2 miles from Sydney. They have no general alignment, and seem to consist inwardly of hard sandstone but to have been transformed superficially into loose, white, dazzling sand by gentle marine action. Processes unknown but perhaps still in operation eventually raised and exposed this [former] sea-bed,[d] which was then colonised by a peculiar assemblage of plants. You find no turf and few herbaceous plants here, but mostly a woody, shrub-like growth 4' to 8' high, much of it covered with creepers. The arrangement of the flowers is generally very striking, and many of the flowers themselves are of unusual beauty. It's still Autumn, though, and Flora is not yet in really rich attire. Innumerable species of insects cover the plants all over, just as the plants have covered the ground. Never have I contemplated the plant world as so crowded a foundation for the insect world. Even this, however, may depend partly on the season. If the Diptera become a burden through their prevalence, it's more on account of the number of individuals in some species, such as the mosquitoes and flies, than of the number of species of Diptera in general. At least, that's how I see it. You've only to ask your friends who've been in Sydney and they'll be quick to tell you wonderful tales of persecution by mosquitoes. It's perhaps because the hotter weather [here] incubates insect eggs more quickly, and even before they're laid, that you often witness the disgusting sight of fully developed maggots being discharged from the ovipositor and dropping on to the meat right before your eyes.—Butterflies are common, but they're not nearly so brightly coloured as those of South America. The Hemiptera exist in unusual diversity, the Orthoptera and beetles in greatest number. The eastern coasts of Australia suit the Mollusca particularly well. In Torres strait and down to 23° S. polyps have built a number of dangerous reefs, on account of which vessels bound for India and China are constrained to proceed southwards around New Holland when making the passage from Sydney to Batavia, Calcutta or Canton. The sea between the reefs is teeming with molluscs, annelida and crustaceans like an aquarium. The variety-hunter stationed there would be sure of a good harvest. The reefs break the force of the swell without impeding the regular ebb and flow of the tides, and the tropical Sun heats the water up to 22°–23° Réaumur, a temperature extraordinarily favourable for the development of marine life. Mr Stewart,[e] the surgeon at the Military Hospital, has paid some attention to the fish of Port Jackson, and, as he is a competent draftsman, he has executed a number of line and wash drawings of unidentified marine fish. He may not, of course, have many new species among them. Since an individual of limited means can achieve his purpose only partially or slowly, and the wealthy people of Sydney are so exclusively concerned with making money, it is not astonishing that so little is yet known about the rich world of marine life. The sea is not accessible to direct observation, or is only partly so. Fishes and molluscs have to be deprived of their freedom for us to examine them, and, as this also means removing them from their native element, it is impossible for us to observe the phenomena of their existence. Were I wealthy I would employ fishermen to construct artificial ponds, like those I saw at Count Rocca Romano's at Naples, which would enable us to increase our scanty knowledge of the behaviour of marine creatures by observing them in their own environment. —Sydney lies on two parallel ridges, which run out as points into the water, and in the gully between them. On the South-West they appear to

Reconnaissance in New South Wales

be joined by a transverse ridge. To the North-East they form the two points between which lies Sydney Cove. The cove is thus the part of the gully just mentioned which has been filled by the sea, and an insignificant stream runs[f] down into it. On the eastern ridge and on the point into which it extends into the harbour, the Government has laid out a park, part of which is reserved for the Botanic Gardens.—The park consists partly of big grass plots surrounded by Eucalyptus trees and Casuarinas, partly of rocky areas of sandstone covered with myrtaceous bushes, and partly of thick bush. Good roads and paths conduce to the enjoyment of this attractive place by carriage, on horseback or on foot. If you enter the Botanic Gardens and go down to the shore you can chip oysters from the rocks yourself, as you can all around this heavenly harbour, and they've a delicate flavour that would remind you of your own English 'natives'. You can amuse yourself looking at tropical plants in a banana grove, or listen quietly to the hissing rattle of swaying bamboo canes, near which the long strands of weeping willows hang over a still pond. On the other hand, should you go strolling under the white-barked Eucalypti in the wilder part of the park, your attention will be caught, now by the rustling sound of lizards (at least two species) darting away, now by the striking colours of birds, pretty well all of which make up in richness of colour for what they lack in song. Reptiles are common around Sydney but there are not many birds. The ornithologist must go to the Blue mountains or to Hunter's river, where his harvest will be abundant. Indigenous mammals, even to the phalangers, are now rare around Sydney. White youngsters have taken their place.

If only we could go walking in this park together on some moonlit night! The pleasure-loving streak in human nature might for the moment restrain even your zealous efforts on behalf of the welfare of your fellow Londoners. You'd look the full Moon in the eye; you'd hearken to the sound of the cicadas and the crickets; your eyes would sweep over the blue mirror of the water to the dark masses of trees that frame it; in this mild weather your whole body would respond to a deep sense of well-being. You'd shout out 'Here let us build a Tabernacle!'[g] And if the Moon is not shining, there's the clear vault of the sky over us, full of brilliant constellations. Every night the Centaur, the Southern Cross, the ship Argo with Maia Placida and Canopus, Canis Major and Minor and Orion, Taurus and Gemini march across the sky; and every night the Magellan Clouds arouse my sense of wonder.—It's very hot during the day but the air is so dry that the sweating skin dries quickly. This in turn keeps the skin cool and makes the heat far easier to bear than the moist heat of China or of the East or West Indies. This rapid cooling, on the other hand, has the disadvantage of causing chills and attacks of rheumatism, especially amongst people who have just arrived from Europe. They are greatly inclined to move briskly, as they would in our colder North; a heavy sweat breaks out through every pore and soaks into their clothing; they sit down without changing their clothes, and the rapid evaporation brings its consequences. The rapid changes of temperature have similar effects. Whilst the Sun expands the mercury column to its maximum at about 2 pm. the clear night sky so facilitates the radiation of heat absorbed by the Earth's surface that the cooling off usually precipitates a heavy dew. The difference between maximum and minimum, according to observations being made at South head (where the tempering effect of the sea must, however, be more marked than in Sydney), is nevertheless only $6°-8°$ Fahrenh. (2.30 pm. and 9 pm.). During the Summer it is probably greater. Amongst the physical relationships that are the reverse in New Holland of what they are in Europe, is the occurrence, confirmed by observation, of rainfall with increasing barometric pressure. I was disposed, at first, to doubt the validity of the interpretation, but I have myself seen it confirmed several times and in a very striking way.

Do excuse me if I have taken up so much space in telling you things you know; for you may have heard and read all that I've been describing, long ago. But I just wanted to let you see how I'm gradually settling down in this new world. Every one of us has his own peculiar way of viewing things and he interprets them accordingly. In two months I may have more to tell you. And I would certainly be delighted should I hear from you in 6 or 7 months. Questions from you would direct my attention to things I've not considered, and you might find the answers useful. Keep an ear cocked all the time, and don't miss a single word about New Holland. And I have one request to make: Letters to and from Germany must be sent through England. Do permit me and my relatives to correspond through you. My German correspondence is so infrequent, I regret to say, because there is no postal agreement between Prussia and England, and I must refrain from putting my people to expense—which means that you shan't be troubled with a letter more than once in two months. Good-bye, my friend; and please remember me to your dear wife and children, and do not forget

Yours most faithfully
Ludwig Leichhardt

⟨[P.S.] My address is Mr Ludwig Leichhardt, Bligh Street at Mr Marsh's, Sydney. If you see Owen[h] give him my compliments so that he doesn't forget me again like he did before.⟩

52 [TO MARK NICHOLSON, ESQ., Port Phillip, New Holland]

[Sydney,] 10 April[, 1842]

My dear friend

I had despaired of receiving a letter from you, as I supposed, that you had left Pt Philipp for England. Imagine my joy, when I found a letter to *Ludovic Leichhardt* on the table—the first letter which I received per post in this colony.—I told you, that I had the intention of becoming tutor to a family or of uniting 6–8 pupils to teach them classics mathematics and modern languages. I commenced to do so! At the same time I followed up my science and I saw soon, that these two occupations clashed against each other. I had one pupil. He was so lazy, so tiresome, that I could not bear it any longer and I sent him off! Do not laugh do not critic! I felt, that to me teaching comes immediately after being a shepherd and that I must have run through the whole series of occupations before returning to the office of schooling boys! I made frequent excursions gathered plants, studied their organisation, gathered insects and everything particular which came my way. I worked as hard as ever and my eyes were very much affected these last days in consequence of my continually examining plants through a powerful magnifying glass.—Sir Thomas Mitchell invited me to dinner party, where I was kindly treated, but were [*sic*] my prospects were little bettered. I was even introduced to Deas Thompson[a] the colonial secretary, to a Dr Nicholson, a very rich young man, but very different from his namesake. I know a number of influential men and I have even the intention of seeing the governor, which I shall be able to effect by Colonel Barney,[b] to whom I am especially recommended. A museum did once exist in Sidney; but it is now in such a state of confusion, that I was not able yet to catch a glimpse of it, in fact I believe nobody knows at present where it is to be found. If it was the intention of the governor to establish a national institution to appoint a curator of the museum who [....ed] at the same time the obligation of giving public lectures—I should make the application and I have the chance of obtaining such a situation. Here is a botanical garden, well situated, but of very little use; the gardener[c] himself knows very little about botany and does not think of giving public instructions. Dr Nicholson told me, I might have a chance of being appointed Director to this garden!— You see there are many possibilities, but few realities.—There is another very important point, which has rendered this colony very interesting to me. I am desperately in love! Think of little D. and be silent! I came out with a family, which consisted of very nice people and of an aimiable [sic] young girl. This girl has taken lodgings in the very centre of my heart and as my present situation is rather unsettled and I would by no means entangle a girl in my own difficulty I can only spread my passions to the winds or cut a mournful tale in the bark of a gum tree, which will give me at least a gummy tear of compassion. ⟨Though this inclination does not interfere with my study, as I have self control enough—at least for the present—to keep it at a certain level, it makes me certainly uneasy about my situation and I wished to employ myself in one way or the other, to fix it. [*From margin:* Ich habe ihn nichts von meiner wundervollen wandelvollen Neigung getheilt. Ich kenne mich von Tag zu Tage selbst nicht und ich will mich nicht vor ungütigen Bekenntnisse scheuen] (I told him nothing of the wonder and range of my feelings. I hardly recognise myself from one day to the next and unfavourable responses shan't make me lose heart)⟩. Many thoughts have crossed my head and it would be foolish to leave those very founded expectations of which I spoke you before, to enter into a different career altogether. To understand me well I must give you the following explanations. If I am unable to obtain a situation which renders possible for me to give myself entirely to my science, I must try another way to make my living. I could follow up medicine and I assure you that, though not ordained I should make a better Dr than many here in Sidney. But the people know that I have not taken my degree, as I told them so myself and my disposition is of such a kind, that I fear the interest which I take in the patients, the continual excitement which it would produce, would destroy very quickly my health.—On the other hand I was from my earliest childhood particularly fond of horticultural pursuits. I did not only cultivate the vine for obtaining grapes but I pressed the grapes and made wine. This excited great interest for the same occupation wherever I saw the vine cultivated as in France, in Italy and in Switzerland and I gathered interesting knowledge on it in these different countries. Here in Sidney I hear continually spoken of it. I am however convinced that the soil of Sidney is not adapted for the grape, as it requires either a calcareous or a volcanic rock to give wine of superior quality—that however a loamy soil would give good grapes, which fetch a good price in Sidney. I thought of Hunters river where the variation of the soil is greater and where alluvial beds cover the sands, which cover every where the sterile rocks round Sidney. I thought moreover of Pt Philipp, where volcanic rocks and limestone are to be found and where at least the possibility would be given of changing the pattern of the soil by mixture—⟨You see, Love drives me to the wine!⟩ I am convinced that such an occupation would soon give me an independence, though not so large, as might be wished. But as my returning home is by far more distant than yours, the slow increase matters little, if it is only sure. I shall certainly try this, as soon as I see, that there is no hope in Sidney or from the governor. But it is clear, that, as long as these hopes are not destroyed, I prefer to keep them, and yet I feel that my eyes would be more content with the cultivation of the vine than with continual application of the glass.

And have you not found my dear friend, men who took some interest in your fate and whose acquaintance made the separation from home less mournful? I found several men of good education, of much knowledge and of kindness of heart, whose conversations have often comforted me. I was introduced to a countryman of mine, Mr Kirchner, in whose family I was very kindly received. Mr Lynd[d] the Barrack master is my companion in the botanical studies; Captain Marlow of the Royal Engineers treats me like his own son—(in fact he is the father of my little maid).[e] I am introduced to an immensity of persons whose name I am even unable of recollecting, which embarrasses me often enough.—If there was any possibility [sic] means to come to Port Philipp on public expense, I should certainly come to see you. I had the intention of coming with the Thetis, but William thought it better to go directly to Sidney, in fact emigrant ships are, I believe, never bound to both places. If you are seriously ill and if you want a friendly hand to have care of you—write and I'll be there at the return of post even at my expense, for you know that 150£ are in the savings bank. I wished well to leave them there for ever; but I fear, that this will be impossible. You imagine perhaps, that your mode of living would not please me, but you are very much mistaken my dear friend! In the contrary I should delight in it particularly if there are many plants about you, and many singular insects, snakes and lizards slugs and snails. And if all is right and I see a little clear about my business I shall certainly come and stay with you a month or two. Write me at what time of the year the greatest number of plants are in blossom, give me a little description of your place, Marks cottage and Marks hut—tell me about the stones which you find in the fields, whether they are all of that porous or cellular structur, of black colour with whitish substances intermixed[f]—whether you have any limestone, any fossils? I read in the newspapers, that they are arranging a botanical garden and a library in Melbourne. If there is no appointment here, it may be there. However the source of application is always the governor; I shall see him next thursday and I'll tell you of my interview in the next letter.

[Neither signed nor initialled.]

Reconnaissance in New South Wales

53 [TO DR WILLIAM NICHOLSON, Bristol]

dn 17tn Mai 1842

Mein theuerer Freund

Jenen Brief, den du mir nach Cork geschrieben empfing ich 4½ Monat später in Sidney. So trennten uns vielleicht ein bis 2 Tage frühere Abreise um eine lange Zeit. Deinen 2tn Brief erhielt ich vor 14 Tagen. Ich selbst habe an dich bei meiner Ankunft geschrieben und würde bereits einen 2tn Brief geschickt haben, wenn ich nicht, wie du den Verlauf eines Dramas hätte abwarten wollen. Wie ich dir bereits geschrieben, versetzte mich die Neue Natur, welche überall von den Ufern Pt Jacksons anlachte in ein Freudenfieber, aus welchem mich selbst die weitern Excursionen nicht hervor brachten, die mich zu den sandigen, unfruchtbaren, freudelosen Hügeln gegen Botany bay zu führten. Ich mußte mir indessen bald gestehen daß die Zone reicher Vegetationen sich nur gerade so weit erstrecke, als der kühlende Meeresathem die Trockenheit der Luft mäßigte nd der raschen Absorbtion eines losen sandigen Bodens widerstand. Herr Marsh, der in Clifton einst dein Nachbar war, bot mir an mit ihm zu wohnen, da er ein großes Haus theuer gemiethet nd von mir einige Erleichterung hoffte. Er gab mir ein kleines Zimmer, ähnlich dem welcher ich zuletzt in Paris bewohnte nd da ich mich leicht begnügte nd die kleine Wohnung so ordentlich nd behaglich als möglich einrichtete lebten wir recht glücklich zusammen. Indem ich nun dachte, daß ich hier wenigstens für eine längere Reihe von Jahren angesiedelt, fühlte ich mich ruhiger befestigter, heimathlicher, als selbst in Paris. Denn es war ja seit 6 Jahren, daß wir mit einander von Ort zu Ort gewandert, gleich den Zugvögeln nur mit dem Unterschiede, daß wir wie mit einem Weibchen zum Nestbau gekommen waren. Selbst die neue Neigung zu einem jungen Mädchen machte Geist und Körper elastischer, energetischer, während nur die Achtung wohl that mit welcher mich diejenigen behandelten, denen die Umstände mich nahe geführt. Das Streben mich mit einer so fremde Natur vertraut zu machen führte mich nun hinaus in die Busch und während ich draussen mit vollen Armen begierig ein reffte, Pflanzen nd Insekten sammelte un geologische Merkwürdigkeiten mir aufzeichnete, war ich zu Hause mit Pflanzen trocknen, Ordnen der Insekten, ausarbeiten der Noten, Lesen von Büchern in Bezug auf die Kolonie beschäftigt. Diese Beschäftigungen nehmen meine Zeit auf eine so völlständige n angenehme Weise in Anspruch, daß ich mit Widerwillen mich loßriß einem ungezogenen unaufmerksamen Knaben für 2–3 Stunden Unterricht zu ertheilen. Ich trug es indessen eine Weile abr als ich später sah, daß mir die Eltern nicht zu Hülfe kommen konnten ud der Knabe es nicht wollte, gab ich ihn auf, sowie die Idee mit Schulmeisterei mich weiterhin zu beschäftigen nd lebte nun für einige Zeit so recht con amore meiner Wissenschaft in die Arme. Auf der andre Seite hatte ich dem lieben Mädchen den Vorschlag gemacht, ihr französischen Unterricht zu ertheilen, da sich indessen ihr Unterricht nicht von dem jungen Schwestern trennen ließ, mußte ich für Eine gleich dem Guten Erzvater Jacob[a] nicht eine Zweite sondern so gar 3 andere in den Kauf nehmen. Daß ich mich glücklich fühlte, einem holden geliebten Mädchen gegenüber zu sitzen nd sie das Runden der Lippen für das französische u nd eu zu lehren, habe ich dir nicht nöthig zu sagen. Doch leider bemerkte ich bald daß der Eifer mit welchem ich auf fleißiges Lernen drang sie .ßrst unbehaglich machte nd daß ihr, wie den Schwestern die französischen Stunden wenig Freude gewährten. Da ich nun einmal nicht Lehrer für faule Schüler oder Schülerinnen sein kann od. sein will, warf ich trotz der Annehmlichkeit die ganze Geschäfte zur Seite. Während sich dieses im Laufe zweier Monate ereignete, wurde ich bei den bedeutendsten Männern der Kolonie eingeführt, n von einigen recht freundlich empfangen u. behandelt. So hat ein wunderliches Glückspiel einen Dr. Nicholson, einen junge ausserordentlich begüterte Mann zu meinem Gönner gemacht. Lieutenant Colonel Barney un seine Familie sind .ßerordentlich freundschaftlich. Mitchell an welchem ich einen Empfehlungsbrief mitbrachte hat mich zwar einst zum Essen eingeladen, doch hält er sich im Ganzen wahrschl. wegen zu großer Beschäftigung sehr zurück. Vor vier Wochen wollte es nun das Geschick, daß der Superintendent des botanischen Gartens starb und mehrere befreundete u einflußreiche Männer machten mir den Vorschlag mich um die Stelle zu bewerben. Der botanische Garten war früher unter Aufsicht des Botanikers Cunningham[b]; als dieser wegen zu niedrigen Gehaltes nd wegen unfreundlicher Behandlung von Seiten Sir Richard Bourkes, des vorigen Gouverneurs, die Stelle niederlegte, gab man sie einem gewöhnlichen Gärtner, einem ganz unwissenschaftlichen Manne, .lcher nun den Garten vollständig zum Küchengarten von Government House niedersinken ließ. Der Mann stand sich etwa 250£; doch wie er starb, war selbst dieß dsem Gouvernour zu viel, man reduzirte die Stelle auf 120£ und wollte kaum auf eine neue Besitzung derselben eingehen. Ich hatte meinen Kopf voll von wissenschaftl. Möglichkeiten und ich hatte die Stelle vielleicht selbst mit einem niedrigen Gehalte angenommen um einen sichern Stützpunkte zu gewinnen nd von dort aus mich bekannter zu machen. Doch das Interesse eines Hrn. Macleay[c] war stärker als das Meinige. Er war früher Colonial secretair gewesen nd hatte so einen bedeutenden Einfluß gewonnen. Sir Richard Bourke hatte ihn von diesem Posten entfernt nd seinen Schwiegersohn Deas Thompson damit bekleidete. Von der Zeit an zeigte der alte Macleay sich stets in festigster Opposition gegen den Gouverneur. Nun hatte er erfahren daß der Gouv. meiner Bewerbung geneigt war nd überdieß hatte er einem befreundeten Gärtner das Versprechen gegeben, ihm die Stelle zu verschaffen. Da er nun das bedeutendste Mitglied der Committee des Bot. Gartens war, dem die Besitzung der Stelle vom Gouverneur überlassen war, so gewann er natürlich leicht den Sieg nd ich erhielt eine abschlägige Antwort. — Wie ich allmählig aus meinem Natur taumel erwachte fühlte ich immer drängender, daß ich auf diese Weise durchaus nicht materiell produktiv wurde nd daß indem ich auf eigene Kasse lebte, endlich eine Zeit kommen müßte, wo ich vielleicht an money making zu denken hätte. Schon in Paris un England vor meiner Abreise hatte ich voraus gesehen, daß mich eine Unterstützung von deiner Seite vielleicht würde fort studiren machen wie ich es in Paris gethan, während eine vollständige Entblossung mich augenblicklich würde zum Broderwerb wieder getrieben haben. — Obgleich diese Betrachtungen mich vielfach quälten, konnte ich nicht leugnen daß mein Studium und die freie unabhängige Lebensweise mich mit einflußreichen Personen genauer hatte bekannt werden lassen, die allmählig finden, daß ich auch praktische Kentnisse besitze und daß ich mich auf vielfache Weise nützlich machen könne. Dieß giebt mir Ruhe und Hoffnung bald in dem Neuen Boden sichere Wurzel zu fassen nd vielleicht kann ich dir schon im nächsten Brief angenehmere Nachrichten darüber mittheilen. — Ich schrieb zwei Briefe an Mark und ich verzwerfelte, je von ihm Antwort zu erhalten. Endlich kam ein Brief von ihm und meine Freude war, wie du dir leicht denken kannst, nicht gering. Seine Lage in Bezug auf gesellschaftlichen Zustand ist keines weges erfreulich. Ich würde mit Freuden ihn besuchen, wenn ich einerseits über meine Zukunft irgend wie sicher wäre nd anderseits nicht mit eig. Gelde zu Werke gehen müßte. Das Leben im Büsche ohne wissenschaftliche Kentnisse, welche ihm vergönnen die Natur in ihre tiefe Brust wie in den Busen eines Freund's zu schauen, ist ausserordentlich ermüdend — und die Erinnerungen an die fernen geliebten, an die theure Heimath zehren an dem Unglückliche gleich dem Adler an der Leber des an den Caucasus geschmiedeten Prometheus.[d] Ich habe mit mehrere jungen Männer gesprochen deren hohle Augen un eingefallenen Wangen keineswegs eine freudevöllig genußreiche Existenz verräthen, und die Beschränktheit ihrer Ideen, welche stets nur zwischen sheep un cattle schwingen erfüllte mich oft mit .ßrordentlichen Mitleiden. So habe ich in diese Kolonie schon manche andere interessante psychologische Beobachtungen gemacht und so abstoßend viele Verhältnisse sein mögen, das beobachtende Auge findet sich überall wohl, wo es eben nur beobachten kann, sollte es auch gleich unserm alten Plinius^e von Aschen schauere mit ewiger Nacht bedeckt werden. Mit dir mein theuerer Freund treibe ich es fast, wie ich es früher mit dem Andenken an Mädchen trieb, die einen tiefern Eindrück auf mich gemacht hatten. Ich idealisirte mir das freundliche empfangene Bild und da meine Kurzsichtigkeit mich vor neuen stärkern Eindrücken mich schützte, konnte ich Jahre lang vor einer idealisch ausgeprägten Madonna... liebdienen, mein Herz war voll, meine Neigungen hatten eine bestimmte Richtung. Jene Augenblicke, in denen die Hoff‑ nung auf kunftigen Besitz aufgegeben werden müßten waren schmerzlich und die Zeit ohne bestimmte Neigung machte mich kalt unbehaglich nd ziellos. Dieß ereignete sich indessen nur zweimal nd die Zwischenräumen waren gewöhnlich kurz, da es wahr scheint daß das menschliche Herz nicht ohne Liebe existiren kann. Dieß zeigt denn auch dein Beispiel, der du in Paris, Neapel, Rom, Florenz (und die *betweens*?) vielfach angegangen würdest, in dem dein scharferes Auge jeden Reiz schneller erfaßte und die hoffnungslosen Erinnerungen an frühere Reize sich gegen den Neuen weder wehren konnte noch meistern. — Wie die Erinnerungen an ein Mädchen mich von der Neigung, andere zu gefallen zurück hält, so hält mich nun die Erinnerung an dich mein theuerer Freund von andern neuen Freundschaftsbündnisse zurück. Ich lebe mit mehreren jungen Männern vertraulich, doch nur du gehst auch hier in Sidney in mein innerstes Gedanken geben mit ein. Ich habe dein Bild in schönen lederholz Rahmen über meinem Arbeits tische und von jeder Excursion bringe ich für dich die schön‑ sten vollsten wilden Blumenbüchsel zurück. So stehest du jetzt unter dem dichten Schatten von hangende Casuarinas, Acacias, den großen gelben Blüthen cylindern der Banksia, von Persoonias, Leptospermum, Epacris grandiflora, Styphelia tubiflora, Lambertia formosa. Du würdest dich über die poetische Combination gewiß freuen, obwohl du nie große

Neigung zu Blumengewinde verriethest. Oft stehe ich im Abend mit dem Lichte in der Hand vor diesem meinem Penatenaltare nd indem ich die Beleuchtung wechsele nd die Schatten verschieden über das Zimmer fallen lasse, scheint es als wenn du Lebend aus ihnen hervorträtest, um mit dem fernen Freunde über die Vergangenen Zeiten zu plaudern.
Die Aussicht dein Microscop zu erhalten gab mir große Freude. Gerade Dieß vermisste ich ausserordentlich, vielleicht nicht, daß ich mehr damit gearbeitet hätte, denn jeder Augenblick war wohl beschäftigt, sondern wegen der Möglichkeit damit arbeiten zu können. Du verstehst mich! der unzufriedener Geist arbeitet so oft in voraus und ärgert sich in voraus über die möglichen Hemmungen, die ihn erwarten mögen. —

[*Sydney,*] *17th May, 1842*

My dear friend,

That letter you posted to Cork reached me in Sydney 4½ months later, so the advance of departure by one or 2 days very likely put us out of touch for a long time. I got your second letter a fortnight ago. I wrote to you myself on my arrival and would have written again by now, were it not that I, like you, have been waiting to see how the play is going to turn out. As I've already told you, the new kind of world which greeted me from all sides around the shores of Port Jackson induced a state of feverish delight from which I've not emerged in spite of my more extended excursions to the sandy, barren, desolate hills over towards Botany Bay. However, I soon had to admit that the zone of richer plant growth did not extend beyond where the cooling breath of the sea moderated the dryness of the air and counteracted against the rapid absorption of a loose, sandy soil. Mr Marsh, who was once your neighbour in Clifton, took a big house at a high rental and invited me to lodge with him, hoping that it would relieve him of some of the expense. He offered me a small room like the last one I had in Paris. Because I was easily pleased [in Paris] and furnished my room as simply but as comfortably as possible, we were really very happy there. Thinking that I would be living here for much longer than we lived there, I felt more at peace, more secure, more at home even than I felt in Paris. For you and

Reconnaissance in New South Wales

I had been wandering from place to place for six whole years, like migratory birds, except that at nesting time we could only act as if we had mates. A new interest in a young woman has been enough to make me more resilient, more energetic in mind and body than I was in Paris; and the mere consideration with which I have been treated by those with whom circumstances have brought me into touch has done me good. My desire to become familiar with the strange world around me soon sent me out into the bush. Out there, I bagged things by the armful, eagerly collecting plants and insects, and making notes on matters of geological interest. Back in my room I've been busy drying plants, sorting insects, working up my notes and reading books about the colony. My time was so completely and pleasantly taken up by these activities that I resented having to waste 2–3 hours in teaching an ill-bred, inattentive boy. I did stand it for a while, but I soon saw that the parents could not, and the boy would not make things easier, so I gave him up, and with him the whole idea of doing any more teaching. For a time I was quite happily wedded to my studies. However, I had offered to teach French to the girl I love; but, as she could not be taught apart from her younger sisters, I was obliged, like the patriarch Jacob, to accept not merely another,[a] *but 3 others into the bargain. I've no need to tell you how delighted I was to sit opposite this dear and lovely girl teaching her how to round her lips to say the French u and eu. Unfortunately, I soon saw that my insistence on hard work made her extremely ill at ease, and that she, like her sisters, was deriving very little pleasure from the study of French. As I can't and won't stand teaching lazy boys and girls, I gave up the whole scheme in spite of its attractions. During the two months that went in this way, I was introduced to the leading men of the colony, some of whom received and treated me with great kindness. Through extraordinary good luck a Dr Nicholson, a young man who is very well off, has become interested in me. Lieutenant-Colonel Barney and his family have made me very welcome. Mitchell, to whom I presented a letter of introduction, has had me to dinner just once, but has been very reserved otherwise, perhaps on account of pressure of work. Four weeks ago the fates claimed the life of the Superintendent of the Botanic Gardens, and several well-disposed and influential people advised me to apply for the position. The botanic garden was at one time under the care of the* botanist Cunningham[b] but, as he resigned from the position on account of the poor salary and his unsympathetic treatment by the former Governor, Sir Richard Bourke, they appointed an ordinary gardener, a man without scientific knowledge, under whom the garden became nothing but a kitchen garden for Government House. His salary was about £250; but when he died the present governor thought that even this was too much, so it was reduced to £120 and they even temporised about filling the position at all. My mind was teeming with the scientific possibilities and I would have accepted the position even on a low salary for the sake of securing a point of support from which I could make myself better known. However, a certain Mr Macleay's[c] interests proved to be stronger than mine. As a former Colonial Secretary he had acquired a good deal of influence. Sir Richard Bourke had deprived Macleay of this office and had appointed his own son-in-law, Deas Thompson, in his stead. From then on, old Mr Macleay has displayed the most vigorous opposition to the Governor. He found out that the Governor favoured my application; moreover, he had promised a gardener of his acquaintance that he would secure the appointment for him; and, as he happened to be the most influential member of the Botanic Gardens Committee, to which the Governor had delegated the authority for appointment, he of course carried the day with ease and my application was rejected. As my excitement over the new manifestations of Nature began to subside, I became more and more aware of the fact that what I was doing was not bringing anything in, and, as I was living on my capital, sooner or later a time would come when I would most likely have to consider how to make money. Back in Paris, and in England before my departure, I could see that if you continued to support me I would very likely go on studying, just as I did in Paris; and I also saw that, were I deprived of all support, I would have to set about earning my daily bread again at once.—Although such considerations have often made me apprehensive, I maintain that it's my studies, my independence and my freedom that are enabling me to make myself better known to influential people. These people are gradually finding out that I also have practical qualifications and am capable of making myself useful in many ways. This is enough to calm me, and to give me hope that I may soon take firmer root in this new soil. I may be able to say something more

Reconnaissance in New South Wales

promising about it in my next letter.—I wrote twice to Mark, and gave up hope of ever hearing from him. But at last I got a letter, to my no small joy, as you can easily imagine. So far as companionship goes, he's not in a happy position at all. I'd gladly go to see him if, on the one hand, I could somehow be sure of my own future, and if, on the other, I did not have to look at every penny. Life in the bush, without that scientific understanding which enables one to sense the deep warmth of Nature like the heart of a friend, is very exhausting—and the thought of those left behind, memories of home, gnaw at the vitals, like the eagle tearing at the liver of Prometheus chained down on a rock in the Caucasus.[d] *I've met a number of young men whose hollow eyes and sunken cheeks hardly betoken a life of pleasure and indulgence; and the limitation of their minds, without interests beyond sheep and cattle, has often aroused a strange sense of pity in me. I've made quite a number of other interesting psychological observations in this colony, and, repulsive though much of the traffic between people may be, the seeing eye is satisfied anywhere, so long as it can see, even if, like Pliny the elder,*[e] *it be overwhelmed with ashes and overtaken by eternal night.*

As for you, my dear friend, I find myself doing just what I used to do about girls who had made a deeper impression on me. I used to idealise their charms; and, as my short-sightedness insulated me from others even more attractive, I was capable of remaining true to my vision of an exquisitely minted Madonna for years on end. My heart was full and my feelings had direction. The times when I had to renounce the hope of ultimate possession were harrowing, and in periods when I was deprived of such attachment I felt cold, uneasy and aimless. As a matter of fact, this happened to me only twice, and the intervals, as with most people, were short, for it seems to be true that the human heart cannot live without love. Even you yourself are proof of it, for in Paris, Naples, Rome, Florence (and in between?) it was your sharper eye that first noticed the pretty girls, and not even the hopeless recollection of past disappointments could make you proof against new attractions.—Just as the memory of one girl *has been enough to restrain me from courting others, my recollections of you, my friend, are making it hard for me to cultivate new friendship here. I've come to know several young men very well, but even here in Sydney you are the only person who can follow the* innermost workings of my mind. Your portrait is hanging in a handsome frame above my work table, and after every excursion I come back laden with tribute of the finest and fullest inflorescences of the native plants. At the moment, you're standing under the heavy shadows cast by drooping Casurainas, Acacias, the big yellow cylindrical blossoms of Banksias, by Persoonias, Leptospermum, Epacris grandiflora, Styphelia tubiflora and Lambertia formosa. You would certainly approve of their artistic arrangement although you never showed much interest in floral display.—In the evening I often stand in front of this altar to my household gods, holding the lamp; and when I move the light and make the shadows go wheeling around the room, you seem to come to life, as if you were going to have a yarn about old times with me.

I was delighted with the prospect of getting your microscope. It's just what I've been wanting so badly. Not, perhaps, that I would have done more if I had had it, for I've not had a spare moment; but on account of what it will enable me to do. You know what I mean—a restless mind works in advance, and worries in advance over the difficulties that might arise.

54 AN HRN. DOVE[a] in Berlin mit den Wetterbeobachtungen von South Head.

d. 27tn März [for Mai,] 1842

Mein hochverehrter Herr

Als ich m Jahre 1836 in Berlin war, hatte ich das Vergnügen ihren öffentlichen Vorlesungen über Meteorologie bei zu wohnen. Sie lenkten zuerst meine Aufmerksamkeit auf die Veränderungen der Atmosphäre und auf die Gesetze, denen sie unterworfen sind. Auf meinen Reisen in Europa unterließ ich nie nach der gelernten Weise mir die Erscheinungen zu erklären. Wie ich zu den Antipoden nach New South Wales herumschiffte hatte ich vielfache Gelegenheit Neue un ab, weichende Verhältnisse zu beobachten nd hier in Sidney erinnerte ich mich ihre Windtheorie und fand, daß sie wie erwartet auch hier, nur in umgekehrter Richtung gelte. Am Eingange des schönen buchtenreichen Port Jackson hat man

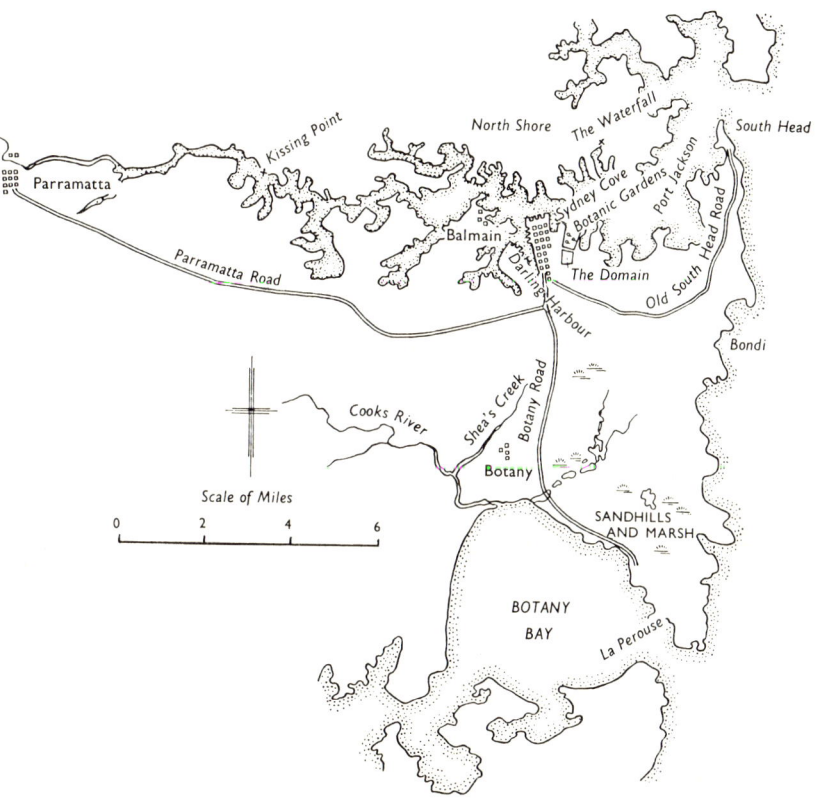

Fig. 6. Surroundings of Sydney, N.S.W. in 1842.

seit längere Zeit [f die Schiffsleute?] meteorologische Beobacht‚
ungen angestellt nd nach den wöchentlichen Beobachtungen
zu ertheilen, die ich zum Theil bewahrheiten konnte, sind
diese Beobachtungen sorgfältig ud deßhalb .ßerst schätzbar.
⟨Ein Englischer Prediger Hr Clarke[b] hat gleichfalls sich mit
der Meteorologie von New South Wales beschäftigt nd er hat
.ßerdem eine Reihe von Aufsätzen im Sydney Herald drucken
lassen, welche zum Theil auf den Vergl. mehrerer Logbücher
beruhen. Die Abhandlung .lche ich ihnen sende ist vielleicht
eine der bessern. Ich muß bekennen, daß der Mann mir zu
[verwirrt?] erscheint, denn in unsere Unterredungen zeigte
sich oft eine .ßerordentliche Ungewißheit nd Unentscheiden‚
heit selbst über Dinge, die er bereits geschrieben. — Er ist mit
einer... Bearbeitung der Natur verhältnisse von Neu Holland
beschäftigt.⟩ *From margin:* Ein Englischer Prediger Hr Clarke
hat sich gleichfalls mit der Meteorologie von New Holland
beschäftigt und hat .ßerdem eine Reihe von Aufsatzen im
Sydney Herald drucken lassen, welche zum Thl. auf dem
Vergl. mehrerer Logbücher beruhen. Er suchte zu beweisen
daß alle Gewitterstürme (Squalls) von S nach W N E
herumgehen während doch der regelmäßige Windwechsel
von S nach O N W ist. Ferner wünschte er dar zu thun daß
die Südost winde welche er als vorherrschend annahm, die
beginnenden Passaten wären. Die Beobachtungen auf South
Head beweisen indessen daß die Südost winde nicht die
vorherrschenden Winde sind, sondern die N Ostwinde. Was
den ersten Punkt betrifft, so ist es wahr daß ein Wind wechsel
von West .ch Nord stets mit Gewitterstürmen oder mit heftigem
unregelmäßigen Wetter verbünden ist; doch glaube ich nicht
daß alle Gewitterstürme diese Bahn folgen, sondern oft von
localen Ursachen abhängig sich ihre eigene Bahn wählen. Für
den 2tn Punkt führt Hr. Clarke die Beobachtung an, daß eine
Menge von Bäumen sich nach NW neigen. Dieß ist indessen
auf keine Weise so allgemein ud wahrschl. von localen
Ursachen abhängig. Ferner wünschte er zu beweisen, daß die
heißen alle Vegetation verdorrenden N Wln. Winde, welche

von Zeit zu Zeit besdrs. im Sommer gewöhnlich auf kurze Zeit in Sidney wehen, die über den Aequator hinüberreich͵ enden von ihre Bahn abgelenkten Nlchn. Passate waren. Ueberdieß ist sein Kopf voll von Strudeln nd Wind[arten] wie man dieß fast aus jener sonderbaren Behauptung schliessen möchte. Er glaubt daß das Innere von Neu Holland keine Wüste sei, da die heißen Winde so elektrisch sind, daß sie selbst die Magnet nadel affiziren, nd da die Bildung von Gasen und Wasserdünsten immer über einem Wasserbecken oder einer mit Vegetation bedeckten Ebene statt finden nd jene Gasbildung die einzige Quelle einer atmosphärische Elek͵ trizität ist, so setzt er entweder einen weiten Salz see oder mit Vegetation bedeckte Flächen voraus. Doch die .ßerordentliche Trockenheit dieser Winde, .lche wunderbar schnell alle Feuchtigkeit absorbiren, streitet gegen den Gedanken daß die Lüfte sich ursprünglich über einem Binnensee erheben nd die Affektion der Magnet nadel wird sich wohl v dem Vorüber͵ ziehen eines continuerlichen warmen Luft stromes erklären lassen. —

Jene heißen Winde machen sich auch in Van Diemensland fühlbar nd zwar in Hobarttown, während Launceston von ihnen frei bleibt. Ebenso hat man sie an einigen Orten der Ostküste von Neuseeland verspürrt, die doch durch hohe Bergketten gegen den West un Nordwestwinde geschutzt ist. Ich glaube daß man es hier überall mit zwei verschiedenen Erfahrungen zu thun habe. Die heißen Winde von Sidney scheinen mir wahre Wüsten winde, welche der Nordwestküste von Neu holland ins Innere hinein wehen die trockenen Lüfte einer weiten Wüste mit s. hinwegführen nd .f der Ostküste v Neuholland hinaus treten. Man hat heiße Winde um ganz Neuholland hin wahrgenommen nd dies ist natürlich. Denn von jeder Seite kann ein starker Wind über ganz Neuholland hinweg wehen. Die heißen Winde in Neu Zeeland sind noch nicht hinlänglich beobachtet; man weiß nicht ob sie [wirk]͵ lich so trocken sind. Es ist möglich daß sie dem sich senkenden zurückkehrenden Equatorial strome angehören. Das partielle

.ftreten des trockenen Windes in Van Diemensld. ist wahrscheinl. von Oeffnungen in den Gebirgsreihen des Südöstlichen Theiles von New holld. abhängig. In Sidney folgt die Nd wst. winde unabänderlich ein heftiger Süd u Süd Westwind, der den losen Sand der Hügel gegen Bot bay in dichten Staubwolken zur Stadt bewegt. Man nennt diesen Wind Brickfielder, da er von den brick fields zur Stadt weht. Er ist kalt nd der Wechsel der Thermometer ist bis weilen .ßerordentlich. — 50°–60° F. *End of marginal additions.*]

Als ich die Bekanntmachung der jährlichen Meteorol. Veränderungen im Sydney Herald fand, glaubte ich dem alten Lehrer eine Freude machen zu können, wenn ich ihm dieselben schnell zu sandte obwohl es wahrschl. ist, daß diese Beobachtungen von der Englischen Regierung an den verschiedenen Akademien des Kontinents versandt werden. Ich habe mich in Europa mehrere Jahre wissenschaftl. vorbereitet, um in den andern Erdhälfte durch Beobachtungen nützlich werden zu können; meine Aufmerksamkeit war indessen mehr auf Zoologie nd später auf Geologie gerichtet. Herr Professor W[ie]gmann wird sicherlich noch meines Namens erinnern. Ich notire indessen sorgfältig jede abweichende Erfahrung nd will es Gott, so kann ich Ihnen in Jahr und Tag vielleicht eine nützliche Arbeit in die liebe Heimath senden. Ich werde wahrscheinl. für mehrer Jahre in Sidney bleiben, nd da ich weiß, wie fruchtbringend mir Winke von den Meistern der Wissenschaft sein können, sehe ich mit Freude Briefen mit Fragen von Ihnen entgegen (kurz vor meiner Abreise von Paris stellte ich mich Hrn. Alexander von Humboldt vor, der indessen zu sehr an den speculirenden Reisenden zu denken u überhaupt zu beschäftigt schien um mir mit That u Wort zu Hülfe zu kommen.)

⟨Meine Beobachtungen die ich gehört, gelesen nd selbst gemacht sind die folgenden: Was die verschiedenen Luftbewegungen in Sidney betrifft, so lassen sich folgende unterscheiden: 1. Der Land un See wind — ersterer zwischen SW W am Morgen bis ungefähr 10 bis 11 Uhr — letzterer zwischen N u Ost von ungefähr 2 Uhr. Dieß sind sanfte Winde welche beim Auftreten stärkerer folgende Strömungen undeutlich werden. 2. Vorherrschende S Ost Winde. Hr Clarke hat mir gesagt daß die hochstehenden Bäume, die er beobachtet sich alle nach NW lehnen. Er selbst glaubt daß diese S Ostwinde den S Ostpassaten angehören, was indessen von Capt. King^c geleugnet wird, der behauptet daß die Südost passate sich von der Ostküste Australiens reflektiren — n dann machen sich die Passat winde weder nördl. noch südl. bis zum 34° herrschend bemerkbar. — 3. Die heißen NW winde, die mit einer außerordl. hohen Temperatur alle Vegetation vertrocknen alle Feuchtigkeit absorbiren. Die luft ist biswelien so ausserordentl. heiß, daß Herr Clarke einst beobachtete, daß das Thermometer im Schatten ebenso hoch stand wie das Thermometer in der Sonne. — Diese Winde scheinen die niedersinkenden erhitzten Lüfte zu sein, welche in der Mitte Australiens oder unter dem Aequator aufsteigen. Doch scheint ersteres wahrschl. da man diese heißen Winde fast um ganz Australien herum wahrnimmt, wie in South Australia, in der Swan River Kolonie. — 4. Der Wind wechselt gewöhnlich von W nach S, O u N. Sehr selten wechselt er von W nach O nd dann sind es Gewitterstürme (squalls). Herr Clarke scheint zu glauben, daß Gewitter‚ stürme (squalls) mit einem Windwechsel von West nach Ost durch Nord verbunden sind.⟩

Wenn es wahr wäre, daß die merkwürdige, heißen Winde von NW erhitzten über dn Tropische Thl. Neu holland aufsteigenden Lüfte sind, so setzt ihre erstaunliche Trockenheit entscheiden eine Wüste im Inner Neuholland voraus. Denn waren es weite Innenseen, so müßten die mit höherer Tempera‚ tur begabten aufsteigenden Luftströme reichliche Wasser‚ dampfe mit sich führen nd diese bei ihren Niedersinken in höhern Breiten unausbleiblich in heftigen Regengüßen aus‚ scheiden (fallen lassen). Dieß ist indessen nicht der Fall, wenigstens liegt nur keine Beobachtung vor, daß heißen Winde mit folgenden Regengüssen verbunden gewesen wären.

To Herr Dove[a] *in Berlin, with the meteorological observations from South Head.*

[Sydney,] 27th [May,] 1842

My dear Sir,

When I was in Berlin in 1836 I had the pleasure of attending your public lectures on meteorology. They drew my attention to the fluctuations of the atmosphere and to the laws which govern them. On my journeys in Europe I never neglected to apply what I had learnt to what I was observing. On my passage out to the Antipodes in New South Wales I was often able to observe relationships that were new, and different from those expected. Here in Sydney I reconsidered your theory of winds, and found, that here too it applies, but, as expected, only in reverse. At the entrance to the beautiful, indented [harbour of] Port Jackson they have been recording meteorological observations for quite a long time, [? in the interests of shipping]. Judging from such of the weekly observations as I was able to verify, they are carefully made, and are of the greatest value. [From margin: ⟨An Anglican clergyman, Mr Clarke,[b] has been making similar observations on the meteorology of New South Wales, and has also been contributing a series of articles on the subject to the Sydney Herald, based, in part, on the comparison of several registers. The article enclosed herewith is perhaps one of his best. I must confess that I think he's too confused in his ideas, because he has often shown extraordinary uncertainty and indecision in his discussions with me, even with regard to things he has already published. He is engaged upon an...elucidation of the natural conditions of New South Wales.⟩ An Anglican clergyman, Mr Clarke, has been making similar observations on the meteorology of New Holland and has also been contributing a series of articles to the Sydney Herald, based, in part, on the comparison of several registers. He has been trying to prove that all thunderstorms move around from the S through W N E, but the regular change of wind is really from S through E N W. Furthermore, he wanted to demonstrate that the South-east winds were the incipient Trade Winds. The observations made at South Head show, however, that the North-easterlies are the prevailing winds, not the South-easterlies. So far as his first point is concerned, a shift of wind from West to North is always accompanied by thunderstorms or

Reconnaissance in New South Wales

by very unsettled weather; but I do not think that all thunderstorms follow this course: some, depending on local conditions, determine their own courses. As for his second point, Mr Clarke draws on his observation that a number of trees are inclined towards the NW. This, however, is by no means the general rule and is probably due to local causes. Again, he wanted to prove that the hot NWly winds, which parch all vegetation, and which now and then—particularly in Summer —blow across Sydney, though not usually for long, were the NE Trades deflected from their course after crossing the Equator. Besides all this, his head is full of [ideas about] vortices and wind-classification, as you may well suppose from his extraordinary deductions. He does not believe that central Australia is arid, because the hot winds are so heavily charged with electricity that they even affect the magnetic needle, and because gases and water vapour are always produced over water basins or over plains covered with vegetation, and this generation of gases is the only source of atmospheric electricity. He therefore assumes the existence either of a great salt lake or of surfaces clad with vegetation. Nevertheless, the extraordinary dryness of these winds, which so quickly dry everything up, is evidence against the belief that the air masses originate over an inland lake; and their effect on the magnetic needle is easily explained as the effect of the passage of steadily streaming, warm air.—

These hot winds are felt in Van Diemen's Land, even in Hobart-town, though Launceston remains unaffected. They have also been detected at several places on the East coast of New Zealand, notwithstanding its protection from Westerly and North-westerly winds by chains of lofty mountains. I think that we are dealing throughout with two different phenomena. The hot winds at Sydney, it seems to me, are true desert winds, which blow inward from the North-west coast of New Holland, pick up the dry air of an extensive desert, and then blow across the East coast of New Holland. Hot winds have been noticed all around New Holland, which is only natural. For strong winds can blow out from across the continent from any side. We lack adequate information about the hot winds of New Zealand, not knowing whether they are really so dry. They may possibly belong to the descending and returning Equatorial stream. The occurrence of hot winds only in parts of Van Diemen's Land may be due to gaps through the mountain ranges of the South-eastern part of New Holland. In Sydney the Nth Wst wind is invariably followed by a strong Southerly or South-westerly wind which carries the loose sand from the hills near Bot. Bay in thick clouds of dust. They call this wind a Brickfielder, because it blows across the brick fields towards the town. It's a cold wind, and the fall in temperature is sometimes extraordinary, 50–60 F. End of marginal additions.]

When I found the yearly weather records published in the Sydney Herald, it occurred to me that my old teacher would be pleased to have them without delay, even if the English Government does very likely communicate them to the various Continental academies. I submitted myself to some years of training in Europe in order to make myself useful as an observer in the other hemisphere, though I paid most attention to zoology, then to geology. Professor W[? ie]gmann would certainly remember my name. However, I am making careful notes of everything that strikes me as unusual, so I hope that some day, God willing, I shall be able to send a useful account of my work home to you all in Germany. I shall probably remain in Sydney for some years, and, as I know how profitable suggestions from masters of their subjects would be to me, I am looking forward with pleasure to receiving letters from you containing questions for me to answer. (Shortly before I left Paris I had an interview with Alexander von Humboldt, but he seemed too interested in travellers engaged in specific investigations, besides being far too busy, to be able to give me any effective advice or help.)

⟨My inferences, from what I have heard, have read, and have myself observed are as follows:

In what concerns atmospheric circulation at Sydney we can distinguish: 1. Land and sea breeze—the former from SW to W in the morning, from about 10 to 11 o'clock—the latter between N and East from about 2 o'clock. These are gentle winds which lose their identity if it begins to blow harder. 2. Prevailing S. East winds. Mr Clarke has told me that all the tall trees he has observed lean towards the NW. He himself believes that these S. East winds belong to the S.E. Trades, though Capt. King^c disagrees with him. King believes that the S.E. Trades are reflected from the East coast of Australia and are then noticeably dominant neither northward nor southward as far as the

34th parallel of latitude.—*3. The hot NW winds, so hot that they shrivel up all vegetation and evaporate all moisture. They attain such a temperature that Mr Clarke once observed that a thermometer in the shade stood as high as one in the sun. These winds seem to be the descending masses of heated air which ascended over central Australia or over the Equator. I think the former more likely, because these hot winds from the interior have been felt all around Australia, as in South Australia and the Swan River Colony. 4. The wind changes direction, as a rule, from W to S, from E to N. It very rarely swings from W to N to E, and when it does we have thunderstorms (squalls). Mr Clarke seems to think that thunderstorms (squalls) accompany changes of wind from West to East through North.*〉

If it be true that the remarkable, hot North-westerly winds are masses of air which have been heated and have risen over the tropical part of New Holland, their astonishing dryness would be decisive indication of a desert in the interior of the country. For, were there an inland sea, air currents ascending as they became hotter would carry away considerable amounts of water-vapour which would inevitably separate out (be precipitated) during their descent in higher latitudes. This, however, is not the case; or at least, there is no record of hot winds having been followed by showers of rain.

55 TO REV. W. B. CLARKE [Parramatta, N.S.W.]

[Sydney,] 27th May[, 1842]

My dear Sir

May I ask you, whether you have ever observed a heavy rainfall immediately after the hot North Westerly winds. Convinced as I am that they are those returning streams, which ascend over the tropical part of the interior of New Holland, I am lead by a simple induction from the law of the capacity of heated air to contain vapours increasing with temperature, to the conclusion that those hot streams of air must rise over a desert. For did they rise over waters they would contain a great quantity of aqueous vapours, which would necessarily fall down in heavy rain, as soon as the stream arrived to a higher latitude.

This is the case in the Northern Hemisphere and is probably distinctly felt in the most Southern parts of South America.— Which are the laws which regulate the rains during the rainy season in Australia? What winds bring particularly rain? I believe that the North Westerly winds... [*not completed.*]

56 TO REV. W. B. CLARKE [Parramatta, N.S.W.]

[Sydney,] 16 June [1842]

My dear Sir,

My time has been so entirely occupied with the arrangement and the study of the different material objects which fell under my observation that day after day went without my having been able to answer your kind letter. As I did foresee the result of my application,[a] I was not vexed by the refusal; at present I must even be thankful for it. In fact my good fortune has made me acquainted with an excellent man, Mr Lynd whom similarity in habits, in disposition and in love of nature had rendered dear to me from the first interview, and whom the intimate connection of daily life has only heightened in my esteem. Soon after your last letter Mr Lynd made me the kind offer to live with him and I accepted it after finding, that a serious consideration had not shaken his resolution. A fortnight ago I left Mr Marsh's and I reentered in that quiet and studious life, which I had enjoyed in Europe. We have made already a very respectable botanical collection, we have a considerable number of shells, birds, some reptiles and if we go on in the same rate will soon create a new Australian Museum. I live in Mr Lynds quarters in the Barracks and I shall be happy to see you, whenever you feel inclined to call on me. I told you of a fossil in the possession of Mr Barney: I found that it is an almost perfect lower jaw of Phascolomys Mitchelli, a drawing of which is given in Sir Thomas Mitchell's work.[b] The fossil itself is now in the possession of Dr Nicholson. I have read your last articles on Meteorology in the Sidney Herald and I hope you will give at the end short recapitulation of the general laws. The tables of

Reconnaissance in New South Wales

the meteorological observations at South Head published some weeks ago seem to me extremely valuable and show by simple ciphers a great number of interesting facts. You had previously excited my curiosity about the prevalence of the South East winds; the observations at South Head give for the year 1840

wind from N − N E for 78 days
from N E − E „ 48 „
N − E   126 days
from S − S E   41 „
S E − S   62 „
      103 „ a difference of 23 days.

In the year 1841 the proportions are a little different

N − N E 55
N E − E 49
    104
E − S E 37
S E − S 63
    100 a difference of 4 days in

favour of the N E quarter.—This would establish the prevalence of the N E winds, which though more frequent seem however by far inferior in strength according to the general law in temperate zones that the strongest winds blow from the coldest quarter.—Could any local circumstance at South Head account for these facts?—

[*From margin* S E − S 62⎫   S E − S 63⎫
    S − S W 45⎪   S − S W 38⎪
       107⎬ 1840   101⎬ 1841]
    S − S W 45⎪   S − S W 38⎪
    S W − W 45⎭   S W − W   57⎭
       90   95

Did you not tell me that hot winds similar to the N Westerly winds in Sidney have been observed all round the coasts of New Holland—even at its N W corner itself?—always blowing from the centre? How could you explain these at the N W coast of New Holland by the deflected or over reaching trade wind of the Northern Hemisphere? I make you several objections to an explanation of the formation of deserts by hot winds between the 15°–30° lat. the heated air ascends at the tropics and is never or only accidentally in contact with the surface of the earth, being separated from it by the cold stream flowing to the equator.—Leopold de Buch suggests, that this returning stream full of moisture lowers as it enters higher latitudes, the vapours condense and fall down in heavy rains. When the sun is at the tropic of Cancer, this stream flows very far to the northward and precipitates its humidity in the form of snow or rain on the North Cape and on the coasts of Norway. As the sun passes the Equator and approaches the tropic of Capricorn, the equatorial stream lowers in lower latitudes [forming?] the rainy season of the respective localities. But the 32°–30° in fact the commencement of the trade winds form the limit of these precipitations—they are wanting between 30° and the equator, moisture is wanting—deserts are the consequence. This theory founded on good observations is extremely ingenious and plausible.—Extensive barren tracts of sand are necessarily more heated, than those covered with vegetation, as the growth of vegetation binds continually great quantities of heat by the formation of vapours and gases. A Desert like that of Sahara will therefore form a centre of rarefaction the cold air rushing from its borders and pressing the lighter and heated air upwards; at the borders of such a desert we may expect a phenomenon equal to the land and sea breeze, the country covered with vegetation representing the sea.—But whenever a stronger wind, produced by another cause sweeps not only to the centre of the desert, but over all the desert, so that it carries the heated air of the centre to the opposite border, we must expect an extremely dry and hot wind like our N Westerly ones or the Sirocco of the Mediterranean. Similar hot winds may happen round the desert, as strong winds may occasionally blow from every quarter. The Sirocco of the Mediterra[nean] itself seems to me of a double nature: during summer it is the real wind of the desert: during winter it is for the greater part the lowering of the equatorial stream, which pours tropical rains on the peuple criard of Naples and covers the distant arc of the Apennines and sometimes the crater of Vesuvius with snow. Whenever the North West Coasts of New Holland will be inhabited simultaneous observations will show that strong winds blow towards the interior when N Westerly winds come from the interior to scorch Sidney. Alas! I fear I am right! for nobody can take a higher interest in the fertility of the interior and in lofty particularly volcanic mountain ranges than I do; (but those winds are not to be satisfactorily explained in any other way.) Have you witnessed the immense fall of rain at the 29th of April 1841 mentioned in the Abstract of the Met. Observations of South Head? It must have been the local descent of a heavy cloud which we call in German Wolken⁄bruch (bursting of a cloud). I am almost inclined to suppose a mistake in the observation.

I remain
My dear Sir
Yours most sincerely
L Leichhardt

I have just read you last paper. Some of the facts admit another explanation.—Are you sure that electricity does affect the magnetic needle? A simple continual current of hot air would produce the same according to the laws of thermo⁄electricity of Baumgartner[c] pag. 645—[?]410. [*From margin*] (in this parti⁄cular case I do not deny that it affects the needle, but I think current would explain in fact). P.S.

57 [GAETANO DURANDO, Rue Copeau, 4, Paris.]

Sidney 23 juin 1842

Mon cher ami

   Bien que je suis déjà quatre mois à Sidney, j'ai toujours tardé de vous écrire en attendant que ma position serait mieux fixée et que je pourrai vous offrir de ma part un point d'appui sur pour l'execution de vos plans ou de nos plans communs.

Quand je suis arrivé ici j'ai pris une petite chambre, pour laquelle et pour le manger j'avais à payer 1£ 10s. (37 francs et demi) par semaine. Si vous voulez paraître un peu en société vous ne pourrez jamais vivre pour moins. Il y a un jardin botanique à Sidney, mais il sert plutôt à l'amusement des habitans et à la cuisine de la maison du gouverneur, qu'à la science ou à l'horticulture. Allan Cunningham était auparavant le botaniste de la colonie et il en tirait une revenue considerable; mais trouvant qu'elle ne suffit pas pour vivre ici aisement, il quitta sa place et l'on la donna à Mr Anderson, jardinier sans connaissance de la botanique, en diminuant au même temps le salaire. Ce M. Anderson mourut peu de temps après mon arrivée et je me hatai sur l'avis de plusieurs amis de faire une application pour la place. Par l'intérêt plus fort d'un des Messieurs de la commission on me la refusa et la donna de nouveau à un jardinier, homme habile dans son art, mais qui ne sait rien de la botanique. — De ce coté mes espérances sont entièrement évanouies et c'est peut être pour le mieux, car on avait réduit de nouveau le salaire à 120£ ce qui est trop peu pour vivre à Sidney, où toutes les nécessités de la vie sont si chères. J'avais fait la connaissance d'un officier anglais Mons. le lieutenant Lynd, qui s'était occupé de la Botanique pour l'amusement et qui avait recuelli un grand nombre de plantes pourtant sans les avoir arrangées et déterminées. Nous fîmes des excursions ensemble, nous arrangeames les plantes et à present nôtre herbarium est très respectable et bien arrangé. Le livre de Robert Brown[a] est excellent, particulièrement quand on est un peu avancé dans la connaissance des plantes indigènes. Nous avons le Prodromus de Decandolle, mais il nous manque plusieurs familles, qui ne se trouvent point ni dans R. Br. ni dans DC; par exemple les Euphorbiacées. Le Bot. Register et le Bot Magazine nous seraient du plus grand avantage. ⟨Nous n'avons pas encore trouvé de novelles plantes mais nous en trouverons certainement, bien que Rob Brown et Allan Cunningham ont bien cherché partout dans les environs de Sidney.⟩ Il est vrai que le botaniste qui cherche à faire une riche recolte de nouvelles plantes ne doit pas venir ici, pour les trouver dans le voisinage de la ville. Sidney est pourtant le centre de la colonie; il existe une communication directe avec les differentes parties de la colonie. On peut se porter facilement d'ici à la Nouvelle Zeelande; mais mon cher ami, il coûte bien d'argent pour faire tout cela. Si vous voulez aller à Moreton Baie, on vous demande 10£ pour aller et 10 pour revenir. On paie presque le même prix pour aller à la Nouvelle Zeelande et à Pt Philipp et si on n'a pas d'amis dans ces pays là, on a encore beaucoup à payer pour y vivre. Il faut avoir *un* ou plusieurs servans, qui demandent encore des dépenses con׳siderables. — Je me serai désespéré déjà longtemps si le nouveau monde, qui m'entoure, les belles plantes qui peuplent les collines sabloneuses de Sidney ne m'avaient inspiré d'un courage ⟨et d'une energie⟩ indomptable, et si de bons amis, comme M. Lynd ne m'avaient pris à la main. Nous avons l'intention de publier une flore de Sidney en anglais et je vais donner un cours de Botanique dans l'école des arts pour exciter l'intérêt des habitans, qui ne pensent qu'à faire de l'argent (make money). M. Lynd m'a offert son logis pour y vivre sans dépense, ce que j'ai accepté avec bien de reconnaissance. Je suis sur, que ma position sera fixée peu à peu et que je serai plus capable d'agir en grand, que je n'ai pu faire jusqu'à present. Chaque chose doit commencer en petit pour ne pas être étiolée après. — Il y aura une expédition dans l'interieur peutêtre en 2 ans — pas plutôt parce que tous les fonds publics sont exhaustés.

Il faut que je vous donne une idée de la localité de Sidney et de la manière d'y vivre. Le port Jackson sur lequel Sidney est situés est extrêmement varié par un grand nombre de petites baies et de petites îles couvertes d'une végétation, qui vous donne l'idée d'un pays riche et fertile. — Les rochers, que vous voyez partout sont composés d'un grès plus ou moins fin et colorés plus ou moins de fer hydraté. — La ville couvre un très grand espace, elle est bien battie et elle a tout׳à׳fait l'aspect d'une ville Européénne ou plutôt Anglaise avec l'exception des églises et des tours qui sont très rares et de peu d'importance. — Toutes les arts et les métiers sont en grande activité et George Street, la rue la plus considérable de la ville vous montre presque le même mouvement que vous voyez dans les rues les plus frequentées de Paris. Ce n'est pourtant qu'une ville com/merciale; car les environs sont stériles et désolés avec peu d'exception. Le port au contraire est rempli de navires, des bateaux à vapeur vont regulièrement à Parammatta 15 miles anglaises de Sidney, à Pt Philipp, à Hunters river, à Pt Stephens et à Moreton baie: des paquets boats vont de temps en temps à Van Diemen, à Norfolk island et à la Nouvelle Zeelande. Sidney est ⟨pour ainsi dire⟩ le cœur de ces colonies, au moins pour le moment: car je crains, qu'il ne perd pas tôt ou tard sa superiorité. La Nouvelle Zeelande est [sic] Van Diemen sont independans de New South Wales, comme la Colonie de South Australia et de Swan River. —

Si vous partez vers Botany Baie, vous vous trouvez bientôt entouré de collines de sable de peu de hauteur couvertes d'arbrisseaux et de petites arbres. Les Eucalyptus et les autres Myrtacées indigènes, les Acacias, les Proteaceae (Pterophila, Isopogon, Lambertia, Grevillea, Banksia, Hakea, Persoonia) forment les parties les plus considerables. Xanthorraea (the Grass tree) donne un caractère particulier à plusieurs localités. Zamia australis ⟨avec son beau cone d'ovules rouges⟩ n'est pas moins frappante. Lampocarya et Gahnia attractent vôtre attention par la hauteur de leurs épis ou panicules brunes dans les marais qui remplissent les depressions entre les collines. Les Epacridées, Styphelia, Epacris, Lysenema, et Sprengelia vous étonnent par l'éclat de leurs fleurs et par leur nombre. Beaucoup de Rutacées ne sont pas moins éclatantes, par example la belle Correa speciosa, Crowea, les Boronias. Nous avons trouvé quelques belles Orchis particulièrement Corysanthes fimbriata, mais il n'y en a peu en fleur à present. Le rivage Septentrionale (the North Shore) est plus riche, le sol est meilleur. Un grand nombre d'éspèces d'Acacia excitent votre attention. Casuarina est plus frequente, bien qu'elle abonde partout en plusieurs

Reconnaissance in New South Wales

éspèces. Les Eucalyptus deviennent plus hautes et la distribution de leur foliage et leur écorce blanchâtre et lisse leur donnent un aspect particulier. Sur le sol vierge le gazon manque entièrement, bien que les graminées sont nombreuses en éspèces. Le caractère de cette flore consiste dans une grande variété de genres et d'éspèces, pour la plupart ligneuses, couvert de fleurs grandes, brillantes, nombreuses ou d'une forme particulière — réunies dans une espace limité. Il me parait quelques fois comme si les points les plus favorables dans les environs de Sidney renfermaient toute la flore dans un diamètre de 1000 metres. Cela n'est pas le cas, mais l'idée seule vous montre cette singulière distribution, ce melange singulier des éspèces; elles sont très peu sociales, comme les hommes qui habitent la ville. Vous savez que Sidney était une colonie fondée par des criminalistes transportés. La plupart de ces hommes, ayant reçu leur liberté sont devenus propriétaires et beaucoup entre eux sont extrêmement riches. De l'autre coté les emigrés libres, arrivés d'Angleterre ont formé peu à peu une société nombreuse, qui ne veut pas s'entre mêler avec les transportés, parce qu'elle se croit plus pur et plus vertueuse. On appelle les premiers Emancipists, les derniers Exclusionists. Il y a donc deux sociétés très differentes l'une riche, l'autre soi disante respectable et l'on doit bien prendre garde dans la choix de ses amis. — Il y a 5–6 journaux, dont le Sidney Herald est publié chaque jour, les autres 1–3 fois par semaine: ils ne valent pas beaucoup. La colonie est à present dans un état très embarassé produit par un esprit de speculation trop grand et par la possibilité d'un credit presqu'illimité. Il s'en suivit l'embarras des maisons riches, elles tombèrent et tirèrent tout ce qui avait des liaisons avec elles dans leur chûte. Au même temps le gouverneur Gibbs demandait pour les terres publiques des prix exorbitans, qu'il ne peut vendre rien; et ayant à payer continuellement pour les émigrés qui arrivèrent journellement, les fonds publics se trouvèrent bientôt exhaustés.

A mon arrivé on n'avait pas eu une goutte de pluie par les 18 mois passés; des milliers de brebis et de bœufs s'étaient morts. Cela augmenta beaucoup la misère, mais pour les 4 mois que je suis ici nous avons eu des pluies abondantes et presque tropiques. — Les émigrés qui arrivent, ont généralement des idées si fausses du bonheur qui les attend, qu'ils se trouvent comme étourdis, quand ils voient ici le véritable état des choses. Un homme qui connait un métier, qui veut se rendre utile, qui peut travailler, trouvera toujours de quoi vivre avec aise. Mais il ne faut pas croire qu'on pourrait trouver de l'argent dans la rue, que l'on n'a autre chose à faire, qu'à se rendre à Sidney pour s'enrichir. Ora et labora vaut ici aussi bien qu'en France. Le climat est charmant. L'air est extrêmement pur et agréablement fraiche pour l'hiver; ceux qui ont vécu plus longtemps dans la colonie le trouvent même froid. — Les vents de mer se font sentir l'après midi. Les levers et les couchers du soleil sont d'une beauté extraordinaire et la lune ne verse jamais des rayons plus purs sur le Golf de Naples, sur L'Amphitheatre de Flavius ou sur le Campanile di Pisa. — Je ne puis dire, que le [ciel] étoilé fasse une plus grande impression sur l'esprit, qu'en France; mais il y a ici une zone extrêmement riche d'étoiles de la première grandeur. Les α et β du Centaur; la croix australe; le navire de l'*Argo* avec Majaplacida et le Canopus, le grand chien avec le Sirius; le petit chien avec le Procyon, le Scorpion, la vièrge, le Bootes, le Lion — pour le moment Jupiter et Venus marchent chaque nuit par leur champs d'Azur. —

Les hommes et les femmes qui sont nés dans la colonie et qui l'on appelle currency lads and currency lasses (celles de l'Angleterre sont Sterling girls) ils sont ordinairement maigres et d'une taille élancée. Les enfans sont très vite dans leur developpement; leurs facultés mentales se forment d'une manière précoce; mais arrivés à l'age de 16-18 ans ils deviennent las et perdent leur energie. Les hommes sont beaucoup donnés aux boissons fermentés et ont des maniéres criardes (boisterous and swaggering).

Dites à Mr Brogniart que je n'oublierai de lui envoyer des empreintes de plantes fossiles. J'en ai déjà vu plusieurs pro-

Reconnaissance in New South Wales

venant des charbonnières de Newcastle. Quand j'aurai terminé mon cours de botanique, j'irai à Newcastle sur le Hunter river pour ramasser tout ce que je pourrai d'échantillons géologiques. Si le Musée du Roi voudra défrayer les dépenses de mes petits voyages, je serai peut être en état de faire plus pour le Musée que chaque autre, qui aurait à payer son voyage d'Europe et son retour. Je voudrai bien, que vous parleriez avec Mons. le professeur Flourens, qui m'a promis son support — sous la condition que l'offre de M. Vieillot n'aurait pas été accepté. Si cet offre est accepté ne parlez pas de la mienne. Vous voyez donc mon cher ami, que l'argent et le A et le O pour faire quelque chose ici. ⟨Je vous écriverai [sic] immédiatement quand mes moyens suffiront d'executer des plans plus larges c'est à dire d'aller dans la Nouvelle Zeelande.⟩ Attendons encore, si la fortune veut nous favoriser. Tenons nous au qui vive, pour prof[iter] de ses graces. Pazienza, pazienza! caro amico mio! cio est [sic] tutto che posso dirvi adesso. Tutto! — Ma posso rendervi mille grazie per vostra amichevole lettera, che [ho] ricevuto si inespettatamente in Cork, dove me sentiva si deserto, il mio ami[co] Nicholson non avendo risposto alla mia lettera. Nei miei escursioni penso spessi volte da voi, sotto questi Eucalypti et tra gli arbusti di Banksia ed i ricchi f[iori?] d'Epacris longiflora, di Styphelia tubiflora immagino quanto voi vi godereste, se una mano magica potrebbe transportarvi della vostra cameretta o del giardino bota[nico] in queste insopitali coste. Sono convinto che ci revedremo un giorno, sia che voi venite qui per percorrere questi paesi e le isole del mare australe sia che io ritorna in Europa e che allora scambino mutuamante nostre impressioni di vaggio. — Fatte mille complimenti a Signo[r] et Signora Witherell et dite loro che non ho ancora trovato M. Bid[will?]]^b

[Drawings of parts of plants, too faint to reproduce.]
[notes]
1 des empreintes accompagnant les feuilles. Couvertes d'une couche noire. Les 2 lignes int. en b ... montrent peutêtre que c'était un fruit rond, comme celui de Persoonia?

2 Une impre[ssion] concave, probab[lement] d'un fruit.
3 6 L'empreinte d'une feuille dans une marne argileuse de Newcastle. — Les feuilles varient en largeur et longeur; mais la bifurcation de ses veines est la même dans toutes: elle est le plus souvent simple.

?7 Ici dans une feuille de la même forme^c on voit clairment l'anastamose, comme dans quelques Hakeas à feuilles plaines.

Bien que je ne puis determiner ces feuilles, je crois que ce sont [des] feuilles de plantes phanerogammes et qu'elles appartenaient peutêtre à la famille des Proteacées. On voit sur le même échantillon des plantes monocotyledones aquatiques à stries parfaitement parallèles, très con ... Les échantillons viennent de Nobby island à l'embouchure de Hunters river. J'ai vu des fougères du grès de la même loc... La surface entre les stries est finement ponctuée. Les points sont élevées, ou la surface est couverte d'une lame noire: mais quand cette matiére noire manque on vo[it] de fines impressions et l'on penserait d'examiner une feuille d'Eucalyptus avec ses glandes. [*Illegible deletion*]

Mille adieux, mon cher Durando, de votre sincère ami

L. Leichhardt

Sydney, 23rd June 1842

My dear Friend,

Although I've now been in Sydney for four months I've been putting off writing to you in the hope that my own position might become secure enough to enable me to provide a firm base for putting your plans, or our joint plans, into operation. When I arrived I took a small room, for which I paid 30 shillings (37 and a half francs) a week, including board. If you want any social footing you can't afford to live more cheaply than that. There's a botanical garden at Sydney, but it's a place of public amusement and the kitchen garden of Government House, rather than a place for the students of botany and horticulture. Allan Cunningham was formerly the Government Botanist here, and was paid a good salary, but, as he found it insufficient to maintain him in comfort here, he resigned from the post. They gave it to a Mr Anderson, a gardener untrained in botany, and they reduced the salary. Anderson died soon after my arrival, and, on the advice of some of my friends, I promptly applied for the position. The influence of one member of the Committee was strong enough to outweigh mine, so I was rejected, and again they appointed a gardener, a man competent within his limitations, but knowing nothing of botany. So you see that my hopes in that direction have been completely dispelled. Perhaps it's just as well, since they've again reduced the salary to £120, which is far too little for Sydney where the necessities of life are so dear. And I've made the acquaintance of an English officer, a Lt. Lynd, who had taken up botany as a pastime, and had made quite a big collection of plants, though he had neither sorted nor determined them. We've been going on excursions together and have been sorting the plants, so that our herbarium is now quite a size, and well classified. Robert Brown's[a] book is excellent, particularly for those who really know something about the native plants. We have De Candolle's Prodromus, but we've nothing on several families that were not considered by R. Br. or De C., such as the Euphorbeaceae. The Bot. Register and the Bot. Magazine would be of great help to us. ⟨So far, we've found no new plants, but we're sure to do so, even if Rob. Brown and Allan Cunningham did make careful searches everywhere around Sydney.⟩ Of course, no botanist who wants to find a lot of new plants should expect to find them anythere near the town. Sydney, however, is centrally situated and communicates with other parts of the colony. From here you can easily get to New Zealand, but it costs money, my dear friend, to do that sort of thing. If you want to get to Moreton Bay they demand £10 each way. It costs about the same to get to New Zealand or Port Phillip, and if you've no friends there, you've the additional cost of living and it's not cheap. You must have at least one servant, and they're a considerable expense.—I would have given up hope long ago, if the novelty of my surroundings, the beautiful plant-life of the sandy hills of Sydney, had not inspired my courage ⟨and energy⟩ beyond defeat, and if good friends like Mr Lynd had not given me a helping hand. We mean to publish a flora of Sydney in English; and I'm going to give a course in Botany at the School of Arts, to encourage people to think about more than making money. Mr Lynd has invited me to share his quarters at no expense to myself. I have accepted the offer and am greatly beholden to him. I feel certain that my position will gradually become secure, and that it will be easier for me to do things in a bigger way than I can manage now. Things should all grow from modest beginnings or they soon wither off.—They may be sending an expedition into the interior in 2 years' time—no sooner, as there's no money left in the exchequer.

I must give you an idea of the situation of Sydney, and of what life is like here. Port Jackson, where the town stands, offers great variety of scenery on account of its numerous little bays and small islands, and the rich vegetation, which gives you the impression that the country must be rich and fertile. The rocks, which are conspicuous everywhere, are formed of more or less fine-grained, quartzose sandstone stained more or less by iron hydrate.—The town covers a good deal of ground, is well built, and looks just like a European, or rather, an English town, except for the churches and towers, which are few and insignificant. All the arts and crafts are busily plied, and George Street, the most important in the town, you'd find almost as bustling as the most crowded streets in Paris. It's no more than a commercial centre, however, as most of the surrounding country is barren and desolate. The port, on the other hand, is full of shipping; steam boats maintain regular service to Parammata, which is 15 miles from Sydney, to Port Phillip, to Hunter's River, Port Stephens and Moreton Bay; and packets leave now and then for Van Diemen's Land, Norfolk Island and New Zealand. Sydney is ⟨so to speak⟩ the heart of the colonies here, at least for the time being, as I'm afraid that sooner or later it will lose the lead. New Zealand and Van Diemen's Land are independent of New South Wales, like South Australia and the Swan River Colony.—If you go out towards Botany Bay you soon find yourself between low sandhills covered with bushes and stunted trees. Most of the vegetation consists of Eucalyptus and the other native Myrtaceae, Acacias and the Proteaceae (Petrophila, Isopogon, Lambertia, Grevillea, Banksia, Hakea, Persoonia). Xanthorrhea (*the grass tree*) gives character to some localities. Zamia australis ⟨with its handsome cone of red ovules⟩⟩ is no less striking. Lampocarya and Gahnia attract attention in the swamps that lie between the hills, on account of their tall spikes or brown

Reconnaissance in New South Wales

panicles. The Epacridae, Styphelia, Epacris, Lycenema and Sprengelia, *are astonishing in the brightness and profusion of their flowers. Many of the* Rutaceae *are just as vivid, for instance, the lovely* Correa speciosa, Crowea *and the* Boronias. *We've found some fine orchids, particularly* Corysanthes fimbriata, *but not many of them are out just now. The North side of the harbour* (North Shore), *where the soil is better, is richer* [in variety]. *Numerous species of* Acacia *excite your interest;* Casuarina, *though several species are to be found plentifully everywhere, is commoner; and the* Eucalypts, *which have a look of their own on account of their smooth, whitish bark and the distribution of their foliage, become taller. The ground* [between the trees] *is bare of turf, although there are numerous species of* Graminae.

The essential features of this flora are, the crowding into a limited space of a variety of genera and species, most of them woody, and the profusion of flowers, either bright, big, or of peculiar form. I've sometimes thought that, in favoured spots close to Sydney, the whole flora was crammed into a circle of about 500 yards in radius. This is not the true state of affairs; but the idea alone is enough to illustrate the peculiar distribution, the odd mingling of species, which are as little given to assembly [of particular species] *as are the people who constitute the town. You know that Sydney was founded by transported criminals. Most of these men, having gained their freedom, are now the owners of property and many of them are extremely well off. On the other hand, the free settlers, who have come from England, have gradually become numerous, and are reluctant to mix with the former convicts on account of their own supposed honour and virtue. The former are called emancipists, the latter exclusionists. So we have two very different elements of society, the one rich, the other, in its own opinion, respectable, and you have to be circumspect in your choice of friends.—There are 5 or 6 newspapers, one of them, the* Sydney Herald, *appearing daily, the others once to 3 times a week. They're pretty poor. At the present time the colony is financially embarrassed on account of excessive speculation and the possibility of almost unlimited credit. The consequence was that the bigger firms got into difficulties, failed, and brought down all their associates with them. At the same time* Gibbs [Gipps], *the Governor,* set such exorbitant prices on Crown lands that they were not selling; and, as the immigrants, who were arriving daily, had to be kept, there was soon nothing in the Treasury.

When I arrived there had not been a drop of rain during the previous 18 months, and thousands of sheep and cattle had succumbed. This only deepened the distress. However, in the 4 months since I've been here the rains have been abundant and almost tropical.—Most of the immigrants have such exalted ideas of the good luck in store for them that they're dumbfounded when they find out how things really stand here. A man who has a trade and is prepared to make himself useful never has any trouble about earning a good living; but it's no use thinking that money is to be picked up in the gutter, or that you'll get rich just by coming to Sydney. Ora et labora *applies here just as it does in France. The climate is delightful. The air is remarkably pure, and just pleasantly cool for Winter. To those who have spent some time in the colony it even feels cold.—In the afternoons we get the sea breeze, and there's an uncommon beauty in sunrise and sunset. The moon never shone more clearly over the Gulf of Naples or the amphitheatre of Flavius or on the campanile at Pisa.—I can't say that the starry heavens are more inspiring here than in France, but there's a zone here which is very rich in stars of the first magnitude:* α *and* β *Centauri, the Southern Cross, the ship in Argo, with Maya Placida and Canopus, the Dog Star and Sirius, Canis Minor and Procyon, Scorpio, Virgo, Boötes and Leo. Just now Jupiter and Venus are traversing their azure realm.—*

Most of the men and women born in the colony have thin features and slim figures. They call them currency lads and currency lasses (those from England are sterling girls). The children develop quickly and are mentally precocious; but at the age of 16 to 18 they flag and lose their liveliness. The men drink heavily and become criards *in their behaviour (boisterous and swaggering).*

Tell M. Brongniart that I shall not forget about sending him impressions of fossil plants. I've already seen a few from the collieries at Newcastle. After I finish my botany course I shall go to Newcastle on Hunter's River, to collect as many geological specimens as I can. If the Royal Museum would be willing to defray my modest travelling expenses I might be of more use to it than anybody who had to pay his

Reconnaissance in New South Wales

fare from and back to Europe. I'd be glad if you would discuss it with Prof. Flourens, who promised me his support; but do so only if M. Vieillet's [Verreaux's] offer has not been accepted. If it has been, say nothing about mine.

So you see, my dear friend, money is the alpha and the omega for doing anything here. ⟨When my means are enough to let me carry out plans for going farther afield, which means New Zealand, I shall let you know at once.⟩ So let us trust Fortune a little longer. But we must keep wide awake so as not to miss what she offers. We must be patient, my dear friend, patient; and that's all I can say just now. But I can at least thank you deeply for the kind letter which I received so unexpectedly at Cork. I was feeling very lonely there as I had heard nothing from my friend Nicholson, who owed me an answer. Often, during my excursions under the Eucalypts, among the Banksia bushes and the bright flowers of Epacris longiflora and Styphelia tubiflora, I think about you, and I imagine how much you would enjoy it could a magic hand whisk you away from your little room, or from the botanical garden, and deposit you on this rock-bound coast. I am positive that we shall meet again some day, either through your coming out to travel in these colonies and in the South Sea Islands, or through my own return to Europe. And what tales of travel we shall be able to tell each other then! Remember me most kindly to Mr and Mrs Witherell, and tell them that I have not yet run across M. Bid[will?][b] . . .

[Drawings of parts of plants, too faded to reproduce.]
[notes]
[1] a and b. Impressions found with the leaves. Covered with a coat of black stuff. The two [inside?] lines in b . . . suggest that it may have been a round fruit; like that of Persoonia?
[2] c Concave impression, most likely of a fruit.
[3] 6 Impression of a leaf in clayey marl, from Newcastle.—the leaves vary in length and breadth, but the bifurcation of the veins is the same in all of them. As a rule, it's simple.

  7 Here, in a leaf of the same shape,[c] the anastomosis can be clearly seen, as in some Hakeas with flat leaves.

Although I can't determine these leaves, I think that they belonged to phanerogams, perhaps of the family Proteaceae. The same specimen

shows some aquatic monocotyledons with perfectly parallel and very . . .
*striae. The specimen came from Nobby's Island at the mouth of
Hunter's River. I've seen ferns, in sandstone from the same loc*[*ality*].
*The surface between the striae is finely dotted; the dots are raised where
the surface is coated with a black film, but where it is free from this film
you can make out delicate impressions, and could think that you were
examining the leaves and fruit of a* Eucalyptus. *Good bye, my dear
Durando,*
from your sincere friend
L. Leichhardt

58 [TO DR WILLIAM NICHOLSON, ?Bristol]

[Sydney, 17 July 1842]

Mein theuerer Freund

Wiederum sind zwei Monate verstrichen und ohne auf deinen Brief länger zu warten, dem ich mit jedem Schiffe entgegen sah, das von England in Port Jackson einlief — gebe ich dir über meinen Lebenslauf neue Auskunft. — Bald nachd. man mir eine abschlägige Antwort auf mein Gesuch um die Stelle im botanischen Garten ertheilt hatte, machte mir Hr. Lynd ein Englischer Offizier un Barackmaster in Sidney das Anerbieten, mit ihm zu leben, um alle Ausgabe wenigstens so lange zu ersparen, als sich irgend eine weitere Aussicht für mein Fortkommen eröffnen würde. Nachdem ich ihm hinlängliche Zeit zur Ueberlegung gelassen hatte, nahm ich diese freundliche Anerbieten an und lebe jetzt mit ihm zusammen in dn Sydney Barracks. Herr Lynd hatte sich seit langer Zeit in verschiedenen Gegenden mit Botanik beschäftigt. Er hatte längere Zeit in Dominica gelebt, war dann nach England zurückgekehrt, um von dort erst nach Ostindien (Bombay) nd hierauf nach Vandiemens land geschickt zu werder. Von Van Diemens land war er nach Sydney versetzt worden. Er hatte zwar überall botanisirt, doch immer nur als Amateur, ohne weder in ernstes Studium, noch in die Anlage einer Sammlung ein zu gehen. Sobald ich mit ihm zusammenkam, setzte ich ihm die Nothwendigkeit einer Sammlung

.seinander, da ohne dieselbe keine Ordnung u k. Fortschritt in der Kenntniß der Pflanzen möglich war. Er sah dieß auch recht wohl ein nd wir machten botanische Excursions mit einander und arbeiteten wacker an ein Herbarium. Dieß wurde nun nach meinem Zusammenleben eifrigst fortgesetzt nd wir sind schon recht weit vorgeschritten. Neue Pflanzen möchten in der Nähe von Sydney nur wenige gefunden werden, da R Brown sich 18 Monate in Sidney aufhielt nd Cunningham Jahre lang hier lebte nd botanisirte. Wie wir indessen mit Sydney fertig sind, werden wir nach Norden un Süden gehen, um die dortigen Flora zu studiren. Ausserdem haben wir die Absicht die Flora von Sidney heraus zu geben. Wir können indessen .tr einem Jahre nicht an die Vollbringung eines solchen Werkes denken. Man ersuchte mich Vorlesungen in der School of Arts zu geben. Ich erklärte mich dazu geneigt und nachdem ich über den Gegenstand nachgedacht glaubte ich, daß Botanik nd die Flora von Sidney vielleicht am meisten das Interesse der gebildetern Einwohner reizen möchten. Zu Ende Juny begann ich meine Vorlesungen nd ich habe es wenigstens von meiner Seite nicht an Eifer fehlen lassen, den Gegenstand so interessant als mögl. zu machen. Meine Zuhörerschaft ist indessen nur ungefähr 25–30, groß genug, wenn ich daran denke, wie leer selbst Brogniarts Vorlesungen zu sein pflegten. Man scheint mit meinem Vorträge zu frieden. Sollte eine Kritik in den Blättern erscheinen, so werde ich sie dir *gut* od. *schlecht* zu senden, damit du wenigstens siehst, daß dein wandernder Freund sich nützlich zu machen strebt. Die School of Arts ist indessen ein Institut von sehr geringer Bedeutung. Die Männer welche die Committee bilden sind so sehr mit ihren eigenen Angelegenheiten beschäftigt, daß sie dem Interesse der Anstalt nur wenig Aufmerksamkeit schenken. Da niemand es sich zur Geschäft macht, die Einwohner zu interessiren, so interessiren sie sich auch wenig nd das Institut wird, so glaube ich, bald seiner Auflösung entgegen gehen.[a]— Man giebt nie regelmäßige Cursus von Vorlesungen, sondern dieser oder jener junger Manne, sobald er glaubt, er habe

Leichhardt's Letters

Fähgkeit genug irgend einen Gegenstand, z. B. über Rhetorik, über Englische Sprache, pp etwas zu sagen erbietet sich diese seine Wissenschaft in einer Vorlesung an den Mann zu bringen. — Die gebildeten Männer der Stadt sprechen sehr bedenklich über viele dieser Lecturers und was ich von ihnen gesehen habe, erhebt sie nicht eben in meiner Achtung. ⟨— Bald nachdem ich mich mit der School of Arts eingelassen sprach mir der Headmaster of Sidney College von Vorlesungen über Natür wissenschaften, welche er von mir in Sidney College gehalten haben wünschte. Ich sprach mit Dr Nicholson darüber und er sagte mir ich sollte das Anerbieten nur gegen einer handsome remuneration annehmen. Wir werden sehen! Ich habe immer den Vortheil, allmählig bekannter zu werden und in diesem sonderbaren Gemeinde leben nach allen Seiten hin Wurzel zu fassen.⟩ — Wie ich dir schon früher gesagt, bin ich fest überzeugt, daß ich vielleicht schneller unabhängig geworden wäre, wenn ich ohne Geld nach Sidney gekommen wäre. Die Schiffsbekanntschaft führte mich in Sidney in einen Gesellschafts kreis, der zwar höchst achtbar ist nd hoch steht, der aber größtentheils aus Regierungs beamten oder solchen Familien besteht, die ohne Interesse für das Neue Vaterland, nach England zurückstreben, die übrige Gesellschaft verachten nd jede Berührung mit derselben soviel als möglich vermeiden. Ein Mann, der durch seine eigenen Kräfte sich nützlich machen, un sein Brod gewinnen will, kann in einer solchen Elite nur geringen Fortschritt erwarten. Es würde mir ebenso interessant gewesen sein die Emancipisten gesellschaft kennen zu lernen, von welcher ich mich jetzt gewissermassen .sge﹎ schlossen sehe. Ich konnte schwerlich mit ihr in Berührung treten ohne von meinen frühn. Freunden mich trennen zu müßen nd vielleicht heftigen Tadel von ihrer Seite zu ver﹎ dulden. Doch wir werden auch hier sehen, was die Zeit mich zu thun gebietet!

[25tn July 1842]

Heut [als den?] 25tn July erhielt ich deinen Brief und der Tod deines Vetters hat mich tief nd schmerzlich bewegt; du

Reconnaissance in New South Wales

hast das Unglück oder das schmerzliche Glück, deinen theuersten Angehörigen die erkältende Hand auf langen Abschied gedrückt zu haben, während ich weder beim Tode meiner Tante, noch meines Vaters gegenwärtig war, nd sicher‧ lich bei dem meiner lieben Mutter gegenwärtig sein möchte. Doch von jenem Lumbago an habe ich mich so fortdauernd mit der Idee des Todes zu befreunden gestrebt, un ich bin so tief von einem freien ungebunden Sein nach dem Tode, von einer fortdauernden Entwicklung überzeugt, daß ich, wie die Alten Tage heranrücken und des Hauses dach sich weiß färbt ruhig hoffend jedem schnellen oder langsamen Hinüber‧ schreiten entgegen sehe. — Ich freue mich zu sehen, wie du dich hervorarbeitest. Du siehst, daß ich gleichfalls thätig bin, obwohl meine Laufbahn weniger klar nd wenigstens weniger gradlinieg sein möchte als die deinige; doch habe ich den Vortheil, im Fall praktische Kenntnisse un selbst Körper‧ werk erfordert werden sollte, damit dienen zu können. — O mein theuerer Freund, du klagst über die Hohlheit nd Herz‧ losigkeit der Menschen! Ich möchte mich gleich Democritus[b] über das Getriebe der großen Kinder, über ihr tägliches Puppenspiel nd über ihr Streben sich gegenseitig soviel als möglich weiß zu brennen, ich möchte mich dar[ü]ber zu Tode lachen! Wenn ich nur eine Seele hätte, die sich theilnehmend mit mir verständigte, die mit mir fühlte, die eben theilnehmend lachte, wenn ich selbst mich kindisch gebehrde. ⟨Wie wir körperlich für die Gemeinschaft mit dem Weibe organisirt sind, so sind wir .tns auch auf ein geistiges Gemein leben mit einer wohlwollenden nachgiebigen mitfühlenden Frauen seele angewiesen — wir suchen und finden und vielleicht in unserem Suchen getäucht, oft, indem uns entweder die [stärkere?] Sinnen Eindrücke vorgeführt, oft oder aber die jetzigen Erziehungs systeme jene Weichheit un Schwiegsamkeit des weiblichen Gemüthes zu ersticken streben. Ich fühle mich oft ausserordentlich einsam; oft möchte ich mich die Länge lang auf der Erde [werfen?], um ihr mein Klagen wie Midas[c] seine Geheimniße ins Ohr zu schreien. Hr Lynd ist ohne

Zweifel ein liebenswürdige Mann: doch einerseits ist er durch langes Alleinleben beständig .f sich selbst hingewiesen gewesen daß es ihm jetzt wohl unmöglich werden möchte aus einer hypochondrischen Selbst beschauung theilnehmend in das Sinnen eines jungen Mannes ein zu gehen: anderseits ist seine Englische Nationalität ausserordentlich schroff .sgebildet nd ich fliehe von [dem Äussern?] eines National dünkels scheu in mich selbst zurück.⟩ [*From margin*. Herr Lynd ist ein liebenswürdiger Mann mit regem Interesse für meine Wissen- schaft nd auch im allgemein wohlgebildet; dennoch ist es druckend für mich, mit ihm zu leben. Ich würde allen sr. guten Eigenschaften viel mehr mich erfreuen, wenn ich als unabhängiger Mann mit ihm verkehren könnte. *End of interpolation*]

Meine Vorlesungen neigen sich zu ihrem Ende; sie waren nicht wenig besucht; man lobte meine Diagrammen, die ich größtentheils nach Australischen Pflanzen mit großem Fleiße gemacht hatte. Ich unterließ indessen nicht, überall die Familien mit welchen ich in Berührung kam, Neigung zu den Natur wissenschaften, ein zu flössen, indem ich mich eines theils als einen Predigend Reisenden ansah [anderseits?] aber den Vortheil erkannte den die Wissenschaft selbst .s dem allgemeinen Beobachten vielen zu gewinnen hatte. — Ich gewann in manchen achtbaren Familien Zutritt. Doch wie ich durch den Menschen zu traulich entgegen ging und wie sie mich auch oft ausserordentlich zutraulich behandelten, immer trieb mich ein Fehlendes Etwas wieder von Ihnen; immer schien es mir unmöglich gegen sie ein tiefstes innerstes Erschließen zu wagen. Und was habe ich denn zu erschließen? Oft kam es mir vor als wenn ich gleich den Freimauerbrüdern mir einbildete un... Geheimniße zu beherbergen, die aber allbekannte Dinge erscheinen, wenn [immer?] ich versuchen möchte, sie aus zu plaudern. — Man zeigte mir gewöhnl. den Garten, die Einrichtung des Hauses, sprach über den jämmer- lichen Zustand der Kolonie und sehnte sich nach England zurück. Dann floß small talk mit ein, Bemerkungen über

Reconnaissance in New South Wales

Freunde nd Feinde un über Kinder. Ueberall strebte ich den engen Gedanken einige Weiten zu geben, nd da mich ein Detail dieser Art anekelte, ohne ihm entgehen zu können, suchte ich wenigstens so viel in mienen Kräften stand an allgemeinen Ansichten an zu knüpfen. — So lernte ich die reichsten Leute der Kolonie kennen, z. B. Hr. Thomas Barker,[d] der als armer Müller begonnen und jetzt weiten Landstrecken besaß. — Eine sehr anregende Bekanntschaft machte ich mit einem Hrn. Robert Scott nd seinem Bruder Walker Scott.[e] Ihre Schwester ist die Frau eines Arztes Hr Mitchell.[f] Sie hatten von frühster Kindheit der Natur Aufmerksamkeit geschenkt und eifrig gesammelt was ihnen merkwürdig schien. Nun war ihre Sammlung von Mineralien un Muscheln am äusserst belehrend für mich. Es ist möglich daß ich mit ihnen in nähere Berührung komme: ihre Besitzungen sind in der Nähe von Newcastle ud Maitland am Hunters river und ich werde mich dorthin begeben, sobald nur immer meine Vorlesungen ganz beseitigt sind. Ich frage mich oft, wann wird endlich die Zeit kommen, in welcher du deine Kenntnisse in praktischen technologischen Beschäftigungen wirst an den Mann bringen? Ich habe einen ausserordentlichen Drang in diese Richtung, besonders da ich fürchte, daß mich anhaltendes Augenwerk blind machen möchte; dennoch ist es mir oft, als ob ein stilles Hausleben mir mehr zu sagte — Wenigstens hat mich lange Gewohnheit ausserordentlich verweichlicht. Du kennst meine Neigung zur Diarrhöe: sie ist hier wie in Paris. Da ich indessen ausserordentlich regelmäßig lebe und augenblicklich Gegenmittel gebrauche, so sind die Anfälle, obwohl für den Augenblick .serordentlich peinig und dennoch nur kurz.

Ich sammle sorgfältig die Materialen, welche späterhin für ein größeres Werk über Neuholland nützlich werden können. Ich würde dir manches mittheilen, wenn ich nicht sähe, daß die Zeit mit andern Beschäftigungen vollkommen angenommen wäre. Einige Bemerkungen möchten dir indessen interessant sein. Deßhalb also einiges Meteorologische. Du erinnerst dich, daß Dove behauptete, der Wind wechsele in der Nördl. Hemisphäre beständig von West nach Nord nach Ost nd Süd. Im Neuholland un vielleicht in der ganzen Südl. Hemisphäre wechselt der Wind regelmäßig von West nach Süd nach Ost ud Nord. Wenn ein anderer Windwechsel statt findet, besonders von West nach Nord n Ost, so ist er ausbleiblich mit gräulichem Unwetter, mit Stürmen nd Regen verbunden. In Sydney und mehr noch an der un mittelbaren Seeküste sind... die regelmäßigen Westl. Landwinde bis 11 Uhr oder Mittag, denn ist für einige Zeit Windstille nd von 2 Uhr treten Ostliche Seewinde ein, die bis nach Sonnenuntergang anhalten. Um 9 Uhr in jetziger Jahreszeit treten die Westl. Landwinde wieder ein, die die Nacht durch anhalten. — Der Süden ist natül. die kälteste Himmelgegend und von S West bis S Ost kommt uns der meiste Regen. Die Nordwestlichen Winde sind im Sommer bisweilen ausserordentlich trocken nd alle Vegetation zerstören — ich habe sie noch nicht gelebt. Es fällt mehr Regen hier, als im feuchten England; die mittlere Regenhöhe ist 72-76″ während es in den verschiedenen Localitäten Englands zwischen 30″-50″ schwankt; doch die Vertheilung des Regens durch das ganze Jahr ist so unregelmäßig daß wir zu einer Zeit fast darin schwimmen, sehr ähnlich den Regengüssen Neaples — zur anderer Zeit dagegen fast vertrocknen. — Der Boden um Sidney ist fast ohne Ausnahme Sandstein; zu Parammata brechen einige schwärzliche feldspathbasige Kegel (Phonolithe)[g] durch die Sandsteinfelsen. Kohlensäume finden sich hier und dort im Sandstein, doch Fossilen existiren hier nicht. Fossilen existiren indessen in großer Anzahl in Illawarra (im Süden von Sydney) un Kohlen mit Pflanzenabdrücken sind häufig am Hunters river (Newcastle u Maitland) un Port Macquarry. Ueber die geologischen Verhältnisse jener Distrikten werde ich später das Interessanteste mittheilen. — Was die Vegetation betrifft — so ist vor allen Dingen Norfolk island, wo Macchonochie[h] die eingefleischtesten Verbrecher auf seine Weise zu erziehen sucht — ausserordentl. reich an Farrenkrauten. Diese sind auch um

Reconnaissance in New South Wales

Sidney häufig nd mannichfach nd in Illawarra haben wir selbst *Farren*bäume. Die Gräser sind zahlreich, doch kein Rasen! ⟨*From margin.* Am Hunters river ist guter Rasen un um Newcastle sind sehr gut[e] Downs. Newcastle, Octbr 18⟩ Eigenthümlich sind die Restiaceae, welche sich hier sehr zahlreich finden nd von denen du im Botanischen Garten dir einige Vorstellung machen kannst. Die offene braune Scheide ist ihre karakteristische Kennzeichen. Von Palmen habe ich nur die niedrige Corypha[i] gesehen. Sehr zahlreich sind die Orchideen, welche gegenwärtig überall hervortreiben besonders schön ist die Rocklily Dendrobium speciosum welches auf den kahlsten Sandsteinfelsen vortrefflich gedeiht. — Zamia spiralis ist ein merkwürdiges Gewächs, sein Fruchtstand ähnelt einer Ananas, die Form des Embryo und seine Befestigung sind sehr interessant. Casuarina mit artikulirten Zweige hast du in Neapel un in Paris gesehen. Die Proteaceae sind ausserordentlich sonderbare Gewächse; ihr Bluthenstand und ihre Frucht weichen so sehr von unser Europaeischen Ideen ab. Die Epacrideen oder Native Heath welche sich von Ericeen nur drch einfächrige Antheren unterscheiden sind schöne Gewächse von sehr mannichfachen Formen. Die Myrthengewächse gleichfalls — die Eucalypten arten sind ausserordentlich zahlreich: eigentl. Myrthe fehlt hier. Wir haben wunderliche Schwämme gefunden, von denen ich dir einige Formen sende. — Wenn es dir möglich ist, bei Leuten meines Faches oder meiner Fächer über das zu sprechen, was ihnen besonders wünschens werth ist, so will ich mich beeifrigen ihre Wünsche, so viel als möglich befriedigen. —

Ich habe an Little einen langen Brief geschrieben, und so an Schmalfuß nd Durando. Es würde mir sehr angenehm sein, wenn du an Schmalfuß ein deutsches Briefchen schriebst und ihm auffordertest an mich unter deiner Addresse zu schreiben. Ich schrieb ihm, er möchte entwd. seine Briefe an dich oder an Little addressiren. Ich würde mich ausserordentlich beruhigt fühlen, wenn du ihnen bisweilen einen Brief schriebst. Im ersten müßtest du aber wohl bemerken, daß Schmalfuß in lateinischen Lettern zu schreiben habe. Ich sehne mich von meiner lieben Mutter zu hören die vielleicht über Nachlässigkeit von meiner Seite klagt. Von Hallmann hast du sicherlich gehört! Vergiß nicht mir ein Wissenschaftliches Journal zu zu senden. Sobald ich Geld zu verdienen anfange, werde ich dich um Bücher angehen. — Nun mein lieber Freund lebe wohl nd grüße alle die deinigen von deinem herzlich dich liebenden Freunde

Ludwig Leichhardt

[*Sydney, 17 July 1842*]

My dear friend,

Another two months have gone by, so, without waiting any longer for the letter from you that I've been anticipating by every ship that has entered Port Jackson from England, I'm going to give you news of what has been happening to me. Not long after they informed me that I had not been appointed to the position at the Botanic Gardens, a Mr Lynd, an English officer who is barrack-master at Sydney, invited me to share his quarters free of all cost to myself, at least until my prospects improve. I gave him time enough to re-consider his kind proposal before I accepted it, and am now living with him at the Sydney barracks. Mr Lynd has been interested in botany for a long time in different parts of the world. He had spent more time in Dominica than elsewhere, but was recalled thence to England, whence he was sent first to the East Indies (Bombay) and then to Van Diemen's Land. From there he was transferred to Sydney. He has been botanising everywhere, but always as an amateur, without regard either to technical study or the organisation of a collection. As soon as I met him I explained the necessity of a collection, without which no orderly progress in our knowledge of the plants was possible. This he readily understood, and we went on botanical excursions together and worked hard on a herbarium. We've continued to do so with great enjoyment since I came to live with him and by now we've made considerable progress. New plants are rarely found near Sydney, as Robert Brown stayed here for 18 months and Cunningham lived here for some years, and both of them collected.

Reconnaissance in New South Wales

However, as we've done all we can at Sydney, we shall proceed both to the North and the South, to study the flora there. Besides this, we are considering the publication of a flora of Sydney, but hardly think that we could complete a work of that kind in less than a year. I received a request to deliver lectures at the School of Arts. I agreed to do so, and, on thinking the matter over, decided that botany and the flora of Sydney might perhaps be the most attractive subject for the [more] educated [of our] citizens. I began my lectures at the end of June, and, at least on my part, no pains were spared to make them as interesting as possible. My audience, however, numbers only about 25 to 30, big enough though, when I remember how poorly attended even Brongniart's lectures used to be. They seem to be satisfied with mine. If I get any notice in the papers I shall send it to you, good or bad, just to let you know that your restless friend is trying to make himself useful. The School of Arts, however, is an institution of small account. The members of the Committee are too busy with their own affairs to have much time to devote to those of the School. As not one of them makes it his business to arouse public interest, the public itself shows very little. I doubt that the School will survive for much longer.[a] They never arrange for regular courses of lectures, but some young man or other offers his wares in a single lecture, as soon as he feels that he has something to say on a subject, any subject, like rhetoric or the English language, and so on. The better educated people here are very critical of many of these lecturers, and, from those I've seen, I've no very high opinion of them.—⟨Soon after my engagement at the School of Arts the Head Master of Sydney College approached me about lectures on natural science which he wanted me to deliver at the College. I consulted Dr Nicholson about it and he advised me to accept the offer only on terms of 'handsome remuneration'. We shall see! I would have the advantage of getting better known, and of making connections on all sides in this oddly constituted society.—⟩ As I said before, I'm pretty sure that I would most likely have become independent the sooner, had I come to Sydney without any means at all. Fellow passengers in the ship introduced me to a set of people who are, in fact, highly respected and of good standing; but they are mainly Government officials or families that have no interest in their new country, want nothing so much as to get back to England, despise the rest of society and avoid all possible contact with it. A man who is making his own way and is earning his living by his own efforts can expect little encouragement in such an exclusive circle. It would have been just as interesting to me to have gained entry into emancipist circles, from which I think I'm to some extent excluded at present; but I could hardly have done so without having to break with my older friends and to suffer their severe censure. Any way, here too we shall see what time decides for me.

[25 July 1842]

To-day—the 2[5th?] of July—I received your letter. I was deeply and acutely affected by [the news of] your cousin's death. You have had the misfortune—or the melancholy satisfaction—of comforting your dearest relations during their last hours on Earth. I, on the other hand, was not there either when Aunt or Father died, but I can't tell you how badly I want to be there when my beloved Mother dies. Ever since that attack of lumbago I've been striving to reconcile myself to the idea of death; and I am so firmly convinced of our untrammelled survival, of our higher and higher development after death, that I contemplate my passing into the next world, be it quick or slow, calmly and hopefully as old age draws nearer and locks begin to get snowy.—I'm delighted to hear how well you're getting on. You'll notice that I'm not idle either, although my course may not be set as clearly and directly as yours. But I have the advantage of being able to turn practical ability and even physical strength to account in case of need.—My good friend, you complain of the shallowness and heartlessness of mankind! For my part, I could almost die with laughter, like Democritus,[b] at the childish motives of grown men, at the puppet show of their daily lives, at the way in which they find all possible excuses for each other, were it not that I have a soul which takes part in my understanding, in my feelings, in my very laughter, even when I behave childishly myself.—⟨As our bodies are fashioned for union with the other sex, so too are [most of us?] dependent on spiritual union with a kind, indulgent, sympathetic wife. We seek and we find, and in our quest we may discover that we have been misled, either through the stronger sensual impressions coming to the fore, or because the educational systems of the present day lead to the repression of the [natural] tenderness and reserve of the feminine disposition. I often feel extraordinarily isolated, and want to throw myself prone to Earth to complain aloud in her ear, the way Midas told her his secrets.ᶜ Mr Lynd is certainly an amiable fellow, but, through living a solitary life, his attention has been so steadily focussed on himself that a hypochondriac attitude towards himself may by now have made it well nigh impossible for him to enter into the workings of a young man's mind. Besides this, his English bluntness of manner is very pronounced, and I shrink back into my shell before any show of national arrogance. For my part, I feel strongly impelled to make myself useful, but the opportunities for doing so are severely limited by several considerations.⟩ [From margin. Mr Lynd is an amiable fellow and is showing a lively interest in my scientific work, his general education having been very good; but I'm finding that life with him has its trials, and I would enjoy his many good qualities much better if I were able to meet him on an independent footing. End of interpolation.]

My lectures are coming to an end. They were not very well attended. My diagrams, which I had prepared with great care, were mainly of Australian plants. I did my best all round to encourage an interest in science in the families I met, since, on the one hand, I looked upon myself as an itinerant preacher, and saw, on the other hand, the advantage to science itself of wide-spread observations made by numerous persons. I myself won acceptance into a number of better-class homes. And yet, although I have been able to show people the same easy confidence with which many of them have been so kindly treating me, there's always been something lacking, something that holds me back from them. It seems impossible for me to risk declaring myself openly and fully to them. And what, you ask, have I to declare? I have often felt as if I were pretending to be the guardian of... secrets, like the masonic brethren, but that they would prove to be nothing but matter of common knowledge were I to choose to divulge them.—People usually showed me the garden, conducted me through the house, talked of the pitiful state of the colony, and told me how much they wanted to get back to England. After that it was just 'small talk', remarks about friends and enemies and children. I've always tried to get them to raise their level of interest, and, as trivialities of this kind disgust me, and there was no escape from them, I tried my utmost to find refuge in generalities.—This is how I got to know some of the richest men of the colony, like Mr Thomas Barker,[d] who began as an humble miller and is now the owner of broad acres.—I've entered into a very stimulating association with a Mr Robert Scott and his brother Walker Scott.[e] Their sister is married to a doctor named Mitchell.[f] These people have been interested in natural history since their childhood, and have been keen collectors of everything that seemed worthy of remark. I have certainly found their collection of minerals and shells instructive in the extreme. It is possible that I may come to know them a good deal better. Their properties are near Newcastle and Maitland on Hunter's River, and I shall hie myself thither just as soon as my lectures are right out of the way. I often wonder how long it will be before I shall be able to turn my practical, technological qualifications to account. I'm strongly impelled in that direction, especially since I've been fearing that constant use of my eyes might reduce me to blindness. All the same, I often think that a quiet domestic existence might suit me better—any way, my settled habits have made me extraordinarily soft. You know how easily I get diarrhoea. It's been like it was in Paris. However, as I live a very regular life and take something for the complaint at once, the attacks don't last long, though they gripe me acutely.

I'm carefully accumulating material for use later on in an extensive work on New Holland. I would tell you much more about it, only I can see that the whole of my time is going to be taken up in other ways. However, a few remarks may be of interest to you. Let us take something on meteorology. You recall that Dove maintained that the wind invariably changes from West to North to East, and then to South, in the Northern Hemisphere. In New Holland, and perhaps all over the Southern Hemisphere, it changes regularly from West to South to East, and then to North. If it changes in another direction, especially from West to North, then East, it invariably brings gloomy, unsettled weather with rain and storms. In Sydney, and more markedly on the neighbouring sea-coast...the regular westerly land wind blows until 11 a.m. or noon, when it becomes calm for a while, until the easterly sea-breeze sets in at 2 o'clock. The latter holds until after sunset, and at about 9 pm at this time of the year the former returns, to last through the night.—The South is naturally the coldest quarter, and most of our rain comes from the South-west to the South-east. The North-westerly winds are sometimes extraordinarily dry in Summer and ravage all vegetation.—I have not yet experienced them. More rain falls here than in damp England, the mean rainfall amounting to 72–76", whilst at various places in England it lies between 30 and 50". But its distribution over the year is so irregular [here] that at one time we're practically swimming in it—very like the downpours at Naples—and at another we're nearly desiccated.—The ground around Sydney consists of sandstone nearly everywhere, but at Parramatta there are several humps of a blackish rock with a feldspathic base (phonolite)[g] which have broken through the sandstone. Coal seams occur here and there in the sandstone, but the rocks contain no fossils. Illawarra, to the South of Sydney, is rich in fossils, however; and coal containing impressions of plants is abundant on Hunter's River (Newcastle and Maitland) and at Port [for Lake] Macquarie. I'll be able to give you a most interesting report on the geology of those districts later on. As for the vegetation, Norfolk Island (where Maconochie[h] is applying his principles of reform to the most inveterate convicts) is outstandingly rich in ferns. They are common around Sydney too, and in Illawarra there are even tree-ferns. There are numerous grasses, but no turf! ⟨[From margin.] There's good turf on Hunter's River, and around Newcastle there are very good downs. Newcastle, Oct. 18.⟩ The Restiaceae are characteristic, and very numerous here. You can form an idea of them from the botanical gardens. The open, brown sheath is their most distinctive feature. The only palms I have noticed are low Coryphas.[i] The orchids are very numerous. Just now, they're springing up all over the place. The rock-lily, Dendrobium speciosum, which grows wonderfully well on the barest sandstone cliffs, is particularly beautiful.—Zamia spiralis is a remarkable plant. Its fruit looks like a pineapple, and the form and attachment of the embryo are very interesting. Casuarina, with its jointed needles, you saw both at Naples and Paris. The Proteaceae are the strangest plants, bearing inflorescences and fruit quite unlike anything we know in Europe.—The epacrids or 'native heaths', to be distinguished from the ericas only by their simpler anthers, are beautiful and are very diversified in form. And so are the myrtle family—there is an extraordinary number of species of eucalypts, but the true myrtle is not found here. We've found most extraordinary fungoid plants, of which I'm sending you some examples. If you have opportunities for asking people who share my interest—or my range of interests—if there is anything they particularly want, I will do my very best to get it for them.—I've written at some length to Little, and also to Durando and Schmalfuss. I would be very pleased if you would write Schmalfuss a few lines in German, asking him to correspond with me through your address. I told him to send his letters through either you or Little. I'd feel so relieved if you were to write to him now and then. But you must say clearly in your first letter that he will have to answer you in Italic script. I'm longing to hear from my old Mother, who may be complaining of my neglect. I hardly suppose that you've heard from Hallmann! Don't forget to forward me a scientific journal. As soon as I have money of my own to spend I'll go into the question of books with you.

Now, my dear friend, good-bye, and remember me to all the family.

Your affectionate friend,
Ludwig Leichhardt.

59 [TO SIR WILLIAM JACKSON HOOKER, The Royal Botanic Gardens, Kew, Surrey.]

Septr. 5 Sidney [1842]

My dear Sir

I take the liberty of sending you the drawing of a fungus,[a] which was found by Lt Lynd Barrack Master in Sidney in the Government Demesne. Should it be new to you and perhaps not yet a described species, I should feel great pleasure in naming it after Mr Lynd, as he is one of the *very few*, who also take an interest in scientific pursuits here and as he was the almost only one not biassed by national prejudice, who would support me and receive me in his house, when even the venerable Alexander M'Cleay acted against me. I cannot expect, that you will recollect my name; a year ago I brought a letter of introduction to you from Mr Webb in Paris. Though you kindly invited me, to call again on you, my departure from London occupied my time and my attention so much, that I found it impossible to do so. During my stay in Sidney I have done everything in my power to excite an interest for botany amongst the money making population: I have given public lectures, in which I particularly endeavoured to illustrate the most striking families in the neighbourhood of Sidney by diagrams; I made public botanical excursions with the more advanced boys of the Australian and Sidney College and induced them to form collections. With the assistance of my friend Mr Lynd I hope to publish a flora of Sidney, at least of the Phanerogamous plants, as the want of an English work on the local flora was a continual drawback to any desire of studying botany here. Knowing your disposition to promote science, I might perhaps not call in vain for your assistance. We have the prodromus of R Brown[b] and the 7 volumes of DC Prodr.,[c] but we are without any book on those families, which have not yet been published in those works. We want very much Jussieu on the Euphorbeaceae,[d] for which we have written to England. You might suggest perhaps better the necessary works, as you know, what has been published.—We have of course the general works on Botany—Lindley's Nat. families,[e] his introduction, Richard[f] pp.—

Mr Webb will give you the necessary information on my studies in Paris, would you be only so kind as to mention my name to him in one of your letters. You will find, that I am not unworthy of your confidence, should you be disposed to assist me in my scientific pursuits.

I remain my dear Sir
Your most obedient
Ludwig Leichhardt

60 DEN MUSIKLEHRER / HERREN / C. SCHMALFUSS / Wohlgeb. zu / Kottbus / Prussia

Sidney den 6ten September 1842

Meine theuerste Mutter

Es ist nun ein Jahr, daß ich von Euch keine Nachricht erhalten habe. Von Paris, von England und von Sidney kurz nach meiner Ankunft habe ich Briefe an Euch geschrieben, die doch wohl sicherlich Euch zu Händen gekommen sind. Jeden Sonntag, wenn ich mich still den Erinnerungen an die Vergangenheit überlasse, befinde ich mich in Euerem Kreise, ich sehe Euch, höre Euch und denke mit Euch; doch dann schiebt sich der ganze Erdball wieder zwischen uns; Ihr kehrt zurück zu der lieben Heimath, während ich in der Fremde mir selbst überlassen, in mir selbst Trost, Ruhe, Befriedigung suchen muß. Ich fühle mich nicht unglücklich! denn seit so langer Zeit habe ich mich gewöhnt mit der Natur zu leben ud in ihrer Betrachtung und Erforschung Freude zu finden. Der einzige Schmerz, der mir bisweilen das Herz beklemmt, ist eben daß ich von Euch getrennt über die Erde wandere, daß ich mit Euch nicht häufiger, selbst auch nur schriftlich verkehren kann. Ich fühle, daß auch Ihr mich sehen möchtet, daß du meine liebe Mutter eben so sehnlich wünschest, deinen Sohn ans Herz zu drücken, wie dieser dich — und doch! und doch kann ich für die Gegenwart keine Hoffnung für die baldige Erfüllung unserer Wünsche rege machen. Als ich noch zu Hause war und meiner Armuth wegen nie hoffen könnte, aus zu führen, was ich jetzt ausführe, glaubte ich, daß ich alles leicht opfern könnte, um den Drang in die Ferne zu befriedigen. Der Himmel erfüllte meine geheimen Wünsche: mir wurde zu Theil, was oft dem Reichsten nicht zu Theil wird; ich fand überall Unterstützung und konnte mich meiner Neigung zum Studium der Natur unbesorgt überlassen; doch eine andere Sorge ließ mich nun nicht loß: ich wurde stets unterstützt, aber ich war nie unabhängig. William gab mir Geld, nach Neuholland zu gehen; hier in Sidney angekommen fand ich bald gute Freunde, welche mich bei sich aufnahmen und mir alle Kosten in einem sehr theuren Orte ersparten, in welchem ich allein für ein einfaches Logis *wöchentlich* mehr als 10 Rtl. zu bezahlen haben würde. Ich kann hier leben, hier studiren, aber ich kann nicht von hier gehen, ohne Jemand zu finden, der mich mit sich nimmt. Auf der andern Seite habe ich mit schätzenswerthen Familien Bekanntschaft gemacht; oft habe

Reconnaissance in New South Wales

ich einige Neigung zu Mädchen empfunden, ja ich bin tief verliebt gewesen, doch meine abhängige Stellung hat mich stets von ernsten Schritten nd Offenbarungen zurück gehalten. So siehst du deinen Sohn von mannichfachen gefühlen beherrscht, von den frühern Erinnerungen, von der stets regen Liebe zu seinem Naturstudium, von den Eindrücken des Augenblicks, welche vielleicht für ihn die gefährlichsten sind nd voreilig sein wanderndes Schifflein zum Anker bringen. Doch wenn auch mein Gemüth oft unruhig sich bewegt, so kann ich dir doch versichern, daß ich glaube stets besser geworden zu sein nd daß ich mich noch ebenso unschuldig fühle, wie damals, als du mich zum letzten Male in deine Arme schloßt. Und ich danke dir dafür! denn wenn ich zur Quelle meiner moralischen Grundsätze zurück spüre, so komme ich zur kleinen Kammer in unserem alten Hause, mit seinem kleinen Fensterchen, in welchem du uns unsere Morgen un Abendgebete lehrtest nd zuerst mit unserem Himmlischen Vater bekannt machtest.

Vor 52 Jahren war auf dieser Stelle, wo jetzt 42000 Menschen mit voller Europaeischer Bequemlichkeit leben, ein wildes ödes Buschwerk, durch welches kaum wilde Horden streiften; jetzt ist hier eine mächtige Stadt nd die ganze Küste ist mit werdenden Städten besetzt. — Die Wilden sind entweder ausgestorben oder zu 50-60 Meilen landeinwärts zurückgedrängt und nur selten kommen sie nach Sidney. Obwohl ich überzeugt bin, daß sie nie Europaeische Kultur annehmen werden, so zeigen sie doch viel natürlichen Scharfsinn nd viel Anstand und Gewandheit. Die freie Haltung mit welcher ein wilder Mann durch die Straßen von Sidney schreitet, setzte mich in Verwunderung: die Weiber sind gewöhnlich zu großem Beschwerden ausgesetzt, indem die Männer sie fast wie Lastthiere behandeln. — Sidney ist von Sandsteinfelsen nd Sandhügeln umgeben, welche mich oft an die sandige Mark Brandenburg erinnern; die Gewächse sind keineswegs so frisch nd grün wie bei Euch und die Landschaft erscheint eigenthümlich matt nd graugrün. Tiefe Waldung nd hohe

Bäume fehlen hier fast ganz, obwohl sie in andern Gegenden der Kolonie sich finden; der Busch wird von niedrigen Bäumen und Gesträuchen gebildet, welche häufig sehr auffallende, große schöngefärbte Blumen haben. In der That giebt es wenige Stellen auf der Erde, wo in einem beschränkten Raume so viele schöne Pflanzen beisammen wachsen. Schlangen giebt es in Menge hier; ich selbst habe schon zwei in diesem letzten Monate getödtet; viele hält man für sehr giftig. Papageyen von allen Farben sieht man hier in großen Zügen. Ausserdem finden sich hier einige auffallend schöne Vögel mit prachtvollem Feder un Farbenschmuck. Euer Sommer ist unser Winter nd der Winter hier ist mild mit vielem heftigem Regen. Obwohl der Winter milder ist als bei uns oder in England, so empfinden wir doch die kleinen Veränderungen, die Nasse Kälte oft sehr unangenehm. Der Boden, obwohl sandig, ist fruchtbar sobald nur die nothwendige Feuchtigkeit vorhanden ist; doch im Sommer vereitelt oft lang anhaltende Dürre jede Bestrebung des Landmannes. Diese Dürre macht das Land zum Landbau wenig geeignet nd die größere Menge der Einwohner hatte sich besonders früher der Schaaf nd Viehzucht zu gewendet. ⌜Doch früher hatten sie freie Arbeiter.⌝ Die Verbrecher welche man von England hierher brachte, waren gezwungen für die Kolonisten zu arbeiten. Gegenwärtig werden keine Verbrecher mehr hierher gebracht nd die freie Einwanderer lassen sich für ihre Arbeit sehr theuer bezahlen. Die Folge davon ist, daß die Schaafzucht wenig Profit abwirft nd daß die früher so reichen Besitzer entweder bankerott werden oder doch ihr Vermögen bedeutend geschmälert sehen. — Besonders jetzt ist deßhalb die Kolonie in sehr bedrängten Umständen nd eine Menge von Leuten welche hierher kamen mit der Hoffnung schnell ein großes Vermögen zu erwerben, sahen sich bitter getäuscht. Doch der Mann, welcher arbeiten kann und will, findet schnell Beschäftigung nd hat reichliche Nahrung für sich selbst und für Weib nd Kind. Die Künstler haben weniger Hoffnung, ebenso die wissenschaftlichen Männer. Alles hier ist nach Körperwerk! —

Der hiesige Landmann nd Arbeiter lebt, was Nahrung betrifft, viel besser als die Landleute bei Euch. Er hat Fleischspeise so viel er will, Gemüse ist weniger reichlich; die Kartoffeln sind keineswegs so gut. Die Wohnungen sind indessen nicht so behaglich. Oft sind es sehr einfache Hütten. Doch das Klima ist mild. Ofen sind nicht nöthig. Dann sind die Besitzungen weit von einander entfernt. Es fehlt die behagliche Gesellschaftskeit eines Dorfes. Entsprungene Verbrecher machen die Straßen unsicher nd berauben die Reisenden, welche indessen selten viel Geld bei sich führen. Ich habe Euch früher geschrieben daß in Van Diemensland nd in Neu Zeeland, wie im Süden von Neu holland reiche menschenreiche Kolonien existiren; daß Europaeische Sitte nd Kultur sich hier völlig heimisch gemacht hat; daß man die Hauptstraßen von Sidney kaum von manchen Straßen Londons unterscheidet; daß Sidney 8 mal so groß ist, wie Kottbus (ich glaube Kottbus hat 5000 Einwr.); daß Dampfschiffe alle Punkte mit einander in regelmäßige Verbindung setzen. Gebt deßhalb alle Idee von Wildniß auf! Die reichen Leute, ja selbst die Armen haben nd erfreuen sich größere Bequemlichkeit, als die Bewohner der Mark Brandenburg nd Ihr selbst! In Sidney werden öffentliche Blätter (Zeitungen) gedrückt! Nachdem ich einige Zeit in meinem getrennten Logis gewohnt hatte, machte ich die Bekanntschaft eines Englischen Offiziers, welcher mich einlud mit ihm zu wohnen, um die so bedeutenden Ausgaben zu ersparen. Der Name dieses Ehrenmannes ist Lynd. Ich lebe nun schon mehr als 3 Monate mit ihm zusammen und ich habe Ursachen, ihn nur noch höher zu schätzen. — Ich versuchte, eine Anstellung im botanischen Garten zu erhalten, doch war ich noch zu unbekannt nd zu kurze Zeit hier nd deßhalb gelang es mir nicht. Ich hätte nicht nur ruhig fort studiren können, sondern man würde mir selbst noch dafür bezahlt haben. ⌜Das war sicherlich mit zu nehmen.⌝ Ich habe indessen die Bekanntschaft vieler angesehenen Leute gemacht und ich hoffe bald eine sichere Stellung zu gewinnen. — Ich habe die ganze Zeit über unablässig gearbeitet nd gesammelt; allmählig, wie sich meine Materialen mehr ordnen, werde ich Euch Sachen zum Druck nach Hause senden; nd wenn Schmalfuß bei einem angesehenen Buchhändler, ⌜z. B. bei Reimer⌝ in Berlin anfragte, könnte ich vielleicht für Euch von hier aus nützlich werden, indem ich Euch den Vortheil des Honorars überliesse. Es scheint mir am besten, eine solche Veröffentlichung in Brief form zu fassen, wie zum Beispiel Raumers[a] Briefe über England oder Italien. Ich bin überzeugt, daß ein solches Buch, von einem anständigen Buchhändler herausgegeben sich wohl bezahlt machen würde. Ich überlasse dieß Euch. ⌜Ihr müßt nicht zögern an mich zu schreiben. Ich habe William gebeten mit Euch in briefliche Verbindung zu treten, da der Verkehr zwischen mir nd William viel leichter ist, als zwischen mir nd Euch. Auch kann ich niemals die Ausgabe übersehen welche Euch meine Briefe verursachen!⌝

In wenigen Tagen beabsichtige ich Sidney für einige Zeit zu verlassen nd nach Newcastle (am Hunters river) zu gehen, welches ungefähr 20 Meilen nach Norden liegt. Ein wohlhabender Besitzer hat mich nämlich eingeladen, ihm zu besuchen nd da die Gegend sehr interessant ist, hoffe ich meine Zeit dort sehr nützlich an zu wenden. Es ist in der That eine Gunst, welche man den einsamen Bewohnern des Landes erzeigt, welche selten nur fremde Gesichter oder überhaupt Menschen bei sich sehen. Es wäre vielleicht möglich, daß ich dort dauernd mich aufhielte, indem mir einige Vorschläge gemacht sind, über die ich selbst noch nicht klar bin, um entscheiden zu können. — Ich fühle mich wohl; doch bin ich sehr zu Diarrhöe geneigt, welche mich bisweilen ausserordentlich erschöpft. Dies ist gewöhnlich mit allen der Fall, welche hier ankommen, bis sich die Natur vielleicht an Klima, Speise oder Wasser gewöhnt hat. Ich war indessen diesem Uebel schon in Paris ausgesetzt nd würde wahrschl. bereits gestorben sein, wenn ich mich der Niger expedition ins Innere von Afrika angeschlossen hätte, wie es ursprünglich meine Absicht war.

Reconnaissance in New South Wales

Lebt nun wohl meine theure Mutter nd meine theueren Angehörigen. Seid überzeugt daß ich Euch stets in meinem Herzen hege, wohin mich auch immer meine Füsse tragen mögen. ⌐Sollte ich wirklich für lange Zeit hier bleiben, so möchten wir nun vielleicht daran denken, junge Schößlinge unserer großen Familie mir folgen zu lassen. Ich hätte dann verwandte Wesen um mich nd sie könnten ihre jüngere Kräfte zum Geldgewinn gar wacker regen. Ein Deutscher Gerber kam mit mir hierher.[b] Ich verschaffte ihm eine Stelle für 280 Tlr. das Jahr ohne weitere bedeutende Ausgaben. Er kann vielleicht in 10 Jahren in die Heimath als wohlhabender Mann zurück kehren.⌐

Euer herzlich Euch liebenden Sch[wager] ud So[hn]
Ludwig.]

The Music Teacher / Mr C. Schmalfuss / at / Kottbus / Prussia.

Sydney, 6th September 1842

My dearest Mother,

It's now a year since I have had any news from the family. I wrote from Paris and from England, and again from Sydney soon after my arrival. You must surely have received the letters. Every Sunday, when I can sit in peace and review the past, I feel that I am really back with you all, as if I were looking at you, talking with you, and sharing your thoughts. But too soon the mass of the world intrudes between us, you recede into our own dear country, whilst I, so far from home, have to review my own situation, and must rely on myself for peace of mind and spirit, and contentment. But don't think that I'm unhappy. I've long been used to living in the world of Nature, and to finding my pleasure in the contemplation and study of it. The only pang that my heart ever feels is that of your absence, for I am too far away for us to communicate oftener, even by letter. I have the feeling that you, over there, would all like to see me, that my dear old Mother would love to put her arms around me, which is just how I feel about her, but—well, as things stand, I can offer you no hope that it's likely to happen in the near future. Before I left home, when I was too poor to hope that I would ever be doing what I am doing now, I used to think that I could easily give up everything to satisfy my desire to see the world. Heaven has fulfilled my secret wishes. I have had good fortune greater than many a rich man has ever known. Everybody has so helped me that I have been able to follow my bent and study Nature without a care. Yet from care of another kind I have never been free. I have always found support, but I have never known independence. William found the money that brought me to New Holland. Once here, I have found good friends who have taken me under their roof and saved me all expense in a place where things are very dear. Otherwise, I would have had to pay more than 10 rix dollars a week for very plain lodgings. Here [in Sydney] I can live and I can study, but I can't go away from here unless I can find somebody who will take me with him. On the other hand, I have made the acquaintance of families held in great esteem. I have often been drawn to young women, even to the extent of falling deeply in love, but my dependence has always restrained me from making serious advances and declaring myself. Your son, you see, is being swayed by different feelings: his earlier recollections, his ever lively love of Nature-study, his impressions of the moment—and the last are perhaps the most dangerous and those most likely to cause his wandering little barque to cast anchor too soon. Still, although my mind is often troubled, let me assure you that I really think that I've been improving steadily, and that I still feel as innocent as I was when you last took me in your arms. And I have you to thank for it. [Why? Because] when I think of the source of my moral principles what comes to mind is the room with the tiny little window in our old house, where you taught us to say our prayers morning and evening, and made us aware of our Father which art in Heaven.

Fifty two years ago this place, where 42,000 people now live as comfortably as they would in Europe, was a rugged waste covered with bushes and was seldom visited even by the little bands of wandering savages. Now there's an extensive town here, and there are towns growing up all along the coast. The savages have either died out or been driven back 50 or 60 miles inland, and very seldom come into Sydney. I am convinced that they will never accept European culture. They show nevertheless, an inherent sharpness of intelligence, grace in their deportment, and no little skill. The easy carriage with which these uncivilised

Reconnaissance in New South Wales

men stride through the streets of Sydney simply astonished me. The women mostly lead lives of great hardship, as the men treat them almost like beasts of burden.—Sydney is surrounded by rough sandstone country and by sandhills which often remind me of our sandy Mark of Brandenburg. The plants are nothing like as fresh and green as those at home, and the landscape has a peculiar, lustreless grey-green hue. There is hardly any dense forest here, and very few tall trees, though both are to be seen in other parts of the colony. The bush is composed of low trees and bushes, and many of them bear big, extraordinary flowers of beautiful colour. In fact, there are few places in the world that produce such a diverse association of plants within a small area. Snakes are common here, and I've killed two myself already, this month. Many of them are supposed to be very venomous. And great flights of gaily coloured parrots can be seen, as well as a few other birds remarkable for their extraordinary beauty of colour and plumage. Your Summer is our Winter; and the Winter here is mild with a lot of heavy rain. Although we have milder Winters than you have at home or in England, we often find the slight changes in the weather, and the raw coldness, quite unpleasant. The soil, though sandy, is fertile as long as it retains the necessary moisture, but long dry spells in Summer can defeat the farmer's best efforts. Because of drought the country is not well suited to farming so most of the country people turned, particularly in former years, to rearing sheep and cattle. ⌜At that time, however, they had free labour.⌝ The convicts that used to be transported from England were compelled to work for the colonists. Convicts are transported no longer, and free settlers demand high pay for their work, so that sheep rearing yields little profit, and land-owners who used to be well off are now either going bankrupt or are finding that their assets are dwindling. For this reason circumstances in the colony, particularly just now, are oppressive, and many folks who came here hoping to get rich quickly have been bitterly disappointed. Nevertheless, a competent, hard-working man soon finds work, and can provide more than bread and butter for himself and his wife and children. For artists there's little opportunity, and no more for scientists. It all depends on brawn here.—So far as food is concerned the farmer or workman is much better off than our farmers at home. He can eat meat as often as he likes, though vegetables are not so plentiful, and the potatoes are nothing like as good as ours. He is not so comfortably housed as at home, and many of the houses are just huts; but the climate is warm and stoves are not needed. The settlements, however, are far apart and there's nothing like our neighbourly village life. Escaped convicts make the roads unsafe by committing robberies, so that travellers rarely carry much money on them. I have already told you about the flourishing and populous colonies that now exist in Van Diemen's Land, New Zealand and in the southern part of New Holland. I've told you how firmly European customs and culture have established themselves here; how you can hardly tell some of Sydney's main streets from the streets of London; I've told you that the place is 8 times the size of Kottbus (I think Kottbus has 5000 inhabts.); that steamers maintain regular communication between all ports. The rich people here—in fact even the poor people—have and enjoy an easier life than the inhabitants of the Mark of Brandenburg, which means yourselves. So just drop all your ideas about my being in the wilds. Why, papers (newspapers) are printed in Sydney!

After I had been in my own lodgings for some time, I made the acquaintance of an English officer who invited me to go to live with him, which would spare me from considerable expense. This gentleman's name is Lynd. I have already been living with him for more than 3 months and my esteem for him has steadily risen, for good reasons. I tried to obtain a post at the Botanic Gardens, but I had not been long enough here and was not well enough known, so I failed. It would have enabled me not only to go on studying quietly, but I would even have been paid for it. ⌜I could hardly say no to that!⌝ However, I've made the acquaintance of a number of respectable families, and hope that I shall soon win secure footing.—All this time I've been studying and collecting without remission. Gradually, as I reduce my materials to order, I shall send articles home to you to be printed. Were Schmalfuß to inquire of a reputable publisher, ⌜say of Reimer in Berlin,⌝ I might, even from this distance, be able to do something to help you, for I [do hereby] make over to the family the benefit of any profit. I think it best to present a publication of this kind in the form of letters, like Raumer's[a] letters from England or Italy, for instance. I'm sure that a book like that, issued by a well-known publisher, would pay well. I leave it all in your hands.

Reconnaissance in New South Wales

⌐Never hesitate to write to me. I've asked William to forward letters to and from you, because communication between him and me is easier than it is between you and me. Nor can I ever overlook the expense that my letters are to you.⌐

I intend to leave Sydney in a few days, and to spend some time at Newcastle (on the Hunter river), which lies about 20 [German] miles North of here. A wealthy land-owner happens to have invited me to visit him, and I'm hoping to make good use of my time there as the district is an interesting one. You do a kindness by visiting people who live isolated lives out in the country, where they rarely see a strange face if they see any one at all. It's even possible that I may remain there for some time. Proposals have been made to me, but they're not clear yet even to me, so that I can't decide.—I'm feeling well but I'm easily given to diarrhoea and sometimes it's terribly exhausting. This is quite usual with people arriving here, until Nature becomes inured, possibly to the climate, the food or the water. However, I was already suffering from this weakness in Paris, and would most likely have been dead by now had I gone on the Niger Expedition to the interior of Africa, which was my original intention.

Well, good-bye Mother and all the rest of you. Never doubt that my heart beats the better for the thought of you all, whithersoever my footsteps may lead me. ⌐Should I really remain here for a long time we might well think of getting some of the younger offspring of our big family to follow me out. I would then have kindred about me, and they could put the energy of their youth into making money with a will. There was a German tanner who came out with me.[b] I got him work at 280 dollars a year [and keep?]. In 10 years he could return home very well off.⌐

Your affectionate brother-in-law and son,
[Ludwig.]

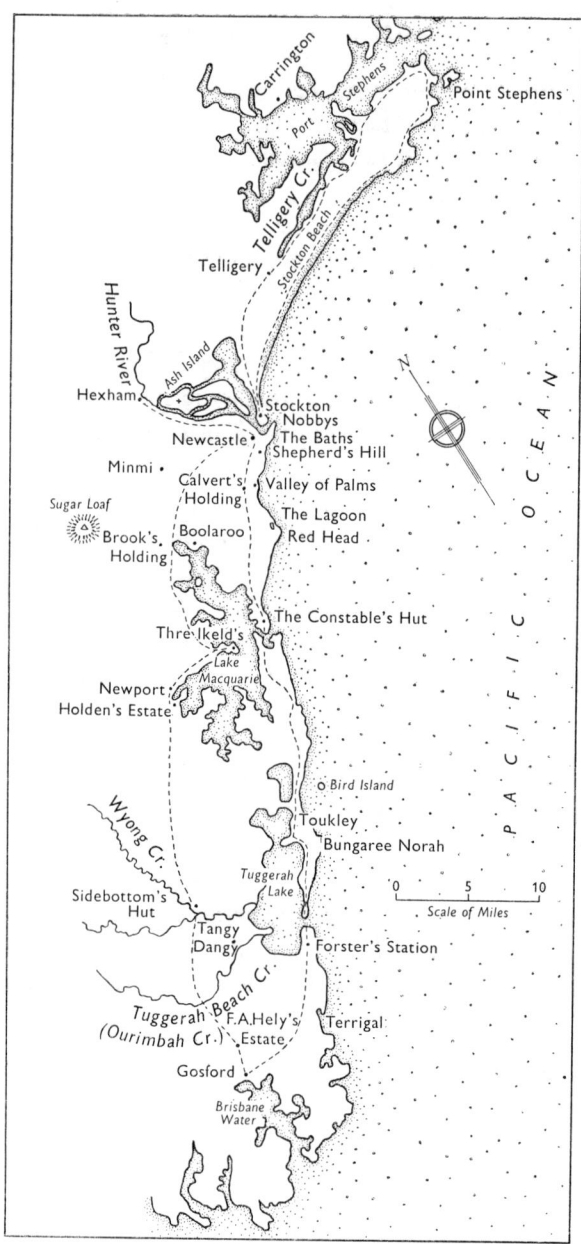

Fig. 7. Journeys on foot from Newcastle, N.S.W. September–December 1842.

B. AROUND NEWCASTLE, ON FOOT

61 [LT. ROBERT LYND, Military Barracks, Sydney.]

[Newcastle, N.S.W., 26 September 1842.]

My dear friend,

I have to keep my promise and to give you an account of the first week at New Castle. We arrived safely Tuesday morning and as soon as we had taken breakfast I accompanied Mr Scott to his establishment at the North Shore called Stoketon [*sic*]. This is very extensive for a private individual and shows that Mr Scott does not lack enterprise and activity. The salt works are the most considerable part of it. The arrangement of the branches for gradation is vertical instead of being horizontal, which would certainly expose the fluid more to the atmosphere [*alt*. the branches are disposed vertically instead of being horizontal, though a horizontal position would certainly expose the fluid more to the atmosphere], as it had continually to drop from one branch to the other.—the pans seem by far too large; it seems to me that smaller pans would be more economical.— Besides the saltworks there is a foundery a forge, there are carpenters and sawyers, who execute the orders given to Mr Scott by the People of Newcastle and of Sidney. The first day was miserably cold and rainy; I tried however to gain some [lenias?], in which I succeeded a new Loranthus and Leucopogon margarodes, which is very caracteristic, the base of its carpels being surrounded by a berry like swelling. Leucopogon Richei is in full fruit and its current like white berries have an agreeable taste.—Wednesday we made another excursion in a SWest direction. I found a new species of Baekea (B. diffusa) in great abundance covering the ground with its prostrate branches and with its fine numerous small flowers.—I think I have a new species of Tetratheca with considerably larger leaves [*from margin* (Hugel[a] calls it however T. thymifolia)]. Prasophyllum elatum and Calochilus poliodosus? Jews beard, as Mr Scott called it, the labellum being covered with long collered hairs.—Melaleuca armillaris is in blossom.—Last Friday we went to Ash island: it is a remarkably fine place, not only to enjoy the beauty of nature, a broad shining river, a luxuriant vegetation, a tasteful comfortable cottage with a plantation of orange trees, but to collect a great number of plants which I had never seen before. I saw at once that you must come to roam with me over the island, which is much larger than I thought, containing full 2000 acres of land. The vegetation seems very similar to that of Illawarra if I can judge by the descriptions I heard of the latter district. Climbing Polypodium, the Aerostichum growing on the trees, a great number of creepers, the nettle Tree, the Caper, the native Olive and many others which we will examine together. When you come here, and I hope you will not delay it long time, we will go together to Ash island and remain there for 3-4 days. Mr Scott has kindly offered us every accomodation we might require.—

62 [LT. ROBERT LYND, Military Barracks, Sydney.]

[Newcastle, N.S.W., Oct. 2nd., 1842]

My dear friend

I am very sorry to hear that your health has suffered again after my departure from Sidney. I wish you could come here and stay with me for some time. Mr Scott has kindly offered us his place at Ash island, as I wrote you last time. My old complaint did trouble me last week, perhaps in consequence of the change of living. Yesterday I commenced seabathing and I feel allready its salutary influence. Last Tuesday I went to the valley of Palms,[a] or rather ferns; for not one single palmtree was to be seen, whilst a great variety of ferns grew everywhere in great abundance and vigour. It is an extremely interesting locality. The valley opens and descends towards the sea, but hills of loose sand have intercepted the communication and it is only during high water and easterly or South easterly winds that the raised waves wash over the sands and form a kind of lagune at the mouth of the valley. The hills slope moderately, they are covered with vigorous Eucalyptus; underwood being formed by Pultaneas Davisias Gastrolobium. The North side is rougher and densely covered with a variety of shrubs tied together by climbers. The first plant I found was Rencalmia paniculata and soon my tin box was filled with many unknown friends. The creeping ferns, which covered either the rocks or the trunks of trees were particularly interesting.—Notholaena pumilis, Polypodium acrostichoides? rupestr., confluens attenuatum and particularly tenellum were in golden plenty. there were the most beautiful specimens of Asplenium Nidus, almost 6' long $\frac{1}{2}$' broad with the finest fructification; they told me that the ferntree attained a heighth of 10 feet; but I did not see it, a fire having destroyed those trees. the ferntree is Alsophila australis, which we found at Balmane [sic]; I saw afterwards a specimen with a trunc about $1\frac{1}{2}$' high.—Pteris tremula and almost all our Sidney ferns were also present; however I do not remember having seen either Osmunda or the Gleichenias. The Proteaceae are very rare in the immediate neighbourhood of Newcastle. I miss that host of Grevilleas which meets our eyes every where at Port Jackson. The Telopea speciosissima is unknown here, Isopogon anemonifolium is dwarfish the Epacridae are in proportion quite as rare; in fact there were only Leucopogon virgatus and Richei, Sprengelia, some Styphelias, Monotoca Lissanthe and these very limited. Of the Rutaceae I found only Zieria and Boronia polygalaefolia (or that Boronia which you brought from Liverpool). Several fine Melaleucas are at present in blossom. They told me of a plant which they call the gigantic lily, of considerable size and purple colour, which is not Telopea as I at first supposed. Have you heard of it?

The geology of Newcastle is not quite as various as I thought, though very interesting. The occurring strata are very regular here and there slightly raised or deranged by the protrusion of volcanic rocks (Trapp) and they extend over an immense space, procuring an inexhaustible source of fuel to the colony.

There are 3 perhaps four beds of coal, of which the lowest is considered the best. They are separated by sandstone or sandy clay: the sandstone between the 2tn and 3tn bed is a hard stone and excellent for building and they quarry it at present for the construction of the breakwater between Nobbys Island and Main land.

Tomorrow I am going with Major Crummer[b] and Mr Bolton (a custom officer) to Major Cr. farm a regular trip in the bush; I hope to be back at Thursday. I intend also to go to the Sugar loaf where Mr Scott has another farm. The Sugar loaf is com⁄posed of Trapp rock and it is possible that the trap has raised lower strata of limestone, which I have not yet been able to find in the neighbourhood of New Castle. It is really very singular that formations of so considerable thickness are so totally deprived of limestone; for there are sections of 3–4–500 feet high and perhaps higher of the 3 constituent parts of this country of Clay sandstone and coal—and a Conglomerate, all rich in Iron.

I found some days ago some interesting [lineas?] on the germination and dissemination of Loranthus. I shall tell you more about it bye and bye: but I am inclined to believe that a snail feeds in the Loranthus and emits the undigested seeds as it is crawling along the branches.—

Write to me when you are coming to Newcastle, that I may be here. Your stay here will cure you from any illness, it will distract your mind and will render you and myself cheerful or at least more cheerful.

63 [W. KIRCHNER, Sydney, N.S.W.]

[Newcastle, N.S.W.,] Octbr. 12[, 1842]

Mein verehrter Freund (an Kirchner)[a]

Ich fürchte ⟨sehr⟩ daß Sie mich der Vergeßlichkeit beschuldigen, da ich trotz meines Versprechens und meines offenbar lebhaften Verlangens, Ihnen bis jetzt weder Raupen noch Schmetterlinge gesandt habe. Hätte ich die Verhältnisse, die Localität gekannt, ich würde bei ebenso starkem Streben

Reconnaissance in New South Wales

mich Ihnen nützlich zu machen, in meinen Hoffnungen weniger sanguinisch gewesen sein. — Herr Scott hat mich ausserordentlich gastfreundlich aufgenommen und hat durch das Interesse, welches er an meinen Studien nahm, nicht wenig zu der nöthigen Geistesfrische und gemüthlichen Heiterkeit beigetragen, welche zum Verfolge meines Werkes vielleicht noch nöthiger sind, als zu dem eines jeden andern. Doch der Wechsel der Lebensweise machte mich für die ersten 14 Tage fast beständig unwohl und obwohl ich mich jetzt an die neue Tages eintheilung gewöhnt habe, so fühle ich doch, daß sie meinem Körper weniger zu sagt, als die frühere. Ich sagte Ihnen, daß die geologischen Verhältnisse von New Castle einen sehr bedeutenden Theil meiner Aufmerksamkeit in Anspruch nehmen würden. Dieß ist denn auch der Fall gewesen. Ich bin Tage lang an den Klippen herumgeklettert um die Verhältnisse der 3 bis 4 Kohlenlager zu erkennen, welche in fast allen Durchschnitten in der Nachbarschaft von New Castle zu Tage kommen. Ich bin überzeugt, daß die Geologie von Australien im Allgemeinen von New Castle ihren Anfang nehmen muß und daß der Geologe sich von hier aus schrittweise nach Norden, Süden und Westen zu wenden habe. Ich habe ausser den gewöhnlichen Farren՚ kräuter Abdrücken noch mehrere andere gefunden, welche alle mit lebenden Gewächsen übereinstimmen und das Jugendliche Alter[b] dieser Bildungen an den Tag legen. Die horizontalen Lagen der Kohlen, des Sandsteines und der Thon letten sind von senkrechten Spalten durchbrochen, welche sich von unten hierauf mit früher feuerflüssigen Gestein massen von ganz andere Natur gefüllt haben. In Nobbys Island ist eine dieser Spalten, welche einem Damme gleich das andere Gestein durchsetzt, recht schön sichtbar. Auf der andern Seite machte ich botanische Excursionen welche mir eine Menge neuer Pflanzen einbrachten. So habe ich während der 3 Wochen meines Hier seins schon 20 Bücher Löschpapier verbraucht. ⌜Ich hoffe daß meine kleinen Freundinnen sich mit mir darüber freuen, denn ich habe sie nicht auf meinen

Wanderungen vergessen.[1] Sie werden von dem Thale der Palmen gehört haben, welches zwar *wenige* Palmen enthält aber desto reicher an andern Pflanzen ist. Es ist eine enge Felsschlucht mit schroffen Felswänden welche sich nach dem Meere hin erweitert: die üppigste Vegetation bedeckt die Gehänge; Bäume und Gesträuche werden von Schlingpflanzen zusammen gebunden und parasitische Gewächse bedecken die Baumstämme. Fluthen welche während der Winter Regengüssen mit unwiderstehlicher Gewalt hier niederströmten, haben starke Baumstämmen niedergebrochen und so erscheint alles im wildesten Gewirre und Leben und Tod halten sich hier auf das innigste umschlungen. Herr Scott war so gefällig mich auf einigen meiner Märsche zu begleiten: Er hatte Geduld genug meinen Schneckengang ruhig ab zu warten, denn meine Kurzsichtigkeit und das Verlangen genau zu sehen zwingen mich überall zu verweilen, überall sorgsam zu vergleichen. Er dagegen strebt als rüstiger Jäger rasch vorwärts zu schreiten und sein scharfes Auge bemerkt wenigstens auffallende neue Pflanzenformen leichter. — Ich war zweimal in Ash island das erste Mal mit einer Damen gesellschaft, welche mich recht[c] sehr in meinem Wissenschaftlichen Treiben hinderte, das zweite Mal in Herr Scotts Geschäften, welche keine Zeit zum Studium liessen. Es ist ein Romantischer Ort, welcher mich auf das lebhafteste anzieht auf welchem ich *vielleicht* befriedigt leben und sterben könnte. Ich sage vielleicht — denn wer kennt sich selbst genug? Wer kann das unaufhörliche Streben nach scheinbar bessern Zuständen aus der Menschen brust verbannen?

Wäre ich in Ash island oder selbst in der Borkenhütte eines Ansiedlers, ich könnte mehr für Entomologie thun, als in Newcastle. Doch selbst in Newcastle läßt sich vieles thun, sobald ich nur erst meine andern Bestrebungen ein wenig befriedigt haben werde. Wenn ich Käfer finde, so sammle ich gewöhnlich eine große Anzahl, damit ich Ihnen späterhin Duplicate mittheilen kann: doch stets finde ich mich an die Schwäche meiner Augen erinnert ⌈und wenn ich manchmal an Blindheit denke, schwindelt mir indem ich mich von fremden Menschen umgeben sehe. Leben Sie wohl mein verehrter Freund und mag dieser Brief Ihnen beweisen, daß ich stets mich Ihrer und meines Wortes erinnere, so wenig ich auch den Versprechungen nach kommen konnte. Grüssen Sie Ihre liebe Frau und Schwägerinnen und meine kleinen Freundinnen deren musikalischen Unterhaltung ich mir gar oft gegenwärtig erwünsche.⌉

[*Newcastle, N.S.W.,*] *12th October*[*, 1842.*]

My good friend,

I've been ⟨much⟩ afraid that you have been charging me with forgetfulness, because, in spite of my promise and my clearly honest intentions, I have so far sent you no grubs and no butterflies. Had I known the conditions and the locality I would have been less hopeful of my ability to do anything for you by no matter what efforts. Mr Scott has treated me most hospitably, and his interest in my activities has done a lot towards maintaining that freshness of mind and serenity of temper which are perhaps more necessary to me in my pursuits than to anybody else. The change in routine, however, had a bad effect on my health for most of the first fortnight. I've now got used to the new time-table, but I feel that it does not suit my system as well as the old. I told you that I would be devoting a good deal of attention to the geological relationships of Newcastle, which has certainly been the case. I go clambering about cliffs all day long, tracing the 3 or 4 coal seams which appear in nearly all the sections that are exposed around Newcastle. I am convinced that the [elucidation of the] geology of Australia must begin mainly at Newcastle, and that the geologist will have to work out steadily from here towards the North, South and West. In addition to the usual [fossil] impressions of ferns, I have found several others, and they all resemble living forms. This reveals the geological youth[b] *of these formations. The horizontal beds of coal, sandstone and various clays are traversed by vertical joints, into which formerly molten rock of quite a different kind has been forced up from below. At Nobbys Island one of these joints, filled by what looks like a wall running up through the other* rocks, is beautifully exposed. On the other side [of my interests] I have made a number of botanical excursions that have brought me a harvest of new plants, so many that I've already used up 20 quires of blotting paper in the 3 weeks that I've been here. ⌈I hope that my little girl friends will feel as pleased as I do about it, as I have not forgotten them in my wanderings.⌉ You'll have heard of the Valley of Palms. It happens that there are very few palms there, which makes it by so much the richer in other kinds of plants. It's a narrow, rocky gully with steep sides, which widens out towards the sea. The slopes are covered with the most luxuriant vegetation; trees and bushes are bound together by climbers; and the trunks of the trees are covered with parasitic plants. Floods, that have swept down with irresistible power during the heavy Winter rains, have uprooted big trees, and have produced the wildest confusion of closely entangled life and death. Mr Scott has very kindly come with me on some of my tramps. He was patient enough to put up with my snail's pace without complaining. For my short sight, and my need to be sure of what I see, oblige me to dwell on things and to compare them carefully wherever I go. He, on the other hand, is inclined to proceed briskly, like a nimble huntsman, and his sharp eyes notice the less conspicuous of the new varieties of plants more easily than mine do. I've been to Ash Island twice. The first time was in company with ladies, who got badly in the way of my scientific activities. The second time I went with Mr Scott, in connection with his interests, which left us no time for studious pursuits. It's a romantic place, which I like well enough to think that—perhaps—I'd be content to live and die there. I said perhaps, for who knows himself well enough to be sure? Which of us could stifle his perpetual longing for what merely seem to be better conditions?

Were I living on Ash Island, or even in a settler's bark shanty, I could do more for entomology that I can in Newcastle. But even in Newcastle there will be a good deal for me to do, when I manage to satisfy some of my curiosity in other directions. When I collect beetles I usually take quite a number, so that I shall be able to let you have duplicates later on. But I'm constantly kept aware of my defective eyes. ⌈If I think of going blind, as I often do, my head swirls at the idea of being amongst foreigners.⌉

Take care of yourself, my good friend, and may this letter assure you

Reconnaissance in New South Wales

that I am ever mindful of you all, and am keeping my word, even if I am falling somewhat short of my promises. Remember me to your wife, your sisters-in-law and my little friends—how often I wish that I were listening to their music!]

64 [TO DR WILLIAM NICHOLSON, Newcastle upon Tyne

Newcastle, N.S.W., c. 26 October, 1842]

Mein theuerster Freund.

Zu Ende September verließ ich Sydney um andere Theile der Kolonie zu sehen. Ich muß gestehen, daß ich zu gleicher Zeit hoffte, vielleicht außerhalb Sydney eine meiner Neigung passende Stellung zu gewinnen. Meine Pläne waren durchaus nicht vollkommen bestimmt — das rein wissenschaftliche Interesse, wie das materielle Streben nach unabhängigkeit beherrschten mich wechselweise und versetzten mich in einen sehr unbehaglichen Zustand fortdauernden Schwankens. So kam ich nach Newcastle, wo Herr Walker Scott mit gast-freundlichst sein Haus anbot. Er hat in Newcastle und in der Nachbarschaft bedeutende Besitzungen, welche nur thätiger rüstiger Arbeiter warten, um ausserordentlich fruchtbar zu werden. — Von Newcastle aus machte ich nach allen Seiten hin Excursionen, sammelte Pflanzen, studirte die Geologie der Umgegend, sammelte Insekten und suchte soviel als möglich Erfahrungen über Colonial akkerbau u Viehzucht zu gewin-nen. — Newcastle selbst besteht aus wenigen Häusern un Straßen; doch da gegenwärtig die Military Barracks vollendet sind, welche ein Regiment fassen, werden überall Häuser gebaut und ich bin überzeugt, daß das Städtchen sich rasch zu einer volkreichen Stadt entwickeln wird. — Der Hunters river an dessen Mündung es liegt ist mehr denn eine Engl. Meile breit und umschließt höher hinauf mehrere Inseln, von denen Ash island, das Eigenthum Hrn. Scott aus 2500 Acres Land besteht. Das Nördliche Ufer ist eine schmale niedrige Landzunge von Meeressanden gebildet, das Südl. Ufer erhebt sich amphitheaterlich. An diesen liegt das Städtchen. Die Hügel sind gegen Westen nd Süden mit dichtem Rasen bedeckt, welcher mich lebhaft an die Downs in England erinnerte. In der Mündung selbst liegt eine schroffe Insel, Nobbys Island, welche wahrscheinl. einst durch unterirdische Kräfte vom Festlande sich loß riß. — Zählt man den obersten Kohlenletten samn. mit so [existiren] hier 4 regelmäßige Kohlenlager, von denen indessen 2, das 3te n 4te nur brauchbare Kohle liefern. Sie werden von Sandstein nd Thonbetten getrennt, sind in ihren obern Schichten gewöhnlich sehr thonig nd nur in ihrem unteren Lagen geben sie eine schöne dichte Glanzkohle, welche die in Newcastle upon Tyne schwerlich an Güte nachsteht. Die Kohlenletten ud Thone enthalten eine Menge von Farrenkräuter abdrücken, welche wahrscheinl. alle noch existirenden Formen entsprechen,[a] obwohl ihre Identität sich nicht nachweisen läßt. Ich glaube daß sich die 3 von mir beobachteten Formen auf Asplenium Adiantum u Nephrodium zurückführen lassen. Man findet .ßerdem Abdrücke von Algen nd Süßwasserpflanzen, .lche sich nicht bestimmen lassen. Diese regelmäßigen geschichteten Gesteine sind vielfach von .trirdischen Kräften [zertrümmert] worden und mehrere Dykes von ausserordentlicher Regelmäßigkeit durchsetzen sie von Osten nach Westen. — Die Beschaffenheit der Kohle, welche beim Verbrennen Asche zurückläßt nd so ihre Ligniten natur verräth, die Farrenkräuter, welche lebenden Formen entsprechen, deuten lebhaft auf das jugendliche Alter dieser Bildungen. Wir werden bald zur Ueberzeugung kommen, daß die Südliche Hemisphäre zwar dieselben allgemeinen geologischen Gesetzen folgt, welche in den Nördl. obwalten, daß es aber unmöglich ist die Stufenreihe beider zu identificiren. Sicherlich werden die Formation[en] je jünger je mehr von einander abweichen. Es scheint mir nun, daß Newcastle wegen der Regelmäßigkeit der Aufeinander lagerung und wegen snr. so karakterische[n] Kohlenlager der Punkt ist, von welchem der Geologe ausgehen muß, um die geolog. Verhältnisse Newhollands zu bestimmen. — Ich bin in Begriff, mir

Reconnaissance in New South Wales

ein Pferd zu kaufen und dann mit meinem Hammer am Sattelknopf, die bewohnten Ostküste zu durchwandern. — Die Umgegend von Newcastle besitzt keines weges eine so reiche Flora, wie Port Jackson, doch erscheinen einige Tropische pflanzenformen, und andere sehr karakteristische Pflanzen Neuhollands, welche nur spärlich un ärmlich in der Nachbarschaft von Sydney gedeihen, treten hier in voller Kraft und Fülle auf. Farrenkräuter sind in einigen engen schattigen Felsschluchten .ßerordentlich zahlreich. Der Grass tree (Xanthorrhoea arborea) und der Farrenbaum (arborescent fern Alsophila australis) bilden Haine oder erreichen ein Höhe von 12 Fuß. Doryanthus excelsa (the gigantic lily) ein Lilien⸌ artiges Gewächs treibt seinen Blumenstiel zu 15 Fuß hoch und die großen liliengleichen dunkelrothen Blumen sind in einen Kopf vereinigt, welcher den eines Mannes oft an Umfang übertrifft. Der Blumenkrone enthält ein halbes Weinglas voll honigsüßen Wassers, in welchem Hunderte von Fliegen ihr Grab finden. Der Honig reichthum der Australischen Blumen im Allgemeinen ist recht merkwürdig. Einige, wie zum Beispiel die Banksias, Lambertia, Styphelia sind .ßerordentlich reich, während einige gänzlich seiner ermangeln. Oft habe ich ermüdet und durstig das Steilende eines Blumenbuschels von Lambertia formosa abgebissen, um das süßesten lieblichsten Honig aus zu saugen.

In Newcastle begegnete ich zum ersten Male dem nakten Wilden. In der Stadt dürfen sie nur bekleidet erscheinen; die Männer tragen alte Hosen, Jacken; die Weiber halten sich in grobe Sackleinwand. Ich ging eines Tages am Ufer des Meeres entlang, als uns ein Wilder in fast völliger Naktheit leichten muntern Schrittes, mit dem Waddi in der Hand mit einer Last auf dem Kopfe entgegen kam. Er hatte ein weißes Band um die Stirn gebunden, ein leichtes spanisches Mäntelchen flatterte um die Schultern, ein haariger Riemen von Opossum fell lag um den Leib. Seine Naktheit genirte ihn nicht im geringsten. Sein Körper war wohl gebildet, die Glieder mager, schmächtig, doch die Muskeln wohl entwickelt, die Brust sehr gewölbt der

Penis wahrschl. in Folge der Kälte .ßerordentlich klein. Die Haut ist keineswegs so sammtartig wie die des Negers; die Haare sind grob, rabenschwarz und lockig, doch keines weges Negerhaft. Die Arkaden der Orbita, besonders der inner Winkel sind .ßerordentlich hervorragend, die Nase eingedrückt, die Nasenflügel breit; der Mund .ßerordentlich breit, die Lippen lang und Pferdartig beweglich. Die Backen Zähne nutzen sich wie die der Egyptisch. Mumien ab, die falschen Molars erschienen mir in einem jungen Manne fast wie die eines Hundes *lobed* [*fig.*]. Obwohl die Frauen .ßerordentlich häßlich sind, so wird dennoch fleischliche Verbundung mit ihnen von den weißen Männern gesucht, wahrscheinl. wegen Mangels weißer Weiber, und die Kinder sind besser geformt nd von bläßlig schwarzer unangenehmen Farbe. Die bastard Knaben werden um 9tn–10tn Jahre getödtet, die bastard Mädchen dagegen verschont man nd es giebt ihrer eine Menge in Newcastle, gewöhnlich mit dem Beinamen Yellow. Die wilden sind ausserordentlich träge. Sie können indessen arbeiten und sind oft sehr intelligent. Vergleicht man die Sclaverei niederer oder weniger gebildeten Raçen mit der Schulzeit der Kinder, so erscheinen diese Wilden im Vergleiche mit den Negern, wie unerzogene Kinder im Vergl. mit Erzogenen. Wäre es von Interesse für die Menschheit, eine so untergeordnete Raçe zu erhalten, so sind diejenige Menschenfreunde in Irrthum, welche Sclaverei bekämpfen. *Zwang* zur Arbeit und strenge Zucht allein könnten diese Wilden bilden un erhalten. Ohne Zwang ist nichts im Stande sie zu dauernde Arbeit zu bewegen. Was ihnen entweder mißverstandenes Wohlwollen oder momentane Arbeit bietet, wird für Brandwein (Botallago) verschleude[r]t. Es ist ein großes Fest für sie Wasser zu trinken, mit welchem man Brandwein fässer spülte. Sie nennen es Bu[ll]. — Bei eine andern Gelegenheit sah ich mehrere Männer Weiber nd Kinder während der Ebbe Krebse fangen und eine Art gestielter Ascidien einsammeln welche Kohlstrünken ähnlich in ausserordentlicher Menge die felsigen Meeresufer zwischen hohem ud niedrigem Wasser bedecken. Sie nennen dieses Thier Congivoi und wissen vermittelst eines kleinen spitzen Stäbchens mit grosser Geschicklichkeit das Thier von der Schaale zu trennen. — Um Krebse zu fangen gehen die Männer bis zu den äussersten mit Wasser bedeckten Klippen hinaus und tauchen nun Kopf über in die tiefen Felsen spalten, wo sie die Krebse bei den Fühlern aus [ihren] Schlupfwinkeln hervorziehen. — Man spricht ihnen hier alle natürlichen Muth ab; Sie wissen nur, so sagten mir mehrere, in der dunkeln Nacht, in welcher sie indessen die bösen Geister fürchten, zu dem Lager ihres Feindes zu schleichen und ihn meuchlings zu ermordern. Hat indessen ein Mitglied des Stammes ein Verbrechen begangen, so wird er einer Art Gottesurtheil unterworfen; ein schmales Holz, in dessen Mitte an der hinter Seite ein Handgriff sich befindet, wird ihnen als Schild gelassen nd alle Männer des Stammes werfen nun ihre Speere nach ihnen. Sie werden indessen selten verwundet oder getödtet, so groß ist ihre Fertigkt, die Speere von sich ab zu lenken. Dieß scheint mir indessen bei der Länge der Speer keinesweges schwieriger als Stoß od. selbst Hiebfechten. [*From margin*: Sie fürchten nd ehren den weißen Mann um so mehr so entschiedener und kürzer er sie behandelt. Selten wagen sie es, ihm körperlich Widerstand zu leisten. Ein Pistol oder ei[ne] Flinte erfüllt sie stets mit Schrecken, sie wissen eine wie tödtl. Waffe es ist, ohne gerade die Gefahr kommen zu sehen. Sie werden dem weißen Manne nie offen begegnen, doch so sie gereizt hinten Bäumen von hinten ihm zu lauern nd ihn rücklings mit ihren Speeren zu durchbohren — a bugery fellow.] —

Die Ansiedler haben die Gewohnheit den stärksten und geehertesten Manne seines Stammes ein messingenes Bruststück zu geben, auf .lchem seine Name [*sic*] eingegraben ist nd ihn mit diesem zum Könige seines Stammes zu mchn. Diese Könige sind indessen um nichts besser als die übrigen; sie sind in großem Miscredit un ein junger Mann, .lchen wir fragten ob er König werden wolle, antwortete, er wünsche es nicht, es waren der Könige schon jetzt zu viele. — Als wir fragten,

warum er sich nicht wasche, da doch das Meer un der Fluß
beständig zu Diensten wären, antwortete er, Sein Volk liebte
es nicht, sich mit dergl. Dingen zu belästigen.
Ich habe schon mehrere Male mein Glück im Büsche
versucht. Obwohl Mark oft seiner einsamen dürftigen Hütte
Erwähnung gethan hat er dir vielleicht doch nicht eine klare
Anschauung dieses einfachsten Ausdrucks menschlicher
Baukunst gegeben. — Ich ging mit Major Crummer dem
Polizei beamten von Newcastle nach Telligerry, wo er eine
Viehstation hat und wo er eine dairy Milchmeierei an zu legen
beabsichtigt. Nach 3 Stunden mühsamen Marsches an dem
Rande eines weiten Morastes (Swamp), in welchem zerstreute
Rinderheerden überall sichtbar waren, gelangten wir zu einem
offenen festen von Bäumen entblößten Waidegrunde, auf
welchem einige Hütten standen, die deinen Anschein nach
nicht unähnlich waren, welche ein Bauer in [Vitnio?] für
seine Schweine oder Gänse baut. — Sie waren ungefähr 15'
lang und 8–9' breit, die Wände bestanden .s Brettern roh
zusammen gefügt, das Dach aus großen Stücken Eucalyptus
Rinde. Die Spalten [wurden] mit einer Mischung von Thon nd
Kuhdünger verschmiert. Der innere Raum der Hütte, besteht
aus der eigentlichen Hütte, welche mit einem holzernen
Bänkchen, einem Tischchen und einigen von Brettern ge‐
bildeten Sideboards mit Teller Tasse dem Teetopf (teapot) nd
Zuckerbüchse meublirt ist, und einem weiten 7' langen, ud
breiten Kamine, über welchem der Schornstein geräumig
emporsteigt. Dieser Kamin ist so groß, weil die Bretterwände
bei der Berührung mit dem Feuer sogleich selbst Feuer fangen
würden. Wir waren unserer Vier: es war ausserordentlich
schwierig uns zu accomodiren. Wir hatten unsere Speise‐
vorrath mit gebracht nd zwei un zwei nahmen nun auf jeder
Ecke der 2 Bänkchen Platz. Wie einer sich ohne vorgängige
Anzeige erhob, schlug der andere mit Bänkchen Teller Tasse
Messer un Gabel über. Dieß machte uns den Cockneys von
Newcastle großen Spaß. — Für die Nacht wurde ein weiches
Lager aus Gras und Farrenkräutern bereitet und wie ich

ausgestreckt da lag und den Widerschein des lustigen Feuers über die Innere Seite des braunen Daches spielen sah, fühlte ich ein Behagen, wie ich es nur immer vor [dem] Zwielicht an einem Englischen Heerde gefühlt hatte. Zwei junge Männer, welche entlaufene Kühe gesucht hatten passirten die Hütte und traten ein. Der breit[rändige] Strohhut mit einem schwarzen losen Bande an das Knopfloch des weißen Staub' hemdes gebunden, das bunte od. auch blaustreifige Hemde, Hosen nach Belieben, dicke Schuhe bilden den Anzug des jungen Hirten. Er hält sich nicht auf Etikette; rasirt sich selten, wechselt sein Hemd nicht beim Zu bett gehen, ißt Damper, trinkt 3 mal des Tages [grünen Thee] mit braunem Zucker und milch wenn er sie hat [*From margin*: Man scheint allgemein im Busche grünen Thee zu trinken], und weiß von entlaufenen Kühen, Ochsen, Kälbern, Pferden zu sprechen und schmut' zige Geschichten zu erzählen. —

Die Ufer des Hunters flußes sind bei weitem fruchtbarer als die Umgegend von Sidney, ein reicher Alluvial boden giebt ohne Dünger in günstigen Jahren ausserordentl. Waizen un Mais erndten. Doch diese günstige Jahre sind selten. Trockene West un Nordwestwinde machen jede Hoffnung des Acker' bauers zu Nichte. Arbeiter fehlen die weiten fruchtbaren Landstrecken zu cultiviren und durch Anlegung künstlicher Taiche dem drückenden Mangel abzuhelfen. Die Arbeit ist .ßerordentlich theuer und man sinnt und spricht hin un her dem Uebel abzuhhelfen. Auf der eine Seite beabsichtigt man in England Geld zu borgen (a loan) um Einwanderer ein zu führen, auf der andern spricht man von der Einführung Indischer Bergcoolies, die Bergbewohner, welche als Arbeiter den Schweizern gleich in die Ebene kommen, um sich etwas Geld zu verdienen un dann in der Heimath zurück zu kehren. — Man sendet diese Coolies nach Mauritius und ist mit ihnen dort recht zufrieden; sie würden sich als Schäfer sehr nützlich machen können. Auch von der Einwanderung von Deutscher hat man gesprochen. Doch Capitalien fehlen. Die Aufhebung des Assignment Systems, wilde Speculationen, zu großer

Kredit, alle Mittel überschreitenden Luxus, Mangel an Arbeitern und Höhe des Arbeits lohnes haben für den Augen, blick diese Kolonie paralysirt. Ein un natüralischer Zustand war eingetreten, eine Art Fieber verrückte die Kopfe. Man wird in die Natürlichen Grenzen allmählich zurück kehren, nd obwohl diese Kolonie vielleicht nicht so rasch un außerordent, lich sich entwickeln wird, wie man voraussetzte, liegen doch viele Lebenskörner in ihr, welche wie nur die Bevölkerung dichter werden wird, nothwendigerweise ans Licht treten müssen.

[*Continued before 28th October.*] Auf meinen Waldexcursionen erregten die Buschfeuer oft meine Aufmerksamkeit. Einige behaupten daß während der heißen Winde die sich reibenden Zweige sich entzünden und beim Niederfallen das Gras in Brand setzen; andere schreiben die Feuer allein den Einge, borenen, den Sägern un Holzfällern bei, die überall zur Bereitung ihrer Speisen Feuer machen, ohne sie beim Weiter, ziehen aus zu löschen. Während der Hitze des Sommers vertrocknet das Gras und fängt deßhalb sehr leicht Feuer, die Blätter der Myrthengewächse voll von flüchtigen Oelen sind in dieser Zeit gleichfalls halb trocken ud so breiten sich denn die Buschfeuer schnell über große Räume aus ohne jedoch für Menschen gefährlich zu werden, da der Wind stets neue Luft nd Sauerstoff zuführt und der Rauch auf der entgegengesetzte Seite die Flamme erstickt, so läuft das Feuer stets gegen den Wind und oft habe ich der rothen oft mehrere 1000 Schritte langen Feuerlinie zugesehen, wie sie die vom Winde in sie hin ein gebeugten trockenen Grashalme begierig erfaßte und in den ihr begegnenden Eucalyptus sträuchern rasch un knisternd emporloderte. Sie zerstört nicht die Bäume, welche sie ihres Laubes beraubt, nicht oder nicht immer und obwohl sie ein schwarzes ödes winterliches Brandfeld hinter sich läßt, so ist sie doch Ursache größerer Fruchtbarkeit, indem nach dem nach, sten Regen feines junges Gras schnell hervortreibt und frisches Laub die entkleideten Zweige wieder deckt. — Man sagte mir, daß die Eingeborenen auf beschränktem Raume Buschfeuer

Reconnaissance in New South Wales

anzünden, um die Wallabis und Kangaroos zu der frischen Weide zu locken. Andere meinten selbst, aber meiner Ueberzeugung nach, irriger Weise, daß sie drch Buschfeuer Schlangen zu tödten suchen, welche sie dann geröstet mit großem Appetite verzehren.

Die Schlangen bilden selbst hier den Gegenstand täglicher Unterhaltung. [Keiner] der aus dem Walde oder von ein[em] Spaziergang zurück kommt verfehlt, zu bemerken, daß er eine .dr mehrere Schlangen gesehen oder getödtet. — Es giebt ihrer in einigen Gegenden .ßrordentlich viele und mehrere Arten, selbst die häufigsten sind sehr giftig, ihr Biß ist ohne schleunige Hülfe tödlich. Als wir nach Telligerry gingen, tödteten wir 5 schwarze Schlangen, einige $4\frac{1}{2}$ Fuß lang und eine Carpet Snake wegen ihrer braun un gelblich gemischte Farbe so gennant. Ich habe indessen viele Excursionen gemacht und habe nie eine Schlange selbst begegnet, obwohl ich mehrer tödtete welche meine erschreckten Begleiter mir zeigten. Die schwarze Schlange hat 2 drch bohrte Giftzähne und eine sehr große, bis hinter die Augen sich erstreckende Gift druse, die Carpet Snake wird für die giftigste gehalten, the Diamond Snake, welche oft 10 Fuß lang wird und den Hühner höfen großen Schaden thut, ist eine Coluber Art und nicht giftig. Ist man vorsichtig und macht man stets mit einem Stabe ein Geräusch zwischen den Gebüschen so wird man nie eine Schlange begegnen. Da sie ein .strordl. scharfes Ohr zu haben scheinen und schleunigst entfliehen. Oft wenn ihnen die Hohle fehlt, gleiten sie ins Wasser um sich zu verbergen; doch können sie nicht lange ohne Luft zu schöpfen in demselben verweilen.

Da ich mich entschlossen habe den bewohnten Theil Australiens zu durchwandern, versuchte ich gestern ein Pferd, welches ich zu kaufen beabsichtigte. Willst du deinen Freund auf einem Solchen Ritte durch den Australischen Busch begleiten? Sitze denn fest im Sattel; denn es ist kein Spaß mit diesen jungen Tollköpfen zu reiten, 'die allzeit auf den Rossen hängen' und durch das Gebusch wie auf Hirschen blindlings hindurch brechen. Man nennt bush gewöhnlich das un cultivirte Land im Allgemeinen, die wilde noch jungfräuliche Erde; dann aber unterscheidet man nach dem Karakter der Vegetation forest land, brush and scrub. Letzteres ist niedriges Gesträuch welches nie oder nur selten bis zur Höhe eines Mannes sich erhebt; [*From margin*: Scrub ein reiches Dickicht auf Alluvial boden]; der bush forest land dagegen besteht aus hohen kräftigen Bäumen mit niedrigen Gebüsch zwischen ein. Dichte Wälde, wie die im Harze und in der Schweiz habe ich hier nie gesehen nd ich glaube nicht daß sie hier existiren: die Baum individuen sind getrennt, wie in vielen Eichenwaldungen in Nördlichen Deutschland. — Wo nun enge Thäler un Felsschluchten ihres Schattens nd ihrer Feuchtigkeit wegen die Vegetation kräftiger machen steigen Schlingpflanzen an den Bäumen hinauf und ranken von einem Baum zum andern, so ein dichtes fast undurchdringliches Gemäsch bildend; Farrenkräuter wuchern in den Felsspalten oder kriechen an den Baum stämmen hinauf und Orkis arten schmücken Fels nd Baumstamme während ihrer Blüthezeit mit oft .ßerordentlich schönen Blüthentrauben. Während der Regenzeit stürzen sich wilde Bergwasser über die Felsen, entwurzeln Bäume nd führen sie mit sich nieder wenn sie nicht von Nachbarstämmen aufgehalten zwischen diesen für allmählige Vermoderung liegen bleiben. La vita e la morte qui si danno l'amplesso d'amor. — Wir ritten zu der Bretter wohnung eines Hrn. Dawson,[b] welches an der südln. Seite eines engen wilden Thales liegt das Hr. Wilton[c] der Prediger von Newcastle valley of palms gennant hat. Die Berg gegend um Newcastle zeigt einen eigenthümlichen Karakter. Man findet überall Kesselthäler (Cockpits) .lche theils zum Meere nach Osten oder zum Hunters flusse nach Westen sich offnen. Oft sind sie völlig geschlossen und ein natürliches Amphitheater ist gegeben. — In einen solchen offenen Keßel oder Mulden thale lag die Bretterwohnung gleich der Sennenhütte auf den Alpen; ein kurzer von der Sonne verbleichten Rasen deckte den Boden; an den Rändern [wuchsen niedrige] Eucalyptus bäume mit sonderbaren gewunden horizontal sich .sbreitenden Zweigen:

Reconnaissance in New South Wales

Man bestellt ein wenig Gartenland zur Gewinnung der nötigsten Gemüse, — 4 Bienenstocke Englischer Bienen standen in der Nähe des Hauses: die Unzahl der Ameisen macht es nothwendig die Pfähle auf .lchen das Bienen haus ruht mit Wasser oder Oel[brettern] zu umgeben, damit jene gefräßigen Thiere den Stöcken nicht schaden können. Man bewirthete mich mit einige Mutton chops Kartoffeln nd Thee. Dieser letztere fehlt, wie schon oben erwähnt bei keinem Mahle. Da man keine Milch hatte schlug man frische Eier in den Thee un machte ihn so recht schmackhaft. Da sie mir erzählten daß Sie seit 18 Monaten schon Bienen besäßen, daß ihnen aber die Mittel fehlten, den Honig .s zu scheiden, erbot ich mich sogleich ihnen den Honig zu verschaffen. Ich nahm einen alten Strohhut durchlöcherte ihn, warf einen Sack über den Kopf, zog zwei Paar Handschuhe an nd begann nun meine Operation. Sie waren über die Honig erndte nicht wenig erbaut. Die Australische Biene, welche ihren Honig in Baumhohlen sammelt hat keinen Stachel und deßhalb spüren ihnen die Eingeborenen nach, indem sie eine Arbeitsbiene fangen und ihr eine feine Feder daune an den Fuß kleben. Die Biene fliegt ihrem Stocke zu nd das scharfe Auge des Wilden folgt ihr dorthin nach. — Doch vor den Stachel führenden Europäischen Bienen hat der Wilde fast ebenso großen Respekt wie vor der Flinte. Wir ritten durch den Busch, um die wilden Kühe nd Ochsen zu sehen. Die Besitzer von Viehstationen pforchen ihre zur Weide bestimmten Ländereien ein und lassen nun im unzäunten mehrer Meilen langen ud breiten Busche das Vieh frei nd ohne Hirten nach Belieben herumstreifen. Alle Jahre fängt man die Kälber und brennt ihnen das Zeichen des Besitzers ein. [*From margin:* Alle die nach 6 Monaten nicht gebrannten Rinder gehören der Regierung, weil sie gewöhnl. auf Regierungs grd weiden.] Diese Freiheit und die seltene Gemeinschaft mit Menschen läßt nun die Rinder in den Zustand halber Wildheit zurückkehren. Beim Nahen des Reitenden Hirten (Stockman) starren sie ihn für einige Augenblicke wundernd an, schaaren sich dann zusammen und traben eilig davon. Der Stockman folgt ihnen auf dem schnelleren Pferde und sucht sie in engen Umzäunung ein zu treiben. — Squatters welche nur einen Run und keine Umzäunung haben, haben viel mehr Mühe das Davon laufen ihres Viehes zu verhindern. Auf den Viehstationen verfolgt man nach der Lage des Ortes und nach der Beschaffenheit des Bodens 3 Absichten: Man strebt die Zahl der Rinder zu ver- [mehren] und die Ueberzahl zu Verkaufen, oder man strebt die Rinder fett zu machen, oder endlich Butter u Käse zu bereiten. Immer ist es in diesen Wasserarmen Lande von höchsten Wichtigkeit das ganze Jahr hindurch frisches Wasser zu besitzen. — In den weiten Swamps in der Nähe des Hunters flußes wachsen besonders Carex Arten nd obwohl sie das Vieh eben nicht fett machen, so scheint es dießselben doch zu lieben und gedeiht wohl auf ihnen. Im Busche wird besonders das feinblättrige Kangaroo gras geschätzt. [*From margin:* Das Cooje gras, eine feine Grasfamilie des Cyper gehöriges Gras.] — Künstliche Futterung hat man in der Nähe von Stadten in der Absicht, mehr Milch und Butter zu gewinnen angewandt. Man hat zu diesem Zwecke Luzern, Gerste, Hafer gebaut, welche letztere 3 bis 4 mal un öfter geschnitten werden. Der Mais ist zu diesem [Zwecke] gleich- falls sehr vortheilhaft. Seine Einsaat erfordert wenig Vor- bereitung doch ist es später von großer Wichtigkeit ihn von Unkraut frei zu halten zu welchem Ende man ihn, wie die Kartoffeln häufelt u hackt.

[Ends with a flourish of the pen.]

[*Newcastle, N.S.W., c. 26th October 1842.*]

My dear friend,

 I left Sydney at the end of September as I wanted to see other parts of the colony. I confess too that I thought I might be able to find agreeable work away from Sydney. My plans, however, were by no means fixed and settled—my purely scientific interests, and my downright need of becoming independent, swayed me in turn and kept me in a most un-

Reconnaissance in New South Wales

comfortable state of indecision, so I came to Newcastle, where Mr Walker Scott has most kindly invited me to stay with him. He has considerable interests in the district, and all they need to make them very profitable is better and busier workmen.—I've made excursions in all directions from Newcastle, collecting plants, studying the geology of the area, collecting insects, and have tried to learn all I can about colonial agriculture and cattle rearing.—Newcastle itself is nothing but a few streets and houses. However, as the military barracks, which hold a regiment, are now built, houses are going up all over the place, and I've no doubt that the little settlement will soon become a thriving town.— The Hunter river, by the mouth of which it stands, is more than an English mile wide. Farther upstream there are several islands in it, of which Ash Island, belonging to Mr Scott, is 2500 acres in area. The North shore is formed by a low, narrow spit of sea-sand, but the South shore forms a kind of amphitheatre which contains the little settlement. The hills to the South and West, which are covered with thick turf, reminded me strongly of the English downs. Right in the mouth of the river there's an abrupt island, Nobby's Island, which was probably sundered from the mainland by subterranean forces ages ago. If you include the uppermost, clay-banded seam, there are 4 evenly disposed beds of coal here, of which only 2, the 3rd and 4th yield a coal at all fit for use. The seams are separated by beds of sandstone and clay, are as a rule very clayey in their upper layers and yield good, dense, bright coal only from their lower bands, but it can hardly compete with Newcastle upon Tyne for quality. The clay-banded coal and the clays contain a number of impressions of ferns which are not well enough preserved for identification though they could all probably be referred to existing genera.[a] *I think that the 3 kinds I have seen myself could be assigned to* Asplenium, Adiantum *and* Nephrodium. *Impressions of algae and fresh-water plants are also found, but it would be impossible to determine them. These evenly bedded rocks have often been shattered by subterranean forces, and several dykes of extraordinary regularity [of form] have burst through them on an East–West alignment.—The constitution of the coal (which leaves an ash when burnt, and so betrays its lignitic nature) and the ferns (which correspond to existing genera), are lively testimony of the geological youth of these formations. We* shall soon be convinced that the Southern Hemisphere has obeyed the same geological laws as the Northern, but that we can't correlate the succession in the one with that in the other. And the younger the formations, of course, the less they will resemble each other. I have come to think that Newcastle, on account of the regularity of stratification here, and because of its very characteristic coal-seams, is the centre from which geologists will have to work outwards, in their study of geological relations in New Holland.—I'm arranging to buy a horse and then, my hammer hanging from the pommel, to ride through the settled parts of the East coast.—*The flora of the Newcastle district is not nearly as rich as that of Port Jackson; but a few tropical plants, and other varieties very characteristic of New Holland, which are quite rare or poorly developed around Sydney, do make their appearance in full vigour and character here. There's extraordinary variety amongst the ferns in the narrow, shady ravines,* [where] *the grass-tree* (Xanthorrhoea arborea) *and the tree-fern (arborescent fern,* Alsophila australis) *form thickets or reach a height of 12 feet.* Doryanthus excelsa, *a lily-like plant* [from margin: *(the giant lily)*] *sends up a flower-stalk 15 feet tall, and its liliform flowers are grouped into a head which often exceeds that of a man's head in circumference. The corolla contains half a wine-glass full of honey-sweet water in which hundreds of flies find a grave. Australian flowers in general are remarkably rich in honey. In some, like the* Banksias, Lambertias *and* Styphelias *there's an extraordinary amount, though in others it's quite lacking. Often, when I've been tired and thirsty, I've bitten off the base of a tuft of* Lambertia formosa *flowers to suck the delightfully sweet honey out of them.—*

It was at Newcastle that I first encountered the naked savage. In the town they have to wear clothes, so the men wear cast-off coats and trousers whilst the women wrap themselves in coarse sacking. One day I was going along the sea-shore when I met an aboriginal coming towards me, walking lightly and briskly. He had a waddy in his hand and was carrying a burden on his head, and he was completely naked. He had a white band tied around his forehead, a light Spanish cape flapping over his shoulders, and a hairy strap of opossum skin around his waist. He was not in the least embarrassed by his nudity. His body was well pro-

portioned, the limbs lank and lean though the muscles were well developed, the chest very deep, and the penis, probably because of the cold, remarkably small. Their skin is nothing like as velvety as that of a negro; the hair is coarse, raven-black and curling but not crinkled like that of a negro at all; the orbital ridges are extraordinarily prominent, particularly at the inner angle; the nose is flattened and the nostrils wide; the mouth is very wide indeed, with long lips as mobile as those of a horse. Their molars get worn down like those of Egyptian mummies, and the canine teeth of a young man looked to me almost the same as those of a dog, lobed [small drawing]. Although the women are extraordinarily ill-favoured white men don't mind fornicating with them perhaps because there are so few white women. The [half-caste] children are better built [than the full-blooded aborigines] but are unpleasantly light in complexion. The bastard boys are killed when they're 9 or 10 years old but the girls are spared. There's a number of them in Newcastle, mostly nick-named 'Yellow'. The blacks are extremely lazy. They can work, nevertheless, and many of them are very intelligent. If we compare slavery amongst the lower or less civilised races with the schooldays of children, these [Australian] savages, compared with Negroes, are like uneducated, as compared with educated, children. If it be in the interests of mankind to preserve so subordinate a people, the philanthropists who oppose slavery are in the wrong; for nothing but compulsion to work, and strong discipline, could save and civilise these savages. Without compulsion nothing will induce them to undertake steady work. Whatever they get from mistaken benevolence or casual work they squander on spirits (botallago). They think it a great treat to drink the rinsings from brandy casks, which they call [bull].

On another occasion I saw some men, women and children catching crabs during the ebb of the tide, and gathering a kind of stalked ascidian which covers the rocky sea-shore between tide-marks in extraordinary numbers and looks like cabbage stalks. They call these creatures congivoi, and are adept at ripping them out of their sleeves with small sharpened sticks. To catch crayfish the men go right out to the outermost, submerged rocks, where they plunge head first into the deep fissures between the rocks and drag the crayfish out of their hiding places by the antennae.

People here deny them all trace of natural courage. All they can do, I've been told, is to sneak up to their enemies in the dark and murder them treacherously.—Yet they're afraid of the dark because of evil spirits; and, if a member of the clan has committed an offence he is submitted to a kind of trial by ordeal. He is allowed, as a shield, a small board which has a handle in the middle of the inner side, and all the men of the clan proceed to hurl spears at him. They are so practised at parrying the spears, however, that they are rarely killed or even wounded. But in my opinion it's no harder than fighting with broadswords or even with foils, as the spears are so long. [From margin: They fear and respect the white man in proportion to his firmness and curtness in dealing with him, and they seldom offer him bodily resistance. A pistol or a gun terrifies them, since they know what a deadly weapon it is, particularly as you can't even see the danger coming. They will never meet the white man in the open, but if resentment be aroused the black man will lurk behind trees, ambush the white, and spear him in the back—a budgery fellow!]

The settlers make a practice of giving the strongest and most respected man of his clan a brass plate to hang from his neck, with his name engraved on it as a sign that they regard him as the king of the clan. These kings, however, are no better than the rest and are looked upon with contempt. When we asked a young man if he would like to become a king he replied that he would not, as there were too many kings already. When we asked him why he did not wash himself as the sea and the river were always there for the purpose, he said that his people could not be bothered doing things like that.

I've already tried my luck in the bush once or twice. Although Mark has often mentioned his lonely little shanty, he may not have given you a clear idea of this simplest of all manifestations of the builder's art.— Major Crummer, a police-officer at Newcastle, and I have been to Telligerry where he has a cattle station and is thinking of setting up a dairy (*Milchmeierei*). After 3 hours of hard going along the edge of a vast morass (*swamp*) where you could see scattered herds of horned cattle all over the place, we reached some firm, open pasture-ground that had been cleared of trees. Here there were a few huts, which would look to you rather like the little sheds the peasants build for their pigs and

Reconnaissance in New South Wales

geese at [Vitnio?]. They were [each] about 15′ long, 8–9′ wide, the walls made of slabs of timber roughly put together, the roof of big sheets of Eucalyptus bark. The cracks were stopped up with a mixture of clay and cow dung. The inside was just the hut itself [was undivided and unlined?], the furniture a small wooden form, a small table and some 'sideboards' [shelves?] made of slabs of wood, with plates, cups, the teapot and sugar canister [on them]. And there was a fireplace 7′ square with a capacious chimney rising above it. The fireplace has to be very big so that the fire can be kept away from the slab walls, lest they themselves catch fire. There were four of us, and you've no idea how hard it was for us to find room. We had brought our provisions with us, and soon sat down, two by two on forms, [one] at each corner [of the table]. If one of us got up without warning, down went the other with his plate, cup, knife and fork clattering after him. We were no end of a joke to the Cockneys from Newcastle. For the night we spread a soft couch of grass and ferns, and, as I lay there stretched out, watching the light from the bright fire playing over the inside of the brown roof, I had such a sense of well-being as I have enjoyed nowhere else but in the dusk at an English fireside. Two young men who had been looking for run-away cows, came in as they were passing by. A wide-brimmed straw hat secured to the buttonhole of a white blouse by a slack, black cord, a coloured or blue-striped shirt, trousers to his liking, and heavy boots, make up the attire of the young stockman. He stands not on ceremony, seldom shaves, sleeps in his shirt, eats damper, drinks green tea 3 times a day with brown sugar and milk if he can get them [From margin: They all seem to drink green tea in the bush], and expresses himself in talk about run-away cows, bullocks, calves and horses and in telling smutty yarns.—

The banks of the Hunter river are much more fertile than the land around Sydney. The rich alluvial soil yields extraordinary harvests of wheat and maize in good years, without any manuring. But good years don't come often. Dry westerly and North-westerly winds [can] dispel all the farmer's expectations; there are not enough hands to till great tracts of fertile land or to construct the dams that would compensate for the insistent shortage of water; wages are extraordinarily high, and some people are thinking and talking of how to overcome the disadvantages.

Leichhardt's Letters

They have in view, on the one hand the idea of raising money in England (a loan) with which to promote immigration, and on the other that of introducing coolie labourers from India, hill-people who go down to the plains to earn money and then return home, like the Swiss. Indian coolies are sent to Mauritius, where they are quite satisfied with them, and they would be very useful here as shepherds. There's also been talk of German immigration. But capital is lacking; and the suspension of the assignment system, wild speculation, easy credit, extravagance beyond all reason, shortage of labour, and high wages, have paralysed this colony for the present. An unnatural state of affairs had come about, and a kind of fever had turned men's heads. Things will gradually adjust themselves to proper bounds, and, although the colony may not grow like a mushroom as it was supposed to do, vitality smoulders within it, and must, if only the population increases, surely burst into flame.

[Continued before 28th October.] During my excursions in the bush my interest in bush fires has often been aroused. Some people maintain that friction between boughs rubbing together in hot wind sets them on fire, and that the falling embers fire the grass. Others ascribe them entirely to the blacks, the timber-getters and the sawyers, who light fires all over the place to cook their food but leave them unextinguished. During the hot Summer the grass dries out and becomes highly inflammable, and the leaves of the myrtaceous plants, which are full of essential oils, also get very dry. The consequence is that bushfires quickly spread over enormous areas, though without becoming a danger to human beings. Because the wind brings a steady supply of fresh air and oxygen, and the smoke extinguishes the flames to leeward, the fire always advances into the wind. I have often watched a red line of fire, which may have been several 1000 paces in length, greedily consuming the dry grasses bowing towards it in the wind, and blazing up with a crackling sound through the Eucalyptus saplings that were in its way. It strips the trees of their foliage but rarely kills them. And, although it leaves a blackened, desolate, wintry, burnt-out area behind it, its effects are quickening, for after the next rains fine young grass springs up, and tender foliage begins to clothe the denuded boughs. I was told that the aborigines burn the bush over limited areas in order to attract wallabies and kangaroos to the fresh pasture. Others even said, though I think

Reconnaissance in New South Wales

wrongly, that they do it in order to kill snakes, as they've a great appetite for roasted snake.

Snakes are the subject of every-day talk here. Nobody who has been in the bush, or has come back from a walk, ever misses telling you that he has seen or killed a snake or two. In some localities they're extraordinarily numerous, and several species, including the commonest, are very venomous, their bite being fatal unless treated at once. On our way to Telligerry we killed 5 black snakes, some of them $4\frac{1}{2}'$ long, and one carpet snake, so called on account of its mixed yellowish colours. Still, I've made a number of excursions on which I've seen no snakes myself, although I've killed several that have been pointed out to me by startled companions. The black snake has two perforated [hollow?] poison fangs and a very big venom gland which extends back to behind the eyes. The carpet snake is considered to be the most venomous. The diamond snake, which can grow to be 10 feet in length and which does a lot of harm in fowl yards, is a species of Coluber and is not venomous. If you're cautious and always make a noise by swishing a stick amongst the bushes, you'll never see a snake, as their hearings seems to be acute and they slither away at once. Often, if they can't find a hole, they'll slide into the water to get out of sight, but they can't remain there long without breathing.

As I've made up my mind to traverse the inhabited parts of Australia, I went yesterday to try a horse that I've been thinking of buying. How would you like to come with me on such a journey on horseback through the bush? Well, just sit firmly in the saddle, as it's no joke to go riding with these young dare-devils who live in the saddle and go charging blindly through the bush as if they were mounted on stags. 'Bush' generally means the uncultivated country in general, the natural, virgin world; but they also distinguish between forest land, scrub and brush, according to the character of the vegetation. The last consists of low bushes that rarely reach a man's height. [From margin: Scrub, a rich thicket on alluvial ground.] The [true] bush (forest land), on the contrary, consists of tall, strong trees with low undergrowth between them. I have never seen dense forests like those in the Harz and in Switzerland here, and I don't think that there are any. The trees [here] stand far apart, like they do in many of the oak forests in northern Germany. All the same, where the shade and moisture of narrow valleys and ravines favour the vegetation, vines cling to the trees and creep from one tree to another, forming a dense, almost impenetrable network; ferns grow luxuriantly in crannies in the rocks or creep up the trunks of the trees; and species of orchids adorn both rock and tree in season, with sprays of extraordinarily beautiful flowers. During the heavy rains torrents of water rush down the ravines, uprooting trees and carrying them away, unless they get caught and held in the tangle of other trees and boughs, there to lie and moulder slowly away. La vita e la morte qui si danno l'amplesso d'amor [Here life and death unite in love's embrace].— We rode to the weatherboard home of a Mr Dawson,[b] which lies to the South of a rugged valley that Mr Wilton,[c] the Newcastle clergyman, has named the Valley of Palms. The higher country near Newcastle has peculiar features. In many places there are basins (cock-pits), some of which open eastwards towards the sea, others westwards towards the Hunter River. Some of them are so nearly enclosed as to form natural amphitheatres. The little wooden house, looking like an Alpine cowman's hut, stood in one of these open pockets, in a patch of grass that had been bleached by the Sun, around which there were low Eucalyptus trees growing whose strangely twisted boughs spread out horizontally. A small patch of ground had been dug for growing the necessary vegetables. There were four beehives with English bees close to the house, and the swarms of ants had made it necessary to surround the posts that supported the apiary with water or with [oiled planks?], so as to prevent these voracious creatures from raiding the hives. I was regaled with mutton chops, potatoes and tea. As I said above, they drink tea with every meal. They had no milk so they stirred fresh eggs into the tea, which made it taste very good. When they told me that they had been keeping bees for 18 months but had no means of extracting the honey I volunteered to extract it for them straight away. I took an old straw hat and made a few holes in it, put a sack over my head, put on two pairs of gloves, and began my operations. They were highly delighted with the amount of honey I got. The Australian bee has no sting and stores its honey in holes in the trunks and boughs of trees. To find the honey the natives catch a worker, attach a tuft of down to one of its feet, and their sharp eyes follow it as it flies towards its hive.—For the sting of the

Reconnaissance in New South Wales

European bee though, they've as much respect as they have for a gun.—
We went for a ride in the bush to see wild cattle. The holders of cattle stations enclose the parts of their property meant for [sheep] grazing, and then let the [horned] cattle roam about untended in unfenced tracts of bush several miles broad and long. Every year the calves are rounded up and branded with the owner's private mark. [From margin: All unbranded cattle more than 6 months old belong to the Government, as they generally graze on Crown land.] In their freedom and comparative ignorance of mankind the cattle revert to a half-wild state. On the approach of a mounted cattle-herd (stockman) they glower at him in wonder for a few moments, then bunch together and trot smartly away. The stockman, who can easily catch up with them on his horse, follows, and tries to drive them into fenced enclosures.—Squatters who have only one run and no fencing have much more trouble in preventing their cattle from running away. On the cattle stations there are 3 things that they endeavour to do, according to the situation and the nature of the ground: to increase the size of herds and sell the surplus beasts; to fatten the cattle; and to produce butter and cheese. In this dry country it's always of the utmost importance to have drinking water throughout the year.
In the broad marshes beside the Hunter River the grass is mostly species of Carex, and, although it's not very fattening, the cattle seem to like it, and they do well on it. The most highly prized of the bush grasses is the fine-bladed kangaroo grass. [From margin: Cooch grass, a family of fine grasses of the Cyper group.] They're thinking of trying prepared fodder close to the towns, to increase the yield of milk and butter. Lucerne, barley and oats have been grown for the purpose, of which the last are cut 3 and 4 times or more. Maize likewise has great advantages for the same use. It needs little preparation for sowing, but has to be kept free from weeds as it grows, so they hoe it and hill it the way you hill potatoes. [Ends with a flourish of the pen.]

65 [TO DR WILLIAM NICHOLSON, Newcastle upon Tyne]

Newcastle, N.S.W., Octbr. 31 [1842]

Mein theuerster Freund,

Am 28tn October 3/4 auf 6 am Morgen, wurde ich durch eine heftige Erschütterung des Hauses .fgeweckt, welche mir durch das Niederfallen eines schweren Körpers gegen das Haus veranlaßt schien. Hrn. Wilton u Crummer indessen hatten es gleichfalls gefühlt, obwohl sie mehrere Tausend Schritte von uns entfernt wohnen nd es blieb keinem Zweifel .trworfen, daß es ein Erdstoß gewesen. Andere Personen hatten um 3 Uhr des Morgens einen ähnlichen Erdstoß empfunden. [*From margin:* Hr. Wilton der Prediger des Ortes sagte mir, daß dieß der 4te Erdstoß sei .lche er während ss. Aufenthaltes in Newcastle empfunden 1837–1841. Hr. Clarke meldet daß Windsor von einem Erdbeben erschüttert wurde 20 Minuten von 6 a.m. Friday 28th Octb. Das Erdbeben wurde gleichfalls in Pt Stephens un Stroud in der Nähe von Pt Stephens gefühlt. In Pt Macquary wurde die Erderschütterung sehr stark gefühlt. *Later, in different ink:* Am Macleay River. Am Paterson. In Van Diemen's Land 1788, 1801, 1803, 1804.] — Daß diese Gegend heftigen Erderschütterungen ausgesetzt gewesen sein muß, geht aus der Gegenwart mehrerer Dykes und tiefer Spalten, welche vielleicht nicht völlständig von flüssig aufstrebenden Massen gefüllt wurden, so wie aus der wunderbar regelmäßigen Zertrümmerung des losen Sandsteins hervor, welcher .tr dem 4tn Kohlenbetter vor Morris Bad und ehe man zur Long beach kommt, das Ufer des Meeres bildet. — (Die Richtung der dykes ist von S Ost by S. nach NW by Nord.) — Man glaubt hier auf einem regelmäßigen Netz oder Mauerwerk zu stehen, welches drch die hervorragenden harten Kanten der Trümmer stücke gebildet ist. In Sidney haben ähnliche stets in einer bestimmten Richtung den Sandstein drch setzende Spalten lange meine Aufmerksamkeit beschäftigt nd es war besonders am Fort Macquary .ter dem Wege zum

Reconnaissance in New South Wales

botanischen Garten und am Wasserfalle[a] am Nördl. Ufer, wo ich sie beobachtete. Das indessen Vulkanische Produkte nicht so nahe waren, glaubte ich zwar, daß Erd[stöße] nd Erd⸗ [wellen?] jene Spaltungen hervorgebracht doch war meine Ueberzeugung weniger innig, als hier wo ich mit meinem Hammer die die Spalten erfüllenden Vulkanischen Gesteine loßbreche. [*From margin:* Concentrirte Stoße bemerkt man recht schon, wo man zuerst zur Ufer niedersteigt. Man sieht ihren Kreis runden Umfang. .s strahlende Spalte vom Mittel⸗ punkt in regelmäßige Zertrümmerung vom Umfange. Alle Hauptspalten haben eine SO by S-NW by N Richtung — die auf ihnen senkrecht stehenden Spalten sind kurz ud krümm nd unregelmäßig.] Die Mauer u Netz förmige Reliefs von Morriss Bad erkläre ich mir folgendermaßen. Erdwellen, .lche stets in einer Richtung wanderten Spalteten die Felsen bis zu dem Heerde vulkanischer Thätigkeit, (welche hier viel⸗ leicht nicht so .ßerordentl. tief liegt) zu eng um flüßige Gestein massen in sich .fsteigen zu lassen, waren es nur die sich ent⸗ wickelnden Gase und besonders Eisen, welche sich erhoben und die Wände der Spalten [ver]härteten nd braun färbten. Wie nun das Meer lose Gerölle über den Sandstein hinwusch, wiederstand[en] die Harten eisenhaltigen Ränder der Trüm⸗ merstücke der zerreibenden Gewalt, während der innere Theil sich zerbröckelte nd allmählig vertiefte.[b]

[*Continued,*] Novembr. 1. Mein theuerster Frd., Gestern brachte mir Hr. Calvert[c] einen Caprimulgus albogularis (the white throated Goatsucker) den er geschossen nd ich machte mich sogl. dabei ihn ab zu häuten. Es fiel mir auf, seinen Magen soweit nach hinten gerückt zu finden nd als ich mit dem Abziehen fertig war, ex[aminir]te ich näher die Lage des Magens nd seinen Inhalt. Er war .ßrordentl. groß un voll un füllte für sich selbst fast die ganze Bauchhölle .s, indem er die Gedärme nach oben n hinten drängte. Der Vormagen zeigte längliche in 3 Reihen geordnete Drüsen, der Mittelmagen war mit dicker Epidermis bekleidet doch die Muskeln waren nur schwach. Er war gepfercht voll von Gryllotalpa, Scarabaeen, einer kleinen Cerambyx un einem den Melolonthiden zu gehörigen Käfer; der Magensaft, welcher die weichen Gryllotalpa umgab nd schon zum Theil verdaut hatte, hatte keine Wirkung auf die harte Bedeckung der übrigen Käfer; doch hatte die Thätigkeit der Magenwände mehrere zusammen gedrückt. — Just als ich mit diesem Vogel fertig war, brachte mir Hr. Bligh der Neffe des früheren Gouverneurs einen andern, blauglänzenden Vogel, .s dessen breiten Schnabel noch Gryllotalpa beine hervorstanden. Die Lage des Magens nd die innere Bildung waren dieselben; .ßer Gryllotalpa nd Scarabaeen enthielt er eine große Menge von Bienen, alle in einem halb verdauten Zustande, doch .ch hier wiederstanden Elytra nd Flügel ud die harte Bedeckung des Körpers dem Einfluß des Magensaftes. — Die Lage des Magens erinnerte mich an die des Kuckkucks und an die Ursache, warum dieser Vogel seine Eier in das Nest anderer Vögel legen soll. Wäre es wahr, daß der Magen ihn am Brüten seiner Eier hinderte, so wäre vor.s zu setzen daß die beiden Capromulgidae dieselbe Unbequemlichkeit fühlen nd dieselbe Abhülfe suchen [würden?]. Es scheint aber, daß der Europ. Ziegenmelker sich selbst ein einfaches Nest baut. Ich glaube dieß .ch vom Australischen nd bezweifele deßhalb die Richtigkeit jener Erklärung. —

[*Continued*] Newcastle dn 5tn November. Mein theuerster Frd., Ich komme noch einmal auf die Dykes zurück, welche den geschichteten Gesteine durchsetzen. Ich hatte stets versäumt ihre Richtung genau zu bestimmen und stieg also Gestern zum Meeres Ufer nieder, un meine Unterlassungssünde gut zu machen. Ich habe dir schon gesagt, daß der Dyke in Nobbys Island so nah als möglich von SSOsten by S. nach NW by N streicht. Ich war deßhalb nicht wenig verwundert zu finden, daß die Richtung des ersten Dyke unter Southhead von S 1/4 W nach N 1/4 East ist, während die des 2tn von SSEast–NNW., die des 3tn vn S 1/4 East–N 1/4 W. — Hiermit verglichen dann die Spalten vor Moris's bade und sie ergeben ein Streichen von S ½ E–N½ W also dem letzten nd nächsten dyke ausserordentlich nahe. [*Marginal diagram.*] Vergleichen wir diese verschiedenen Richtungen, so ergiebt sich nicht eine einfache Linie sondern ein Band welches zwischen S Ost by Süd un Süd ¼ West liegt (3¼ Punkte des Compaß 36° 25'). Ist nun an zu nehmen daß die Fortpflanzung des Vulkanischen Stoßes auf diese Richtung senkrecht steht, so werden wir vielleicht nach Neu Calidonien nd nach den diesem benachbarten Vulkanan geführt. Unter dem Gefängnisse hat man bei tiefe Ebbe Gelegenheit, eine Art von Charybdis[d] zu beobachten. Eine Spalte 24 drch setzt die Felsen, welche in der Tiefe wahrscheinl. durch das beständige Waschen des Meeres .sgehölt sind. Sie weitet sich ein wenig gegen das Ufer zu nd bildet eine Art Trichter. Wie nun das Meer sich zurück zieht, stürtzt das Wasser in den Schlund hinab und läßt die von Seethieren besonders von Balanus bedeckten Felsen wieder zu einer Tiefe von vielleicht 30' sichtbar. Die nächste Welle dagegen füllt die unteren Höhlungen .s und treibt das Wasser schäumend drch die enge Trichter röhre empor. Dieses [regelmäßige] Niederstürzen nd Aufzischen des Wassers erregte meine Aufmerksamkeit .ßerordentlich und ich hätte diesem wilden Spiele Stunden lang zu sehen können. Es giebt gewisse Eindrücke welche für sich selbst unsere Seele erfüllen, nicht indem sie alter Erinnerungen beleben, oder uns zu wissenschaftlich strenger Beobachtung [und zu] möglicher Erklärung reizen. Sie sind gewaltige Anschauungen, .lche unsere Seele übermannen, denen wir uns leidend dunkler großer Empfindung hingeben.

[*Continued*] Novbr. 7. Mein theuerer Freund, Soeben komme ich von einer Buschexpedition zurück. Zwei junge Männer, welche als Steerage Passengers auf demselben Schiffe mit mir von England gekommen waren, erkannten mich hier wieder nd luden mich ein, sie in ihrer Buschwohnung zu besuchen. — Diese befindete sich in einer sehr interessanten Gegend an der andern Seite des Palmenthales in der Nähe von Red Head, welche letztere Localität ich geologisch zu untersuchen beabsichtigte. Als ich über Shepherds Hill ging fand ich Hakea pugioniformis nd dactyloides in Blüthe, welche beide einen angenehmen süßen Geruch haben. Banksia collina bildet neue Schooße. Ich habe schon früher bemerkt daß Banksia serrata blüthe und jetzt junge Früchte zeigt. — Die verschiedenen Glieder einer Familie scheinen drch.s nicht in diesem immer milden Klima an eine Blüthenzeit gebunden. Gesträuche wie z. B. Westringia rosmarinifolia scheinen immer in Blüthe zu stehen. — Wie ich über das flache Felsen ufer hin schritt in welches das Meer durch Spalten überall eindringt, stiegen hier und dort wie größere Wellen gegen die scharfen Klippen anschlungen schäumend schneeweiße Wasser auf, welche der Naturdichter vielleicht nicht verfehlt haben würde als die wehenden Schleier der Nereiden zu begrüßen. Wie der Matrose sich säubert und schmückt, wenn er dem Lande sich nähert, so nahen sich die Wellen gleichfalls im weißen Kopfputz dem Lande und erzählen tosend dem lauschenden Ufer von den fernen Windern, die sie gesehen. — Auch während eines sanften Seewindes treiben ausserordentlich lange hohe Wellen gegen die Ufer dieser Küste; während stürmischen Wetters ist die Brandung .ßerordentlich und man erkannt daß ein weiter Meere vor diesen Ufer sich .sbreitet, auf welchem kein Hinderniß der wachsenden Größe der Wellen entgegensteht bis sie nach einer Wanderschaft von Tausenden von Meilen gegen Neuhollands Küsten anbranden.

Als ich im hohen Kesselthal bei der Romantisch gelegenen Hütte meiner beiden Freunde anlangte, fand ich sie just beschäftigt den Honig von dem Wachse zu trennen. Ich ersparte ihnen die Mühe, die sie sich gaben, indem ich die Waben mit dem noch in ihnen befindlichen Honig über ein gelindes Feuer setzte ud den Honig .slaufen ließ. Nachdem dieß geschehen, wusch ich die Wachsreste mehrere Male .t Wasser nd füllte dieses nun zur Gährung in ein kleines Fäßchen. Bald nachdem unsere Arbeit vollendet war kamen zwei Eingeborene welche sich nicht wenig an der noch in den Wachsresten vorhandenen Süßigkeit erquickten, indem sie

dieselben sorgfältig .ssaugten, in Wasser tauchten nd wiederum
.ssaugten. — Am Sonntag Morgen ging ich nach Great Head,
um zu sehen, ob die Kohlenlager wesentliche Veränderungen
zeigten und ob ich vielleicht noch mehrere dykes finden
könnte. — Es fiel mir auf, daß .f einigen sehr beschränkten
Theile der Küste eine Menge von Muscheln aufgehauft waren,
von denen sich oft wenige Schritte weiter keine Spur fand.
Dieß hängt auf jede Fall von der Natur des Meeresbodens ab,
welcher entweder einer gewisse Gattung von Muschelthieren
reichliche Nahrung bietet, oder aber eine größere Strömung
verursacht, welche alle Muscheln, wie sie die Kraft verlieren
sich in der Tiefe zu halten, sogleich auf das Ufer schleudert.
Hier fanden sich Haliotis Parmophora und Triton in Menge.
— Das 3te Kohlenbett ist sehr entwickelt: Unter demselben im
Sandstein finden sich Eindrücke sehr großer Farrenkräuter un
Equisetums. — Auch hier ist das Felsen Ufer in regelmäßige
rhombische un trapezoidische Tafeln zerspalten und tiefe
Risse, wahrschl. nicht völlig erfüllte dykes, bieten das Schau⸗
spiel hoch aufschäumender und niederstürzender Wasser,
dessen ich schon früher Erwähnung gethan. — Die Richtung
dieser Risse ist von S by East to North by West. In dem
Conglomerate, .lches das Ufer bildete fand ich in Silex
verwandelten Baumstämme, während auch die in Eisenerz
verwandelten nicht fehlten. — Es was Flutzeit und wir konn⸗
ten nur mit Schwierigkeit an der Küste gegen den Endpunkt
von Red Head herumgehen. Große von der Atmosphäre
angefressenen Blöcke lagen unregelmäßig übereinander: das
Gestein war Conglomerate: die Felsen waren senkrecht.

Zurück kehrend fanden wir auf den mit sparsamer Vegeta⸗
tion bedeckten Sanddünen Leucopogon Richei, in dessen
Schatten ein Geranium blüthe. Auch ein Gewachs .lches mir
eine neue Art von Pimelia zu sein schien wurde gefunden.
Wormia prangt noch immer mit ihren goldgelben großen
Blüthen; Ipomaea war dicht über dem Meere. Eine große
Menge von Gewächsen mit fleischigen Blättern scheinen die
Nähe des Meeres zu lieben. Die oben erwähnte Familie,

Tetragonia, Salicornia, eine dem Leontodon verwandte Composita, Senecio.

Vor dem Mittag essen untersuchte ich die geologischen Verhältnisse des Klippen drch schnitts dicht vor der Lagune. Ich werde die nähere Beschreibung in meine geologischen Bemerkungen geben. Wir gingen zum oberen Ende der Lagune nd sahen hier die Wirkung des Buschfeuers, die Lagune hat eine ungefähre Richtung von SO–NWest, wäre also in Bezug .f Lage der Kultur nicht ungünstig. — Die beiden Wilden welche während der Nacht der Hütte genüber ihr Lager aufschlugen waren John Magill [*sic*] der König des Macquary stammes^e und Gorman der König eines andern Stammes. Sie waren weder von Weibern noch von andern wilden begleitet. Magill kam in die Hütte, forderte Feuer un Kessel: Calvert gab ihm die Honigreste, worüber er sich nicht Wenig freute und Mahl mit welchem er wohl verstand Doboys zu bereiten. Es war indessen nicht eben erbaulich, zu sehen, wie er beim Doboy kneten seine Pfeife rauchte. Er nahm den Kessel, in .lchem noch das Waßer war, indem Calvert 2 Hühner gekockt, um darin seine Doboys zu kochen. — So nährten die beide Natur menschen .tr einem dicken Eucalyptus baume ein kleines Feuerchen, neben welches sie sich läßig .sstreckten bis ihre Speisen zubereitet waren. Dann ließen sie nicht ab zu essen, bis sie den letzten Brocken verschluckt hatten und schliefen nun trotz eines etwas regnerischen Nacht bis in den späten Morgen hinein, stets ihr Feuerchen wieder schürend, wie die Kälte sie packte.

[*Continued*] Nov. 8. Mein theuerer Freund, Du weißt mit welcher Liebe ich Alles was auf Weinbau und Kelterung Bezug hatte aufnahm und wie oft ich mit dir über die Möglich⟋ keit sprach, in Neuholland die Reben zu ziehn wie der alter Samierkönig Ankaeos.^{f} Meine Mittel erlaubten mir nicht, Reben von Frankreich mit zu nehmen, wie ich zuerst beab⟋ sichtigte, und dieß war auch recht gut; denn es würde mir hier unmöglich gewesen sein, sie entweder selbst zu pflanzen oder verkäuflich an den Mann zu bringen. Man sprach indessen bei

Reconnaissance in New South Wales

meiner Ankunft viel über die Kultur des Weines^g ud Mons de Ligny ein Franzose machte mehrere Reisen zum Hunters flüsse un zu andern Thln. der Kolonie um zu sehen, ob denn auch der Boden zur Kultur geeignet wäre. Meine Ausflüchte um Sidney, welche mich überall nur über einen armen wasserlosen sandigen Boden führten, liessen mich zur Ueberzeugung kommen, daß sich Port Jackson nicht für die Cultur der Trauben eigne. Ich hatte guten Wein immer nur auf Kalkstein oder vulkanischem Boden gefunden und obwohl ich zu gebe, daß sich ein leichter wohl trinkbarer Wein auch auf Sandsteinboden z.b. auf dem Pagelflur u Molasse bergen der Schweiz ziehen lasse, würde dieser doch nicht die Kolonisten befriedigen, welche einen schönen Wein zur Ausfuhr zu gewinnen trachten, der ihnen zugleich die Einfuhr der Spanischen un Madeiraweine vermindern oder gänzlich aufheben sollte. Mons de Ligny ließ mehrere Aufsätze im Herald einrücken in welchen er zeigte, daß der Boden des Hunters flusses sich sehr wohl für die Kultur des Weinstocks eigne. Er machte sich verbindlich die Aufsicht von Wein bergen zu übernehmen u die Kelterung vor zu stehen, wenn sich Grundbesitzer zur Anlegung von Weinbergen n Weingärten geneigt finden sollten. Er verlegte seinen Aufenthalt auf Raimond Terrace, wo er, wie es scheint, Anstalt trifft seinen Versprechungen nach zu kommen. Als ich nach Newcastle kam, überzeugte ich mich gleichfalls daß dieser Boden, sind nur Arbeiter in hinlängliche Zahl vorhanden, sich gar bald mit Reichen Weinpflanzungen bedecken könnte und daß, obwohl wir keinen Madeira zu erwarten haben, doch ein recht guter Wein sich erziehen lassen würde. Ich sprach mit Hr. Scott, der indessen so sehr verlangt Geld im Großen zu machen, daß er schwerlich einem solchen Unternehmen hinlängliche Aufmerksamkeit schenken möchte. Er scheint indessen doch zu wünschen, daß wir sobald die Zeit der Kelterung heran rückt ein [sic] Versuch zum Pressen machen. Viele Besitzer am Hunters flusse machen indessen recht ernstliche Anstalten. Hr. Captain Scott hat einen Deutschen

mit Namen Luther,[h] mit welchem er den Gewinn theilt.
Hr. Robert Scott hat gleichfalls einen Weingarten mit einem
Französischen Weingärtner, der indessen seinen Platz verlassen zu haben scheint. Der Saame ist ausgesät oder in der
Aussaat, wir haben zu erwarten, wie man die jungen Pflanzen
pflegen nd was für Früchte man gewinnen wird. Meiner
Theorie nach wird man nur auf dem Kalkboden des obern
Hunter, im Wellington valley, auf dem Vulkanischen boden
von Portland Bay (Pt Philipp) gute edle Weine gewinnen.
Dieses Land hat indessen schon in so vieler Beziehung merkwürdige Ausnahmen gemacht daß ich noch nicht wundern
würde, wenn man wirklich bessern Wein auch .f ungünstigem
Boden hervor brächte. — Ich habe in den Engl. Zeitungen
viel von Deutschen Ansiedlern in der Kolonie gelesen, welche
.ßerordentlich gute Weine im Großen keltern sollen. Doch
wen ich auch hier in der Kolonie fragte, keiner wüßte mir zu
sagen, wo sie lebten und von wannen jene großsprecherischen
Nachrichten kommen konnten. Die Weine welche ich in der
Kolonie kostete waren alle fade, obwohl sich einige recht gut
trinken ließen. Hr. Macarthur in Camden[i] besitzt einen
großen Weingarten un keltert seinen Wein mehr als Liebhaber, denn als Wein bauer. Man sagte mir daß er einen recht
guten Wein machte. [*From margin:* Hr. W. Brooks hat am
Westl. Ende vom Macquary see einen Weingarten begonnen.
Der Boden ist gut d[er H]olz gedeiht üppig, doch da es nicht
gehörig verschnitten wurde, hatten nur wenige Trauben
angesezt.

Auf Conglomerate und flachen Erdkrume wird Wein in
Threlkelds[j] Garten gebaut. Die Holzschösse sind schwach
und dürftig un wenige Trauben. Man denkt nicht daran den
Boden um den Stock auf zu locken. In Healys[k] [*sic*] Apfelsinenbaum pflanzung gedeiht der Wein auf sandigem Boden
außerordentl. wohl. Hier sind die strauchartig gezogenen
Stöcke bei weitem kräftiger als die an Spalieren gezogenen.
A. p. ed. Nov. 21.]

Sydney ist die letzte Zeit über in großer Bewegung gewesen.

Reconnaissance in New South Wales

Sir George Gibbs [sic] hat der Stadt eine Municipalität gegeben, hat die Bürgerschaft in eine Corporation vereinigt, ihr die Verwaltung ihrer eigenen Angelegenheiten übergeben nd die town in eine city verwandelt. Der erste November war der Tag an welchem die verschiedenen Bürgerfunktionen beginnen sollten. Mehrere Wochen vorher hatten sich die reichern nd angesehenern Bürger in den verschiedenen Wards zum Platze von Stadträthen (counsellors) gemeldet. Jeder von ihnen wurde von einer Anzahl von Mitbürgern unterstützt, welche jeden Tag ihren Zuwachs im Morning Herald bekt. machten. Es konnte nicht fehlen, daß in einer Bürgergemeinde, wie die von Sidney Männer von sehr bedenklichem Karakter, die durch schlaue Klugheit oder durch Betrug reich geworden, obwohl ihnen selbst die oberflächlichtigste Bildung mangelte, sich zu der hohen Würde von Stadträthen melden und daß eine Menge ähnliche niedriger Genossen sie unterstützen würden. So sind denn anerkannte Schmuggler und Diebe zu Stadträthen erwählt worden, während sehr gebildete Männer wie z. B. Dr Nicholson, der nicht um Stimmen betteln mochte, in ihrer Bewerbung drch gefallen sind.

Man hat vielfach über die Zweckmäßigkeit gestritten, in den jetzigen bedrängten Umständen der Kolonie, den Städten Municipalrechte zu geben. Doch es scheint mir, daß gerade diese Zeit für solche Maßregeln viel geeigneter ist, als die Zeit der Blüthe nd des Reichthums. Ein armer Mann ist gewöhnlich weit zufriedener .t seinem Schicksale, wenn er seine eigenen Ausgaben bestreitet und weiß woher der Mangel an Mitteln komme, als wenn seine Geschäfte von andern verwaltet werden, da es in der Natur der Menschenseele liegt, die Ursachen unglücklicher Verhältnisse niemals in sich selbst, sondern in andern zu suchen und stets an die eigene höhere Vollkommenheit zu glauben. — Folgende war die Art wie die Wähler ihre Wahlen unterstützten. —

Sydney Septr. 1842

To the Electors of Maquary [sic] Ward (of Philipp, Bourke, Brisbane Ward)

Gentlemen. We the undersigned citizens of the above Ward, beg to recommend the following gentlemen as fit and proper persons to represent your interests in the city Council viz. Mr F. Holt pp.
Die so unterstützten Candidaten dankten dann ihren Patronen auf eine kürzere oder längere Weise.
[*Continued*] November 9. Mein theuerer Freund, Da die Erziehung der Jugend jetzt mehr als immer zur Sprache kommt, wird es dir vielleicht ni[cht] unangenehm sein drbr. [*sentence incomplete*]. Obwohl ich nur wenige Theile der Kolonie gesehen habe, so läßt sich doch von der Eigenthümlich zer׳ streuten Bevölkerung schließen, daß gerade die Thle. die ich gesehen, mir alle Materialien zu einem Richtigen Urtheile liefern. — Im Busche ist an regelmäßige Erziehung nicht zu denken: einzelne Familien mögen vielleicht Hauslehrer mit sich nehmen; doch dieß ist selten nd von geringer Bedeutung. Kleine Städte oder Stadtembryonen wie z. B. Newcastle haben kaum Elementarschulen. Hier in Newcastle hält ein Hr. Balmain, ein Presbyterianer eine Privatschule, welche .ch allem was ich sehe u höre dem Lehrer un Unternehmer Ehre macht. Doch so stark ist der Sekten geist der Episcopalen un in der That Aller, daß der arme Mann nicht ganz so tüchtig .tr stützt wird, wie er es sein sollte. Herr Prediger Wilton hat einige Pensionäre doch dieß ist alles, was bis jetzt hier gethan wird. — In Sydney sind drei Hauptschulen: das Australian College, welches Herr Dr Lang stiftete, welchem er indessen wegen seiner Reisen nicht die nothwendige Aufmerksamkeit schenken konnte wor.f es in schlechte Hände gerieth und nun seiner Auflösung nahe ist. Herr Dr Lang hat alle Verbindung .t der Schule aufgegeben. Sydney College ist soviel ich urtheilen kann eine kräftige schülerreiche Schule: sie ist mit keiner Kirche in Verbindung nd scheint mehr un mehr Grund zu gewinnen; dennoch ist der Geist der Schüler nicht gerregelt [*sic*], nicht edel, nicht geachtet — es ist immer noch eine gemeine Bürgerschule. — Mr Rennies Schule, welcher er den tönenden Titel high college gegeben ist eine beschränkte

Privatschule, recht gut doch nach einem wunderlichen Stück‑
werk Plane gelietet, der am Ende die Möglichkeit aufhebt die
Schule auf einen höhern Standpunkt zu erheben. Er ist
Direktor, sein Sohn nd ein anderer junger Mann sind Unter‑
lehrer. Die Knaben sind von 9–16 Jahren. In Sydney College
un Australian College sind sie bis zu 19 Jahr alt.
Da nun das Bedürfniß einer höhere Bildungs anstalt immer
fühlbarer wird, hat Herr Dr Lang versucht eine neue unab‑
hängige Anstalt zu gründen, welcher mir vorzüglich den
Zweck zu haben scheint junge Prediger für die Kolonie zu
bilden: dieß sagt un glücklicher Weise wiederum zu viel. Es
zeigt, daß die Anstalt nur den Presbyterianern offen stehn und
alle übrigen Sekten .sschließe. Diese Kolonie muß bei der
Mischung der Sekten meiner Ansicht nach allgemeine Insti‑
tute zu gewinnen trachten, um die existirende Bitterkeit der
Anders denkenden allmählig zu zerstreuen. Unglücklicher
weise glauben die armen Leute, daß die [Ausrng.?] intoleranter
Gesinnungen Religion ist, und so hat man fast ein Tantalen[1]
Werk vor sich, wenn man gegen diesen Barbarismus anzu
kämpfen trachtet! ['trachtet' ends with a flourish.]

Newcastle, N.S.W., Octbr. 31[, 1842.]

My dear friend,

On the 28th of October at ¼ to 6 in the morning I was awakened by a violent shaking of the house, as if some heavy body had dropped on to it. Messrs Wilton and Crummer, however, had the same experience although they live several thousand paces away from us. There was therefore no doubt that an earthquake had occurred. Other people had felt a similar shock at 3 o'clock in the morning. [From margin: *Mr Wilton, the local clergyman, told me that this was the 4th earth‑ quake he had experienced during the time he had been at Newcastle 1837–1841. Mr Clarke informs me that a shock was felt at Windsor at 20 minutes to 6 on the morning of Friday the 28th of Oct. The earthquake was also felt at Port Stephens, and at Stroud (near Port Stephens). It was felt very strongly at Port Macquarie. Later, in* different ink: *On the Macleay River. On the Paterson. In Van Diemen's Land 1788, 1801, 1803, 1804.*]—That this region must have been affected by strong disturbances is shown by the presence of several dykes, and of deep fissures into which the molten material probably did not ascend so far, as well as by the wonderfully regular shattering of the weak sandstone below the 4th coal seam, which forms the sea shore in front of Morris's Baths before you get to the long beach.—(*The strike of the dykes is from S.E. by S. to N.W. by N.*)—You feel here as if you are standing on a regular criss-cross of low walls which are formed by the hardened, projecting edges of the shatter-blocks. At Sydney I've been interested for a long time in similar joints in sandstone, which all strike in the same direction. I've noticed them particularly at Fort Macquarie on the way to the Botanic Gardens, and at the waterfall[a] on the North Shore. As there were no volcanic rocks in the neighbourhood I concluded that these joints were caused by seismic disturbances; but I am less certain about it since I came here, where I can hammer chips off the volcanic rocks that have filled the fissures. [From margin:] You can see [evidence of] concentrated shocks where you first go down to the shore. You can see their circular effect [in the form of] joints radiating out from a centre and regular shattering from the circumference. All the master joints run from SE by S to NW by N. The joints at right angles to them are short, crooked and irregular.] I offer the following explanation to account for the relief network at Morris's Baths: earthquake waves coming always from the same direction split the rocks right down to the hearth of volcanic activity (which may not, perhaps, lie so very deep here), but did not open fissures wide enough to permit the molten rock to rise high in them. It was the nascent gases, particularly those bearing iron, which escaped through the joints, hardening the walls and colouring them brown. When eventually the sea began to wash loose stones back and forth over the sandstone, the indurated, ferruginous [top] edges of the shatter-blocks withstood the friction, whilst their tops were gradually worn down and deepened.[b]

[Continued,] *November 1. My dear friend*, Yesterday Mr Calvert[c] brought me a *Caprimulgus albogularis* (the white-throated goatsucker) that he had shot, and I set about skinning it at once. I was surprised to find the stomach so far to the rear, so when I had finished skinning the bird I gave closer consideration to the position of the stomach, and I examined its contents. It was extraordinarily big and full, taking up nearly all the room in the abdominal cavity and squeezing the entrails both upwards and downwards. The glandular stomach showed 3 rows of elongated glands; the muscular stomach was covered with a thick epidermis, yet the muscles were weak. It was crammed full of Gryllotalpa, Scarabaei, a small Cerambyx and a beetle of the Melolonthid group. The gastric juice which covered the soft Gryllotalpa and had already partly digested them was not acting on the hard covering of the other beetles, though some of them had been kneaded together by the activity of the walls of the gizzard. I had hardly finished with this bird when Mr Bligh, the nephew of the former governor, brought me another, a bright blue bird with the legs of Gryllotalpa still protruding from its wide beak. The position of the stomach and the internal anatomy were the same. As well as Gryllotalpa and Scarabaei there was a great number of bees, all in a half-digested state, though here too elytra and wings had withstood the action of the gastric juice.—The position of the stomach reminded me of that of the cuckoo, and of the supposed reason why it lays its eggs in another bird's nest. If it be true that the position of the stomach prevents it from hatching its eggs, the two Capromulgidae presumably suffer from the same difficulty, and would find the same way out of it. But the European goat-sucker, it seems, does build a simple nest, and I believe that the Australian bird does likewise, so that I doubt the explanation of the cuckoo's behaviour.—

[Continued,] Newcastle the 5th November. My dear frd., to revert to the dykes which cut through the stratified rocks: I had neglected to determine their strike with precision, so yesterday I clambered down to the sea shore to make good for this sin of omission. I've already told you that the dyke at Nobbys Island strikes as nearly as possible from S.S.E. by S. to N.W. by N. I was therefore rather surprised to find that the strike of the first one, below South Head, is S. $\frac{1}{4}$ W. to N. $\frac{1}{4}$ East, whilst that of the 2nd is from S.S.East to N.N.W., and of the 3rd from S. $\frac{1}{4}$ E. to N. $\frac{1}{4}$ W. I then compared them with the joints in front of Morris's Baths, which give a strike of S. $\frac{1}{2}$ E. to N. $\frac{1}{2}$ W—extraordinarily close to that of the last and the next dyke. If we compare these different alignments [marginal diagram] *we get not a line, but a zone, which lies between S. East by South ¼ West (3¼ points of the compass 36° 35′).* If we assume that the propagation of the volcanic impulse was at right angles to this direction, we are led, perhaps, to New Caledonia and the volcanoes that lie near to it.

Below the gaol, when the tide is very low, you can watch a kind of Charybdis.[d] There's a fissure, 24°, through the rocks, which has been enlarged, deep down, most likely by the incessant work of the waves. It also widens [in plan] slightly towards the shoreline to form a kind of funnel. When the tide is going out, the water withdraws from the cavern, exposing rocks encrusted with a marine growth consisting mainly of Balanus, to a depth of 30′; but the next wave surging back fills up the hollows below, and sends the water foaming up through the narrow neck of the funnel. This regular falling of the water, only to come hissing back, fascinated me, and I could have watched it by the hour. There are certain impressions which are enough in themselves to engross our whole being; not because they revive old memories, or stimulate us towards more rigorous scientific observation and the possible solution of problems; but because they are intimations of power so overwhelming that we surrender to it passively in a spacious darkness of the mind.

[Continued,] *Novbr. 7. My dear friend, I've just come back from a trip in the bush.* Two young men who came out as steerage passengers in the ship that brought me from England recognised me again here, and invited me to go to see them at their place in the bush.—It's in a very interesting district near Red Head, beyond the Valley of Palms. Red Head is a locality that I have been meaning to examine geologically. When crossing Shepherd's Hill I found Hakea pugioniformis *and* dactyloides *in bloom, both of which have a pleasantly sweet smell.* Banksia collina *is putting out new shoots. I've already remarked that* Banksia serrata *was in flower, and now it's showing young fruit.*—The several members of one family do not, on the whole, seem to come into flower at the same time in this continually mild climate. Shrubs like Westringia rosmarinifolia *seem always to be in bloom.*—

As I made my way along the flat, rocky shore, with the sea rushing in through fissures everywhere, fringes of snow-white spray were being thrown up here and there, as the bigger rollers crashed on to the sharp

Reconnaissance in New South Wales

rocks. Your true poet could hardly have seen them for anything else but the waving veils of nereids; and, just as the sailor begins to spruce himself up when his landfall is made, the waves put on their gay white caps as they sweep towards the land, to tell their resounding tale of the world's immensity to hearkening shores.—Even in gentle breezes from the sea the slow rollers that sweep in towards this coast are high and very long. During storms the surf is tremendous. One senses the wide, open Australasian sea that spreads before these shores, where there is nothing, in thousands of miles, to oppose or impede the growth of the waves that surge towards them.

When I reached my two friends' place, which was in a romantic spot in a high-lying pocket, I found them just about to extract honey from the wax. I saved them from waste of effort by warming the combs in which there was still some honey left, over a slow fire, and letting the honey drain out. After this, I washed the wax several times with water and poured the rinsings into a small barrel to ferment. Soon after we had finished our work two natives came along, and enjoyed the last sweet oozings from what remained of the honey comb. They carefully sucked the moisture out of the wax, which they then dipped in water and sucked again.—On the Sunday morning I went out towards Great [Red] Head to see whether or not there was any essential change in the coal measures, and also to see if I could manage to find a dyke or two. I was astonished to find a number of accumulations of shells, each confined to a small area along the coast, with hardly another shell to be found near many of them even a few yards away. In every instance this was due to the character of the sea bottom, which either provided food in abundance for a particular kind of mollusc, or produced an acceleration of currents which swept the shellfish ashore when they had lost the strength to cling to their submerged rocks. In such places Haliotis, Parmophora and Triton were common.—The 3rd coal seam is well developed. In the sandstone below it there are impressions of very big ferns, and of Equisetum. Here too, the rocky shore has been split into regular rhombic and trapezoidal blocks, and deep fissures (probably filled by dykes at some depth below) are the scene of that drama of foaming flood and sinking surface which I've already described.—These fissures strike from S. by East to North by West. In the conglomerate that forms the shore I

Leichhardt's Letters

found the trunks of trees, petrified mostly into silex, though some had been changed into ironstone.—It was high tide and we had difficulty in rounding the point of Red Head along the shoreline. There were great blocks [of stone], brought down by the forces of weathering, lying at all angles, and the cliffs were sheer. The rock was conglomerate.

On the way back, on the thinly overgrown sand dunes, we found Leucopogon Richei, *with a beautiful* Geranium *growing in its shade. We also found a plant that looked to me like a new species of* Pinelia; Wormia *was still making a magnificent show of big, golden yellow flowers; and* Ipomaea *was dense, overlooking the sea. A great number of plants with fleshy leaves seem to like to be near the sea—the families just mentioned,* Tetragonia, Salicornia, *a composite related to* Leontodon, Senecio.

Before we had lunch I studied the geological section exposed in the cliffs close to the lagoon. I shall be giving a fuller account of it in my geological observations. We went to the upper end of the lagoon, where we saw the effects of a bush fire. The lagoon extends approximately from S.E. to N.West, so its aspect is one not unfavourable for cultivation.

The two blacks, who had pitched their camp opposite the hut during the night, were John M'Gill, king of the Lake Macquarie clan,[e] and Gorman, king of another clan. They had neither their women nor any other blacks with them. M'Gill came into the hut and asked for some embers and a kettle. Calvert gave him what was left over from the honey, with which he was highly delighted, and some flour. The latter he knew quite well how to use to make doughboys, though it was hardly edifying to see him kneading the dough and smoking his pipe at the same time. He used the kettle, which still contained the water in which Calvert had boiled two fowls, for cooking the doughboys. The two noble savages went over to the small fire they had lit under a Eucalyptus tree, stretched themselves out lazily beside it until their meal was ready, ate without stopping until they had swallowed the last scraps, and then slept until late the next morning, regardless of the somewhat showery night, but putting more wood on their little fire whenever they felt the cold.

[Continued] *Nov. 8. My dear friend, you remember how interested I used to become, in anything to do with wine growing or wine pressing;*

Reconnaissance in New South Wales

and how often I discussed with you the possibility of taking vines to New Holland, like the old Samian king Ankaeos.[f] My means did not permit me to bring French vines out with me, as I first thought of doing, but it was just as well, because out here I would not have been able to cultivate them myself or to sell them to advantage. But there was a lot of talk about wine growing at the time of my arrival,[g] and M. de Ligny, a Frenchman, made several trips to the Hunter River and other parts of the colony, to find out if the soil were really suitable for the purpose. My outings in the Sydney district, which took me mainly over poor, dry, sandy soil, led me to conclude that grapes would not do well around Port Jackson. I had always found that good wine was the product of soils derived from limestone or volcanic rocks, though I grant that a light but very pleasant wine can be produced on sandstone, e.g., on the Pagelflur or in the Molasse mountains in Switzerland. But a light wine is not what they want here, as they are trying to produce a superior wine that will be suitable for export, and that may reduce and eventually stop the importation of wines from Spain and Madeira. M. de Ligny had several articles printed in the [Sydney Morning] Herald, in which he demonstrated that the Hunter River country was eminently suitable for vines. He declared himself willing to supervise cultivation and to superintend the pressing, should landed proprietors be disposed to lay out vineyards. He has gone to reside at Raymond Terrace, where, it seems, he is trying to live up to his promises. When I came to Newcastle I too was convinced that this part of the country could very quickly become covered with flourishing vineyards, given the necessary labour; and that, although we can hardly expect Madeira, it will be possible to produce a really good wine. I spoke to Mr Scott about it, but he's so bent on making his fortune that he can hardly give the necessary attention to consider such a venture. All the same, he seems to want us to try our hands at pressing wine as soon as the time is right for it. However, quite a number of land-owners are making serious preparations. Captain Scott has a German named Luther,[h] with whom he shares the profits. Mr Robert Scott likewise has a small vineyard and a French vintner but the latter seems to have left the place. The seeds [of the idea] have been sown, or are being sown, and it remains to be seen how the young plants will be tended and what kind of fruit they will yield. My own theory is that wines of superior quality will be produced only on the limestone soils of the upper Hunter, in the Wellington Valley and on the volcanic soils of Portland Bay (Port Phillip). This country has, nevertheless, proved the rule by so many exceptions that I would not be at all astonished if it really were to produce good wines from [supposedly] unfavourable soil.—I've been reading quite a lot in the English papers about German settlers in the colony who are said to be producing extraordinarily good wine on a wholesale scale; yet nobody here can tell me where to find them, or knows who made such a boastful report. All the wines that I've tasted in the colony have been undistinguished, though some of them are quite pleasant to drink. Mr Macarthur[i] at Camden has a big vineyard and presses wine, though more as an amateur than a wine-grower; but I've been told that he makes a very good wine. [From margin: Mr W. Brooks has recently planted a small vineyard at the western end of Lake Macquarie. The soil is good, and the vines have made plenty of healthy wood, but they were not properly pruned so very little fruit has formed. There are vines in Threlkeld's[j] garden, growing on conglomerate and level top soil. The canes are few and weak, with little fruit. They've not considered loosening the soil around the stocks. Vines are doing extremely well in Hely's[k] orange grove, on a sandy soil. The stems here that have been trained to become bushes are much more vigorous than the trellised vines. A.p.ed. Nov. 21].

Sydney has been in a state of continual excitement recently. Sir George Gipps has given the town municipal standing, united the citizens into a corporation, has transferred the management of their own affairs to them and has changed the town into a city. The various municipal functions were to come into force on the first of November. Several weeks before that date [some of] the wealthier and better established citizens of the several wards had stood for office as town-councillors. Each of them was supported by a number of fellow citizens, and every day lists of names of additional supporters were published in the [Sydney] Morning Herald. It was inevitable, in an urban community such as Sydney, that men of doubtful character, men who had made money by sharp practice or deceit, should stand for the high office of town-councillor, though they lacked even the most superficial education; and it was also inevitable that a following of baser associates should support them. So we have known smugglers and thieves accepted as town-councillors, whilst highly educated men like Dr Nicholson, who did not like to canvass for votes, have failed in their candidature.

There has been a good deal of dispute, in view of the present financial embarrassment of the colony, as to the propriety of granting municipal rights to the towns. Yet it seems to me that this, rather than a time of prosperity, is the very time for such expedients. A poor man, as a rule, is more contented with his lot if he is meeting his obligations himself and can see where the money goes, than if others are managing his affairs for him. It's only human nature to blame anybody but ourselves for our misfortunes, and to hold unshaken to our belief in our own superior perfection.—This will show you how the electors supported their candidates:

Sydney, September 1842.
To the Electors of Macquarie (Phillip, Bourke, Brisbane) Ward.
Gentlemen:
We the undersigned citizens of the above ward beg to recommend the following gentlemen as fit and proper persons to represent your interests in the city Council, viz., Mr F. Holt, etc.

The candidates so nominated then thanked their patrons at shorter or greater length.

[Continued,] November 9. My dear friend, As education is now being discussed more than ever, you may perhaps be glad [sentence incomplete] about it. Although I have seen only a few parts of the colony, the population is so peculiarly scattered that the very parts I've seen do, I believe, provide all the evidence I need for forming a sound opinion.—In the bush regular teaching is out of the question. Isolated families may have taken private tutors with them, but that's exceptional and of little significance. Small towns or the beginnings of towns— Newcastle, for example—hardly support even primary schools. Here in Newcastle a Mr Balmain, a Presbyterian, is keeping a private school which, from all I've seen and heard, is a credit to him both as teacher and organiser. Yet the sectarian feeling is so strong amongst the Episcopalians—in fact amongst all of them—that the poor man gets nothing like the support he deserves. The Rev. Mr Wilton has a few boarders, but this is all that has been done here so far. In Sydney there are three leading schools: the Australian College, founded by Dr Lang, but, as his journeys have prevented him from devoting the necessary attention to it, the college has fallen into the wrong hands and is soon to be closed. Dr Lang himself has severed his connection with it. Sydney College, so far as I can judge, is firmly established and well attended. It is not denominational, and seems to be gaining ground; but the behaviour of the boys is undignified, disorderly, and fails to command respect. It is still nothing but an ordinary middle-class school. Mr Rennie's school, which he pompously calls a 'high college' is a restricted private school, and a very good one; but it's conducted in accordance with a wonderful patchwork of planning, which will effectively prevent it from rising into a higher grade. Mr Rennie is the principal, with his son and another young man as assistant masters. The boys are aged from 9 to 16. At Sydney College and the Australian College they're up to 18.

As there is increasing need to provide for higher education here, Dr Lang has been trying to found a new and independent institution which, as I see it, is to serve especially for the training of young clerics for the colony; and this, in its turn, unfortunately says too much; for it shows that the institution would be open only to Presbyterians, to the exclusion of all other sects. The colony must, in my opinion, strive, through the mingling of all sects, to establish general institutions, in order to dissipate, if only gradually, the bitterness between those who hold to conflicting beliefs. Sad to relate, the poor people think that the expression of intolerant convictions is religion. And so, whoever strives to fight against this kind of barbarism has something like a work of Tantalus[1] before him!

66 HERREN MUSIK LEHRER / C. SCHMALFUSS / Wohl-geb. / Cottbus / p. Berlin

Newcastle den 10tn November 1842

Meine liebe theuere Mutter

Sollte ich den heutigen Tag vorüber gehen lassen, ohne an dich zu denken, ohne mit dir zu sein und meine Glückwünsche mit denen aller deiner Kinder und Enkel zu mischen, welche den Vortheil haben, sich in deiner Nähe zu befinden und deine

Reconnaissance in New South Wales

Hände zu drücken und an deinem Halse zu hängen! — O meine Mutter — mich trieb die unablässige Begierde, zu lernen von dir und obwohl die Heimath stets in meiner Erinnerung frisch blieb, so waren es doch nur Erinnerungen; ich beklagte oft mein Schicksal welches mir den Vortheil, das Vergnügen raubte, mit Euch zu leben, während es mir die weite Welt eröffnete. Doch du hast wenigstens ein Vergnügen, welches nicht allen zu Theil wird; du hast eine liebende Seele an der entgegen gesetzten Seite der Erde, welche von dort her um dein Glück täglich an den himmlichen Vater bittet. —

Nachdem ich ungefähr 6 Monate in Sydney gelebt und die Umgegend der Stadt, die Bewohner, ihre bürgerliche Verfassung, ihre Art nd Weise zu leben beobachtet hatte, wünschte ich andere Theile dieser Kolonie zu sehen. ⌈Erinnert Euch, daß Sydney eine Stadt von 42000 Einwohnern einen Raum bedeckt, der vielleicht Berlin an Größe übertrifft, indem die Häuser gewöhnlich niedrig sind und nur von Einer Familie bewohnt werden. Also solch eine Stadt in einer jungen Kolonie kaum 52 Jahr alt, mit allen Bequemlichkeiten Europaeisch. Lebens nd mit einem bei weiten lebendigern Handel als Berlin!!⌉ Von dieser Stadt ging ich auf einem Dampfschiffe nach Newcastle⌈, welches ungefähr 18 Meilen nördlicher an der Mündung des Hunters flußes liegt⌉. Das Dampfschiff verläßt gewöhnlich Sydney um 10 Uhr in der Nacht und ist am Morgen um 7 Uhr in Newcastle.ᵃ Dieß ist ein kleines, sich eben erst entwickelndes Städtchen, dessen zerstreute Häuser rings an den Abhängen der zum Flusse sich senkenden Hügel liegen. Ein reicher Landbesitzer, dessen Bekanntschaft ich in Sydney gemacht, bot mir seine Wohnung während meines Aufenthaltes an und ich lebe mit ihm nun schon diese 7 Wochen. Ausser den Pflanzen nd Gesteine nd Thieren, die ich hier sammelte, sah ich Einiges vom Australischen Akkerbau. Den Hunters fluß hinauf findet sich ein reiches vom Flusse im Laufe der Zeiten angeschwämmtes Land, welches ohne Dünger reiche Erndten giebt; doch unglücklicher Weise von heißen Winden und von der Sonnenhitze so ausgetrocknet wird, daß der Akkerbau immer nur in 3 Jahren auf eine gute Erndte rechnen kann. Diesem Wassermangel könnte vielleicht durch künstliche Taiche abgehelfen werden; doch sind so wenige Arbeiter hier, und die wenigen verlangen so ausserordentlich hohen Arbeitslohn, daß man an große Unternehmungen der Art jetzt noch nicht denken darf. Ausser Weizen baut man besonders Türkischen Weizen welcher zur Futterung benutzt wird. Ihr werdet indessen leicht einsehen, daß die Landbesitzer wenig Neigung haben ihr Land zu bestellen, da das Mehl und Getreide, welches von Van Diemens land nd Südamerika eingeführt wird, wohlfeiler ist, als sie selbst es gewinnen können. Sie schenken deßhalb der Vieh nd Schaafzucht weit größere Aufmerksamkeit. Die Schaafheerden sind ungefähr 100–200 Engl. Meilen (25–50 Meilen) landeinwärts; die Rinder läßt man frei in weiten ungezäunten Waldstrecken herum laufen, nachdem man jedem das Zeichen des Besitzers eingebrannt. Jedes Jahr werden die jungen Kälber eingefangen und gebrannt. Die Kühe melkt man nur in der Nähe der Städte, in welchen man Butter und Milch verkaufen kann. Man strebt gewöhnlich, die Heerde zu vermehren und die fetten Stücke an die Schlächter zu verkaufen. Die Thiere werden fast wild, da sie so wenig Gemeinschaft mit Menschen haben. Die Hirten, welche nur von Zeit zu Zeit Musterung halten sind alle zu Pferde und oft ist Roß und Reiter den Angriffen der Kühe und Bullen ausgesetzt. — Mit den Schaafen ist es verschieden. Hier müssen regelmäßige Hirten gehalten werden, welchen man jährlich 220–350 Thr. zu bezahlen hat. Jeder Hirt hat nach der Beschaffentheit des Landes von 500–1000 Stück Schaafe. Diese sind den Angriffen des wilden Hundes ausgesetzt, welcher der Verschmitztheit des Fuchses mit der Stärke eines guten Schäferhundes vereinigt. Die jungen Männer, welche von England hier her Kommen, sind gewöhnlich von guter Familie und haben gewöhnlich etwas Geld. Sind sie arm, so treten sie für einige Zeit als Oekonomien oder Inspektoren in den Dienst der angesiedelten Besitzer. Nachdem sie sich einiges Geld erspaart, gehen sie

Reconnaissance in New South Wales

landeinwärts in die noch nicht in Besitz genommenen Theile und grunden auf eigene Rechnung eine Vieh oder Schaafstation. Sie packten gewöhnlich von der Regierung Strecken von 1000-2000 Morgen Landes, welche ihrer Heerde hinreichende Nahrung gewährt nd bezahlen für dieß den festgesetzten Preiß von [?50 Thr] jährlich. — Sie sind indessen nicht Eigenthümer und wenn das Land zum öffentl. Verkauf kommt, hören ihre Ansprüche daran auf. Wenn der junge Mann mit seinen Hirten an den gewählten Ort angekommen ist, baut er sich eine Hütte von gespaltenem Holze, welches er mit Baumrinde bedeckt. Die Hütte besteht aus dem Raume, in .lchem der Besitzer ißt nd schläft und aus dem Kamine, welcher sehr geräumig ist, um das Verbrennen zu verhindern. Er hat einen ⌜zinnernen⌝ Thee topf, einige Teller, eine Tasse, ein Paar Löffel. Er lebt ausschließlich von Fleisch und einem schweren ungegährenen Weizenbrode, welches man Damper nennt. — Man knetet Mehl mit Wasser in einen dichten Teig, macht ein großes Feuer mit trockener Baumrinde, nd wenn es .sgebrannt, legt man den flachen Teigkuchen in die heiße Asche. Thee ist fast das einzige Getränk. Zum Frühstück, zum Mittag nd Abendbrod wird Thee getrunken, welchen man mit braunem un raffinirten Zucker süß macht. Hat man keine Milch, welches gewöhnlich der Fall ist, schlägt man ein Ei in jede Tasse. Wenn Eier fehlen muß man sich ohne Milch zu behelfen wissen. — So lebt der junge Mann, der vielleicht in England an jeden Luxus gewöhnt war, und auf dessen Winke im Vaterhause vielleicht 2-3 Bedienten warteten, in tiefster Einsamkeit, rings vom Busche umgeben; alle seine Aufmerksamkeit richtet sich auf seine Heerde. Oft sitzt er vom frühen Morgen bis zur späten Nacht zu Pfe[rde,] entlaufene Kühe oder Schaafe zu suchen. Wenn er müde in seine Hütte z[urück] kehrt hat er mit eigener Hand sein Feuer zu schüren, seinen Thee zu bereiten sein Beefstake oder Hammelkeule zu braten. Dabei kann er sich oft nicht auf seinen Hirten verlassen, denn da die Arbeiter so selten sind, muß man oft diese ausverschämten Menschen mit großer Zuvorkommenheit behandeln. Doch trotz dieser harten Lebensweise hat diese Einsamkeit für das junge kräftige Gemüth viele eigenthümliche Reize. Er sieht mit Verwunderung, wie wenig der Mensch bedarf um seinen Körper zu befriedigen, und wie viel Kraft er selbst besitzt, jedem Mangel ab zu helfen. Oft haben mir die jungen Männer versichert, daß diese Selbstkenntniß, welche sie vielleicht in England nie gewonnen haben würden, ihnen mehr Nutzen gewähre, als Reichthum, und einige kehren nach England zurück, völlig zufrieden, gelernt zu haben, wie man wohlfeil leben könne. — Die Leichtigkeit mit welcher sie auf ihren an den Wald gewohnten Pferden den Busch durch streifen, das Gefühl der Unabhängigkeit welches ihnen dieses sonderbare wilde Leben giebt, macht ihnen ihre einsame Borkenhütte theuer und oft erinnern sie sich mit stillem Behagen derselben, wenn sie entweder zu den Städten oder nach England zurück kehrten. Wenn ich Euch von einer elenden Brettenwohnung spreche, drch welche Wind und Wetter überall eindringen, so müß Ihr nicht vergessen, daß wir hier in einem milden gesegneten Klima leben, in welchem das Thermometer nie bis zum Nullpunkt sinkt. Fast 8 Monate hindurch ist der Himmel immer heiter, oder nur für Tage getrübt und die Sonnenhitze ist oft ausserordentlich. Während der heißen Nordwest winde hat man das Thermometer bis zu 39° R (120° F.) steigen gesehen. Die gewöhnliche Wärme ist 16°-20° R.

Während des Winters (April Mai Juni July) haben wir gewöhnlich viel von heftigen Regengüssen zu leiden; denn obgleich eine ausserordentlich Dürre die Erndte vereitelt nd oft 1000 von Rindern un Schaafen vor Durst sterben läßt, so ist doch die Regenmenge hier fast doppelt so groß, als in England oder Deutschland. Könnte man den Regen der Erdoberfläche gegen Abfließen nd Verdünstung schützen, so würde in einem Jahre eine Wasser masse von 6 Fuß Tiefe Australien bedecken, während in Deutschland kaum 3 Fuß Wasser fällt. Doch in Deutschland ist es über das ganze Jahr vertheilt, während es hier wie in Wolkenbrüchen in wenigen Wochen nieder

Reconnaissance in New South Wales

strömt. — ⌜Wahrscheinl. sind 6 Monate vergangen während Ihr diesen Brief erhaltet; doch seid überzeugt, daß ich ebenso lebhaft an Euch denke, wenn Ihr diesen Brief lest, als jetzt wo ich ihn schreibe.⌝ Lebe wohl theuere Mutter. Tausend Küße, tausend Glückwünsche von

Deinem wandernden Sohne Ludwig.

[*P.S.*] ⌜Ich habe seit meinen Abreise von Paris in August 1841 noch keine Nachricht von Euch erhalten. Ihr wißt Williams Addresse. Er versprach mir an Euch zu schreiben. Ich wünschte, ich konnte Euch das Postgeld erspaaren.⌝

C. Schmalfuss, Esq., / Music Master, / Cottbus, / near Berlin

Newcastle, 10th November 1842

My dear old Mother,

How can I let this day pass without having you in mind, without feeling that we are together, and without adding mine to the greetings of all those children and grandchildren who will be enjoying the privilege of holding your hands and throwing their arms around your neck! You dear old thing, it was my insatiable appetite for learning that drew me away from you, and, although my recollections of home are as clear and fresh as ever, it's true that they're only recollections. I've often upbraided Fortune for offering me the world, but at the cost of everything I would enjoy by remaining with you all. Still, you have one thing that not many can claim to have—the daily concern, and the daily prayer for your health and happiness, of somebody living on the other side of the world.

After spending about 6 months in Sydney, where I studied the surrounding districts, the townspeople, the management of public affairs, and the general way of life there, I wanted to see other parts of the colony. ⌜*Bear in mind that Sydney has 42,000 inhabitants and covers an area greater, perhaps, than that occupied by Berlin. The reason is that most of the houses are not built high, and there's only one family in each house. Just think of such a town, in a young colony founded only 52 years ago; but it has all the conveniences that are enjoyed in Europe, and leads a much brisker business life than Berlin!*⌝ *I came from there to New-*

castle⌐, *which is about 18* [*German*] *miles north of it at the mouth of the Hunter River,*⌐ *by steamer. The ship leaves Sydney, as a rule, at about 10 o'clock at night and arrives at Newcastle at about 7 o'clock the next morning.*[a] *The place is very small, and only just beginning to grow. The scattered houses lie around the encircling slopes that fall from the hills down to the river. A wealthy landowner whom I met in Sydney had invited me to stay with him when I should visit Newcastle, and I've been his guest now for the last 7 weeks. Besides collecting plants, rocks and animals, I've been seeing something of Australian agriculture. Along the Hunter there is rich land, made of soil deposited by the river during the long course of time. It yields heavy harvests without manuring. Unfortunately it becomes so dried out by hot winds and the heat of the Sun that farmers can count on good yields only one year in three. They could very likely make up for this lack of moisture by building dams, but there are so few labourers in the country, and the few demand such high wages, that any undertaking on such a scale is as yet quite out of the question. Besides wheat they grow a lot of Indian corn, which is used as fodder; but you'll easily understand that landholders are hardly disposed to farm their land whilst flour and grain can be imported from Van Diemen's Land and South America more cheaply than they can be produced here. This is why they are far more interested in rearing sheep and cattle. The sheep stations are 100–200 English miles (25–50 miles) farther inland. The cattle are allowed to run wild in unfenced tracts of bush, after they have been branded with the owner's mark. The new calves are rounded up and branded every year, and the cows are milked only in the neighbourhood of the towns, where there's a market for milk and butter. The usual practice is to let the beasts multiply and to sell the fat stock to the butchers. The animals become like wild beasts as they see so little of human beings. The stockmen, who round them up only from time to time, are all mounted, and both horse and rider are exposed to attack by cows as well as bulls.—It's otherwise with sheep. Regular shepherds have to be employed, and are paid from £30 to £50 a year. Each shepherd has to tend from 500 to 1000 sheep, according to the nature of the country. The sheep are liable to attack by the wild dogs, which combine the cunning of the fox with the strength of a good sheep⸗ dog. The young men who come out here from England belong as a rule*

Reconnaissance in New South Wales

to good families, and most of them have some means. If they happen to be poor they begin by taking positions as managers or overseers on established stations. When they've saved a little money they go farther out, into the unsettled districts, where they take up sheep or cattle stations of their own. They generally lease from 1000 to 2000 acres of land from the Government, which is enough to support their stock, and they pay the fixed rent of [? £7] a year for it. The land, however, is not their property, and when it is offered for public sale they forfeit their claim to it. When a young man and his stockmen get to the place he has selected, he builds a hut out of split timber and roofs it with bark. The inside of the hut consists of a free space where the owner eats and sleeps, and a fire-place, which has to be large if the hut itself is not to catch fire. He has a ⌈pewter⌉ teapot, a few plates, a cup and several spoons. He lives entirely on meat and a heavy, unleavened, wheaten bread which he calls 'damper'.—First you knead flour and water into a stiff paste, then you make a big fire out of dry bark, and when the fire's burnt out you lay the flattened cake of dough in the hot ashes. Tea is about the only drink, and they drink it at breakfast, luncheon and supper, sweetened with unrefined brown sugar. If there's no milk, as is usually the case, you break an egg into each cup. If there are no eggs you have to put up with plain tea. There are young men here, living like this, who were used to every luxury at home, where they were waited on hand and foot. Here, they lead solitary lives away out in the bush, giving all their time to their cattle. Any one of them may be in the saddle from early morning until late at night, on the tracks of strayed sheep or cattle. When he gets back to his hut tired out, he has to get the fire going, make his tea and grill his steak or roast his leg of mutton himself. And he can never rely on his stockmen. Labour is so scarce that squatters often have to treat impudent stockmen with extraordinary forbearance. Yet notwithstanding the hard life, this isolation may prove to be very attractive to a stout-hearted young man. He is astonished to learn how little a man needs for his bodily satisfaction, and to discover how much he can do for himself. Many a young man has assured me that this knowledge of himself, which he would perhaps never have acquired in England, has been of more use to him than money; and some of them go back to England quite content just to have learnt to live modestly.—The ease with which they can roam far and wide through the bush on horses that are used to the work, the feeling of independence that comes with this strange life in wild country, make them fond of their bark huts; and often, when they return to town or go back to England, they find themselves thinking of their quiet contentment in the bush. When I talk of draughty, leaky little huts, don't forget that we live in a favoured climate here, where the temperature never falls to zero. For nearly 8 months on end the sky is clear, or overcast only for a few days at a time. It can be extraordinarily hot in the sunlight. During the hot North-west winds the thermometer has been known to rise to 39° R (120° F.). It's usually between 16° and 20° R.

During the Winter (April, May, June and July) we generally have to endure heavy rains; for, although severe drought may ruin the harvest and leave thousands of sheep and cattle to die of thirst, we get nearly twice as much rain here as falls in England or Germany. If we could prevent the rain that falls on the Earth's surface from running off or evaporating, what falls in a year would cover Australia with water 6 feet deep; but over Germany there would hardly be 3 feet. In Germany, however, it falls at all times of the year, whilst here it pours down in a few weeks as if the clouds had burst. ⌜—It may take six months for this letter to reach you, but believe me when I say that you'll all be as much in my mind when you come to read it as you are now as I write it.⌝ So good bye, my dear Mother. With kisses and all the best of good wishes from   Your wandering son Ludwig.

[P.S.] ⌜I've not yet had a word from any of you since I left Paris in August 1841. You know William's address. He promised me to write to you. I wanted to save you the cost of postage.⌝

67 [TO LT. ROBERT LYND, Barrack Master, Sydney, N.S.W.]

[Newcastle, N.S.W.,] 11th November[, 1842.]

My dear friend,

I am exceedingly obliged to you for the paper you send me. The greater part of our plants are dry, except that Ophioglossum which you found on Ash island and some other ferns and succulent plants. I wish not to charge Mr Scott with any message but I'll send you some specimens though not yet dry, that you may show them to Mr Robertson. When ever you find out names of plants which we have seen together pray let me know them. Some few new plants have been added to my collection; but I am sorry to say that many of our Ash island plants are mildewed and not well dried, as I had not sufficient paper to separate them. So many plants together entered even in 24 hours in a state of fermentation, which made the leaves fall and inclined the flowers to become mildewed, though I took every pain to save them. I shall try to gather other specimens as soon as I go to Ash island.

I am sorry to hear that your income is so considerably shortened; should you want money, consider that which I have your own. Do not think that these are empty words. I should be so glad to assist a man whom I esteem so highly and also who has acted so kindly towards me when I was a perfect stranger to him.—I send you an order for 30 pound, of which you may send me 25. I shall want 20£ for a horse and 2£-3£ for saddle and bridle. The mare which you saw and which is excellent costs only 16£ but she would involve me into the law business as it is very possible that the creditors of Mr Dawson will claim her. Could I ascertain that this would not be the case I would immediately buy her. I would save 4£ and she would sell well [at?] my return.—In my account with the bank you will find that I took out 50£ in August. I lent this money to the father of my little haunchback friend,[a] a poor but industrious man, who has been enabled by it to carry his business on in a more actif and profitable manner.

Mr Scott has spoken to me about a plan, the execution of which would be perhaps very advantageous to me. I'll give you the outlines of it. Mr Scott will clear 10 acres in Burwood,[b] near the parsons glebe, where we were together. He will build a cottage, establish a vineyard, make all the necessary expenses.— I have either 100£ per annum, and no share in the profit, or no salary whatsoever but half the profits. I have however the permission in the latter case to draw on Mr Scott to a certain amount.—The situation of the spot is certainly favourable: You will remember the parsons glebe which has not the advantage of the moist sea breeze which is of very great conse/quence. The place is perfectly sheltered against the cold winds. —If I accept the certain salary, I have no risk, it is true; but I am convinced that the share in the produce of 12500 vines will be by far the higher, having besides the advantage of establishing a kind of kitchen garden and keeping cows, which will provide me with plenty of milk.—I am sure that I could make a good thing of it to myself or to Mr Scott. I have a certain home, the continual gnawing feeling of my dependence from the kindness of others would cease; the surrounding nature is beautiful and interesting; avoiding society and being economical of my time, I could even continue my studies—perhaps not quickly, but more agreeing with my health and particularly with my eyes.—

But before I settle down in this colony, I must and will see other parts of it. My plans of going to New England are not damped. I set off next Monday on a walking excursion to Brisbane water, to find out if possible if the coal of Newcastle is above or below the sandstone of Sidney. As soon as I return to Newcastle, I sett off for Glendon and Patricks plains.

Now let us suppose that the mentioned plan is a bubble. What have I to do when I come to Sidney. Mrs Mitchell[c] wishes that I may become tutor to her children. I really do not know whether this is still her wish, for Mr Mitchell has never mentioned a word to me about it. Could I perhaps obtain the place of teacher of natural philosophy and chemistry at Sidney college for a small salary—for these are the only immediately useful and practical branches of the natural sciences, which it is advisable to teach in a school like Sidney college. You have perhaps opportunity of speaking with Mr Braim[d] generally on the subject. Think about all this my dear friend and write to me when you have leisure. My tour to Brisbane [Water] will last about 10 days.

Reconnaissance in New South Wales

According to all I hear, my expenses will be trifling in the [up?] country. I shall give for small amounts orders upon you and I think that it is not necessary to put the money in another bank. I am going to Glendon, to the burning mountain, to Patricks plains. I cross the Liverpool Range and so to New England. Make my compliments to Mr Macdonald and poor Capt. Marlow and family and believe me

Your affectionate frd. L.

68 [TO DR W. J.] LITTLE[, London.]

Newcastle on the Hunter 12 Novbr. 1842

Mein verehrter Freund,

Aus meinen Briefen werden Sie ersehen haben, welchen Eindruck diese junge Colonie auf mich hervorbrachte und wie ich allmählig an dem ersten Enthusiasmus zu ruhigerer Anschauung zurück kam. Ich fand mehr Theilnahme als ich erwartet hatte, wahrscheinlich weil ich meine Erwartungen nie zu hoch spanne. Ein wohlwollender botanisirender Barrack master, Herr Leutnant Lynd, nahm mich in seine Wohnung auf und theilte brüderlich sein kleines Einkommen mit mir während wir botanische Excursionen machten und nach allen Richtungen hin die Umgegend von Sydney durch suchten. Die Bestimmung der Pflanzen, die mir alle neu waren, war ein Tantalisches Werk. Doch Fleiß und Ausdauer ließen mich die Hauptschwierigkeiten besiegen. Ich gab Vorlesungen über Botanik in der School of Arts, welche nicht eben zahlreich besucht waren; doch war man mit meinem Vortrage und meinen Auseinandersetzungen wohl zu frieden. Ich machte öffentliche botanische Excursionen mit einigen Schülern des Sydney und Australian College[a] und versuchte überall der Wissenschaft Freunde zu gewinnen.—Nachdem ich nun 5 Monate in Sydney gelebt, wünschte ich andere Theile der Colonie zu sehen und folgte daher der Einladung Herren

Scotts eines reichen Landeigenthümers, der mich seit 6 Wochen hier in Newcastle recht Gastfreundschaftlich bewirthet. — Newcastle erhielt seinen Namen, weil es gleich Newcastle upon Tyne von reichen Lagern guter Kohle umgeben ist. Diese Kohlenlager haben eine außerordentliche Ausdehnung und treten überall im Becken des Hunters auf von seiner Mündung und vom Lake Macquary bis zur Liverpool Range 200 Meilen landeinwärts. Newcastle ist gegenwärtig ein kleines Städtchen doch seine Lage an der Mündung des Hunter flußes und seine Kohlminen, welche von der Australian Agricultural Company bebaut werden, sind seiner raschen Entwicklung außerordentlich günstig. Maitland, welches 19 Meilen von Newcastle entfernt liegt, ist viel bedeutender mit vielleicht 9-10000 Einwohnern und bildet jetzt moralish die Hauptstadt des Beckens des Hunters. Ueberall hin sind Städtchen in Werden, wie z. B. Raimond Terrace und Morpeth und Scone, alle Keime eines fruchtbaren und einst gewiß sehr bevölkerten Landes. *Die Colonie ist jetzt in einem sehr bedrängten Zustande. Früher gewährte das Aßignment system den Landbesitzern soviele Arbeiter als sie wünschten, und diese kosteten ihnen sehr wenig. Die Einwanderer brachten Capitalien ins Land und da das Preiß des Landes sehr niedrig war, konnten sie sich durch Kauf oder durch Schenkung (grants) große Länderstrecken erwerben. Ihre Bedürfnisse steigerten sich, der Handel belebte sich außerordentlich und die Colonie bot das Schauspiel einer in weniger denn 50 Jahren aus vollkommenen Wildniß hervorgetretenen Bürgerlichen Gesellschaft, welche allen Anschein von Civilisation besaß, der indessen wissenschaftliche und moralische Basen fast ganz fehlten. Alle Aufmerksamkeit lenkte sich auf die Erzeugung von Wolle als Ausfuhr artikel, und von Schaafen und Rindern, um die Ueberzahl an die Einwanderen zu verkaufen, die mit ihren Heerden unter Convict Hirten zu 300 Meilen landeinwärts Berg und Ebene über schwemmten. — Um eine moralische Basis für die Colonie zu gewinnen, vereinigten sich mehrere

* Copy in *Tgb.* begins.

reiche Besitzer und trugen darauf an, die Einführung von
Verbrechern und ihre Vertheilung an den Colonisten auf zu
heben — und anstatt dessen die Einwanderung freier un,
bescholtener Familien zu begünstigen. Daß jene Männer nicht
wußten, wie viel die Colonie materiell bis dieser Veränderung
für den Augenblick einbüßte, geht daraus hervor, daß sie
selbst in allem dem Luxus fortlebten, welchen ihnen die
unbezahlte Verbrecher Arbeit gestattet hatte. — Freie Arbeiter
kamen; doch sie forderten ausserordentlichen Lohn. Hirten,
welche früher nichts gekostet hatten, verlangten jetzt 30–50 £
jährlich und so jeder Arbeiter im Verhältnisse seiner Brauch,
barkeit. — Man lebte wie man früher gelebt, ohne Reduction
im Haushalte; man borgte, indem man auf bessere Zeiten
hoffte, man gerieth in Schulden; denn Credit wurde leicht
erlangt, indem die Kaufleute immer noch an die frühern
Verhältnisse dachten. Eine Unmasse von Gütern wurden von
England eingeführt und alle Speicher waren voll. Auf einmal
zeigten sich Lücken; einige bedeutende Häuser für Zahlung
gepresst, pressten ihre Schuldner; diese konnten nicht bezahlen
und nun fiel alles über den Haufen. Mehr denn 500 Bankerotte
fanden in erstaunlich kurzer Zeit statt und jeden Tag lesen wir
im Sydney Morning Herald, daß Landeigenthümer, Kaufleute,
Arbeiter, Handwerker, Doctoren, kurz aus allen Klassen
'filed their schedule'. —

Die Colonisten wissen jetzt sehr wohl, wo ihnen der Schuh
drückt und man sinnt und spricht darüber hin und her, wie
dem Uebel ab zu helfen sei. Jeder schreit nach Arbeit und
nach wohlfeiler Arbeit. Da der land fund gänzlich erschöpft
ist, um die Einführung von Einwanderer zu bezahlen, so
beabsichtigt man in England eine Anleihe zu machen. Andere
wünschen die Bergcoolies von Ostindien ein zu führen, welche
sich in Mauritius sehr nützlich machen: noch andere wünschen
Deutsche Einwanderer herbei zu holen. Da den Privaten Geld
fehlt, wird wohl nur das erste Mittel in Ausführung kommen.
Man hat sehr allgemein sich in seiner Ausgaben zu beschränken
angefangen und während früher hunderte von gigs und

Carossen am Nachmittag ihre Promenade durch Government garden machten, sieht man jetzt nur hin und wieder ein bescheidenes Fuhrwerk.* Ich habe schon eine recht bedeutende Pflanzensammlung gemacht und die Umgegend von Newcastle hat mir viele Blätterabdrücke aus den Thonlagen über und unter der Kohle geliefert. In wenigen Tagen bin ich Willens nach New England zu gehen, um die Geologie jener Gegenden zu studiren. Meine Stellung macht mir mit unter große Sorge. Ich werde zwar freundlich aufgenommen und unterstützt, doch sehe ich noch nicht, wo und wie ich eine sichere Stellung gewinnen kann, um mir mein eigenes Brod zu verdienen, welches ich bei der Beschränktheit meiner Mittel nothwendiger Weise thun muß. Ich habe zwar einige Vorschläge; doch diese binden mich wieder zu sehr und so bin ich beständig schwank‧ end welches mir oft die für den Verfolg meiner Wissenschaft nöthige Heiterkeit raubt. Ich habe die Absicht, mit Herren Lynd eine Flora von Sydney heraus zu geben und habe an Sir William Hooker geschrieben, ob er mir dabei helfen wolle und könne. Vielleicht könnten Sie mir durch ihre Bekannt‧ schaft mit Lindley oder andern Botanikern da[bei] nützlich werden. An eine Expedition ins Innere ist wegen der Erschöp‧ fung der Colonial Kasse für lange Zeit nicht zu denken. Sir Thomas Mitchell [und?] Sir George Gibbs [sic] sind keine guten Freunde, und es fragt sich, ob Sir Thomas der Führer einer andern Expedition werden möchte, indem die Colonisten ihn zwar als Surveyor schätzen, doch sein Talent als Entdecker neuer Länderstrecken sehr bezweifeln. Die Squatters haben vielleicht mehr gethan als er, indem sie mit ihren Heerden weit über Mitchells äußerste Puncte hinaus gedrungen sind. Das Leben im Busche behagt mir außerordentlich und ich vergesse leicht bei dem Thee, Damper und Salzfleisch in der Borken‧ hütte des Settlers und Squatters alle Luxus artikel unter welchem sich ihre Londoner Mittagstafeln beugen. — Schreiben Sie mir nur wenige Zeilen, wenn Sie nicht Zeit

* Copy ends.

haben, mehr zu schreiben. Denken Sie sich, daß ich noch in Gloucester Street wohnte. Leben Sie wohl, mein verehrter Freund und grüßen Sie Ihre liebe Frau von ihren ergebensten Freunde,

L. Leichhardt.

Newcastle on the Hunter, November 12th, 1842

My esteemed friend,

My letters will have given you some idea of my first impressions of this young colony, and of how my enthusiasm gradually gave way to a balanced appreciation of things. I have found people more sympathetic than I expected, probably because I never pitch my hopes too high. A well disposed barrack-master who is interested in botany, a Lt. Lynd, has taken me into his quarters and has been sharing his modest income with me like a brother, whilst we have been going on botanical excursions together in all directions out of Sydney in search of plants. The determination of our plants, which were all new to me, has been a tantalising effort, but patience and perseverance have enabled me to overcome the main difficulties. I gave some lectures on Botany at the School of Arts. I could hardly call them crowded, but the audience was very interested in everything I had to say. I also conducted some public botanical excursions with some of the boys from the Sydney and Australian Colleges.[a] *I also made general efforts to attract people to scientific study.—By the time I had been 5 months in Sydney I wanted to see other parts of the colony, so I accepted an invitation from Mr Scott, a wealthy land owner, who has been entertaining me here in Newcastle most hospitably for the past 6 weeks.—Newcastle was named because, like Newcastle-upon-Tyne, it is situated on extensive deposits of good coal. The coal seams crop out all over the place in the basin of the Hunter and the coalfield extends from the mouth of the Hunter to Lake Macquarie [along the coast] and for 200 miles [inland,] to the Liverpool Range. Newcastle at present is nothing but a small settlement; but its position at the mouth of the Hunter, and its coal mines, which are being worked by the Australian Agricultural Company, are highly favourable for its rapid development. Maitland, which is 19 miles from Newcastle, is a much more important place, of perhaps 9–10,000 inhabitants, and is now the* civic focus of the basin of the Hunter. But small towns like Raymond Terrace, Morpeth and Scone, are beginning to spring up at many points, like seeds germinating in what is bound some day to be a fruitful and populous countryside.* Just now the colony is very hard pressed for money. Under the Assignment System land owners were able to count on having all the labourers they wanted, and it cost them very little. The immigrants had been bringing capital into the country, and, as the price of land was very low, they were able to acquire extensive tracts of country by purchase or by Government grant. The increase in their wants led to an extraordinary expansion of business, and the colony presented the dramatic spectacle of organised society, with all the trappings of civilisation, appearing in what had been an absolute wilderness, in less than 50 years. But it was a society organised with an almost total disregard for scientific and moral principle. Nothing mattered but the production of wool for export and the breeding of sheep and cattle for sale to immigrants. And they, with their convict stockmen and their herds of beasts spread out like a flood over mountain and plain as far as 300 miles inland.—To infuse morality into public life certain wealthy land owners united in proposing that the transportation of convicts and their assignment to the colonists be suspended—and that the immigration of free families, of honourable antecedents, be encouraged instead. That the proposers of this measure had no idea of what the colony stood to lose whilst the change to the new system was in progress, is shown by their own behaviour. For they went on living with the extravagance they had been able to afford when they had unpaid convict labourers.—Free workmen came [out here]; but they demanded very high wages. Stockmen, formerly unpaid, now wanted £30–£50 a year, and so [it went] with all working men, according to their skill.—People went on living as before with no reduction of their current expenses; they borrowed money, hoping for better times; they ran into debt, credit being easily given by merchants who still saw things in terms of the past. Goods in immense amounts were imported from England and all the warehouses were full. All at once the extent of indebtedness became apparent. Leading houses, pressed for payment, pressed their debtors. The latter could not pay, and everything collapsed. More than 500 bankruptcies occurred in no time,

* Copy in Tgb. begins.

and we read every day in the Sydney Morning Herald *of landed proprietors, merchants, labourers, artisans, doctors, in short, of* [men of] *all classes having 'filed their schedule'.—*
By now, the colonists can feel just where the shoe pinches, and they've been thinking and talking up hill and down dale about how to deal with the trouble. Everybody's clamouring for labour, and for cheap labour. Since the Land Fund, which was used to finance immigration, has been spent to the last penny, some people are talking of raising a loan in England; others want to bring out hill coolies from India, as they've been a success in Mauritius; and still others are recommending German immigration. Since private persons lack means, only the first of these suggested measures is likely to be taken. Quite a number of people have begun to economise, and instead of the hundreds of smart gigs and carriages that we used to see being driven in the Domain every afternoon, there's only an occasional very modest turnout.*
I've already assembled a respectable collection of plants, and the surroundings of Newcastle have yielded a number of impressions of leaves, from the clays that lie above and below the coal seams. I'm ready to set out for New England in a few days' time, to study the regional geology. I sometimes get very worried about the situation I'm in. I'm being received and entertained very kindly indeed; but so far I can't see where or how I might establish myself to earn my daily bread. But do this I must, as my means are so slender. I've several proposals in view, of course, but they would put new ties on my freedom, so I'm living in a state of indecision which deprives me of the serenity of mind that I need for my scientific activities. I intend to publish a flora of Sydney jointly with Mr Lynd, and have written to Sir William Hooker asking him if he is willing and able to help in the matter. Perhaps you yourself could help, through your acquaintance with Lindley and other botanists. The Colonial Treasury being empty, an expedition into the interior will be out of the question for a long time. Sir Thomas Mitchell and Sir George Gipps are not the best of friends, and it is questionable that Sir Thomas would want to lead another expedition; for, although the colonists appreciate him as a surveyor, they certainly question his ability to find new country. The squatters have probably done more than he has,

* Copy ends.

because they've pressed on, with their herds, far out beyond the farthest point reached by Mitchell. Life in the bush suits me extraordinarily well, and when I'm drinking tea and eating damper and salted beef in a settler's or squatter's hut I soon forget all the rich dishes under which your London dining tables are groaning.—Write me just a few lines if you've no time to write more, as if I were still living in Gloucester Street. Good-bye, my friend, and give your dear wife the greetings of

Your devoted friend
L. Leichhardt.

69 [TO LT. R. LYND, Barrack-master, Sydney, N.S.W.

Newcastle, N.S.W., 24th November, 1842]

My dear friend,

A day before yesterday I returned from my arduous journey to Brisbane water. I was accompanied by the post master of Newcastle a Mr Flood, who went down to Gosford to collect the quit rents for Government. We went the first day to lake Macquary, which we forded the next morning; the second day we visited the coal mines of Mr Threlkeld's; the third day we arrived at a new township Newport, which however is only planned on paper; the fourth day we travelled during the most oppressive heat to Tukkerah Beach Creek and the 5th day we arrived at West Gosford near Brisbane water. We were generally received in the most hospitable manner by the settlers, who treated us with milk thea and damper and gave us a 'shake down' for the night. In West Gosford, which is prettily situated we lodged in an inn, which made us pay most dearly.— The flora was very uniform; the journey through the bush without change rather fatiguing. At the west end of lake Macquary at Mr William Brooks in [for is?] a good place for botanising; the finest place however and superior to any I have seen till now in the colony is between Tukkerah Beach Creek and Brisbane Water. Vegetation is here extremely rich and various. The Seaforthia, the cabbage tree, fern trees to 18′ and higher, a great number of unknown trees grow here most luxuriantly. As I was walking on foot I was not sufficiently provided with paper to gather good specimens; But I hope to go down on horse back a second time, as the Rev. Mr Rogers[a] has invited me to stay with him, in case I return.—The Geology between Newcastle and Gosford is equally uniform. The ranges of Newcastle, all the surface around lake Macquary down to Tangiganji are a iron coloured conglomerate or pudding sometimes interupted by a loose sandstone.—Towards Brisbane water the sandstone becomes general and offers a good material for building being harder and denser, than that of Newcastle. We returned from Brisbane water by the beach walking the first day to the south point of Tukkerah Beach lake to the station of Mr Foster, and having refreshed ourselves with a cup of tea and a piece of bread we forded the entrance of the lake, which will be shut by and by by the sands and walked along the beach for about 24 miles, till we were opposite Birds Island. The sun had set, the night became dark, the beach rocky and wild and having no place to go to we were compelled to light a fire and to remain for the night at the sea side. I slept well enough for the place, though my stomach felt rather empty. Next morning we had a most fatiguing journey through the scrub, over hills and gulleys without any food except the honey which the flowers of Lambertia yielded us. At three oclock in the afternoon we arrived at lake Macquary, where a kind constable and an hospitable fisherman restored our exhausted strength by a cup of tea and by some cold fish. Next day, after having examined the sandbar of the lake, we returned to Newcastle when the next days were taken up by writing my notes out and arranging the collected geological specimens and plants.—

70 [TO DR W. A. NICHOLSON, Newcastle‑upon‑Tyne.

[Newcastle, N.S.W., 24tn November, 1842]

Mein theuerster Freund

Von einer ermüdenden [Fuß] Reise nach Newcastle zurück gekehrt fand ich deinen lieben Brief und die Trauer botschaft aus der Heimath. Ich habe nicht nöthig dir zu sagen, wie tief mich die letztere erschütterte, da ich ein kräftiges junges lebens‑ frohes Wesen, das just hatte angefangen recht glücklich zu sein aus diesem ihrem iridischen Himmel, aus den Armen eines liebenden Mannes fortgerissen sah.[a] Dein Brief trug nicht wenig dazu bei, durch die lebhaftere Erregung meiner Erin‑ nerung an dich, das bekümmerte und niedergedrückte Gemüth zu beruhigen. Mein Schwager wußte gleichfalls die alten Bilder der Kindheit so lebhaft wieder .f zu frischen, daß der tiefe Schmerz bald einer ernsten resignirenden Trauer Platz machte. Meine Fußreise zum Lake Macquary nd nach Brisbane water haben meine Erfahrungen über die Kolonie wesentlich erweitert. Ich glaube die Verhältnisse in .lchen die Kohlenbetten von Newcastle zum Sandstein von Sydney stehen besser erkannt zu haben: sie scheinen gleichzeitigen Bildungen,[b] die indessen unter verschiedenen Verhältnissen natürlicher Weise einen verschiedenen Karakter zeigen, indem die Kohlenbetten von Newcastle einer weiten Fluß mündung ihr Entstehen verdanken, während der Sandstein von Sydney eine Meeresbildung ist. — Das Wandern durch den Busch ist auf die Länge sehr ermüdend. Eucalyptus bäume mit ihrer weißlichen Rinde, Casuarinas mit ihren articulirten, den den Fichten Nadeln fast ähnlichen Zweigen, stets sich gleich‑ bleibende niedere Gebüsche begleiten dich über Hunderte von Meilen. Das Terrain ist wellig und hüglig, doch selten gestatten größere Bergmassen weite freie An oder Aussichten. Lake Macquary ist indessen eine sehr ansehnliche Wasserfläche, welche wegen der Menge ihrer Buchten und der springenden bewaldeten Hügel[n] viel Mannichfaltigkeit zeigt und den

Reconnaissance in New South Wales

Zürich see an Schönheit weit übertreffen würde, wenn eine frischere Vegetation das matte Grün des Australischen Busches ersetzte. Brisbane water ist noch malerischer, indem höhere Berge einen bessern Hintergrund bilden. Ueberall wurden wir von den Kolonisten gastfreundlichst aufgenommen und mit Thee un Milch nd Damper bewirthet, welche uns besser schmeckten, als die wohlbesetzte Tafel in Newcastle. Da die Kolonisten sehr getrennt leben und nur selten mit der Stadt, bevölkerung in Verbindung kommen, sehen sie den Reisenden gern und schwatzen sich wieder einmal recht herzlich aus. Ich hatte Gelegenheit eine junge wohlerzogene Dame in der Buschhütte zu finden. Man konnte nicht leugnen, daß sie allmählig jenes feine Gefühl für strenge Ordnung un Nettigkeit verloren hatte, und daß sie sich nicht sehr genirte nur Schuhe aber keine Strümpfe zu tragen. Die jungen Männer, welche in den Busch gehen, gewöhnen sich gleichfalls sehr bald an dieses ungenirte leben: sie verbauern! Doch Gastfreundlichkeit scheint in demselben Grade zu zu nehmen, und sie theilen gern mit dir, was sie haben. Am schlechtesten ist es bei verheiratheten Personen mit der Erziehung der Kinder bestellt; diese wachsen gleich wilden Thieren auf und ich hatte Beispiele, daß sie als sie unsere fremden Gesichter erblickten nach einigem Anstarren gleich wilden Kälbern brüllend davon liefen. In den kleinern Städten wie z. B. Newcastle wird für die Erziehung ebenso wenig gethan. Die jungen Damen ergeben sich hier mehr gesellschaft. Vergnügen. Sie reiten aus, faulenzen oder spielen und lassen sich denn wohl zur Veränderung die Kur machen. Es ist wirklich bedenklich eine Currency lass zu heirathen; doch wandern jährlich so viele Sterling girls ein, daß man auch hier ein sichtbares wohlgebildetes Mädchen haben kann, wenn man nur nicht gar zu genau auf die Mitgift sieht. — Es ist indessen für den jungen Buschmann schwierig, in achtbare Familien eingeführt zu werden, oder es würden gewiß nicht so viele un verheirathete junge Männer sich Ausschweifungen ergeben, welche ihre Gesundheit zerrütten.

Ich glaube dir schon früher geschrieben zu haben, daß ich

Hrn. Clarke kenne; er ist Prediger in der Nähe von Paramatta und Secretair des Museums. Der Mann thut sein Bestes und hatte gute Erfahrungen gemacht; doch ist er in seinen Erklärungen nd Verallgemeinungen oft sehr unglücklich. Herr Alexandr Macleay war früher Secretair der Linnean Society und kam nach Newholland als Colonial Secretary unter Governor Ralph Darling. Als Burke Gouverneur wurde verabschiedete er Macleay nd setzte Deas Thompson, .s Schwiegersohn in .s Stelle. Der alte Macleay hat eine ausserordentl. schöne Insekten sammlung. Sein Sohn Sir William Macleay[c] ist der eigentl. Naturforscher un er hat .s drch einige allgemeine Aufsätze über [Ossificationen?] bekt gemacht. Dr Nicholson versprach mir, mich bei ihm ein zuführen doch verzog sich dieß bis mein Anhalten um die Stelle im Bot. Garten dazwischen kam; ich ging zu Macleay er versprach mir zuerst seine Hülfe, doch zeigte er .s später als mein größter Widersacher, da der Gouverneur dessen personl. Feind er [war?] mein Gesuch begünstigte. Ich habe einen Dr Welsh kennen gelernt, der hier ein reiches Mädchen geheirathet hat, doch weiß ich nicht, ob er derselbe ist, von welchem du sprichst; er scheint ein junger Mann von ungefähr 30 Jahren: ich habe wenig von ihm gesehen.

Ich wundere mich daß du noch nicht meine Briefe von Sydney erhalten hast, welche ich in Februar an dich schrieb. Ich freue mich, daß du nun eine briefliche Kommunikation mit den Meinigen eröffnet hast. Mein Bruder Adolph hat mein Vaters Gut übernommen und so es dem Leichhardtchen Namen erhalten.

Ich kann dir nicht .s drücken, wie tief mich deine freundschaftlichen Versicherungen bewegen. Ich danke dir viele tausend Male. Mein ganzes Streben geht darauf hin, dir wieder zu erstatten, was du an mir gethan hast. — Und solltest du durch Umstände gezwungen werden, mir zu folgen, so soll meine Erfahrung, die .s täglich vermehrt zu deiner .sschließlichen Disposition stehen. Ich kann nicht leugnen, daß ich mich gegenwärtig in einer etwas voreiligen Stellung finde, die mir meine ursprünglichen Pläne etwas verwirrte; doch kann auch dieß vielleicht mir nützlicher werden, als ich glaube. Lebe wohl mein theuerster William n grüße alle deine Angehörigen von deinem herzlich dich liebenden Freunde

L.L.

[*Newcastle, N.S.W., 24th November, 1842*]

My dearest friend,

On returning to Newcastle after a fatiguing journey on foot, I found your good letter, and the saddest news from home. There's no need for me to tell you how badly shaken I was by the latter, for it meant that a strong young creature at the height of her vigour, who had barely entered into life's fulfilment, had been borne away from the arms of a loving husband and from all her joy of life on Earth.[a] *But your letter contributed in no small measure to the calming of my troubled and depressed spirits, through the way in which it aroused my memories of you yourself. My brother-in-law too, was able so to revive my old memories of childhood that the sharpness of grief soon gave way to sorrowful resignation. My tramp to Lake Macquarie and Brisbane Water has added substantially to my first-hand knowledge of the colony. I think that I've a clearer idea of the relations between the Newcastle coal measures and the Sydney sandstone—they seem to be contemporaneous formations, whose differences in character correspond naturally to differences in their manner of deposition.*[b] *The Newcastle coal measures were deposited in a vast estuary, whereas the Sydney sandstone is a marine formation.—Wandering about in the bush becomes very tiresome in the long run. Eucalyptus trees with their whitish bark, casuarinas with their articulated leaves that are so like pine-needles, and unchanging undergrowth, are with you for hundreds of miles. The ground is undulating and hilly, but you very seldom catch sight of, or enjoy a view from, an outstanding eminence. Lake Macquarie, nevertheless, is a very imposing sheet of water. Its numerous bays and abrupt, wooded hills present great diversity, and it would far exceed the lake of Zurich in beauty, could a livelier green replace the subdued green of the Australian bush. Brisbane Water is even more picturesque, its grander* shores providing a more impressive background. The colonists received us very kindly everywhere. They regaled us with tea and milk and damper, which all tasted better to us than excellent fare in Newcastle. As these colonists live far apart and seldom have any intercourse with townspeople, they like to see travellers, and they enjoy the opportunity for having a good long talk again at last. I happened to come across a well-bred young woman in a hut in the bush, and could not help noticing how she had gradually lost her fine sense of order and neatness, and how little she minded being seen wearing shoes but no stockings. The young men who take to a life in the bush are the same—they soon get used to the free and easy existence and become quite countrified. But their hospitality seems to be all the readier for it, and they'll share whatever they have with you. Things are hardest for married people with children to rear. For the children grow up like wild animals. There were occasions when bush children, on catching sight of our strange faces, just stared at us for a few moments, then turned and ran off bellowing, like wild calves. In the smaller towns like Newcastle not much more is being done for education [than in the bush]. The young women in such places are more concerned with social pleasure. They go out riding, they idle the time away, play cards, or, just for a change, they take a cure. It's not considered the right thing to marry a currency lass, but so many sterling girls come out here every year as immigrants that a fellow can find a presentable, well brought up girl even here, provided that he's not too fussy about the dowry.—All the same, young men from the bush find that respectable families are slow to take them up, which is why so many single young men give way to excesses that ruin their health.

I think I've already told you that I know Mr Clarke. He's a clergyman who lives near Parramatta, and is [also] Secretary of the Museum. He is doing his best, poor man, and has acquired a great deal of first-hand knowledge; but he has often been very unfortunate in his interpretation of it, and in his generalisations. Mr Alexander Macleay, who was formerly Secretary of the Linnean Society, came out to New Holland as Colonial Secretary under Governor Ralph Darling. When Bourke became Governor he dismissed Macleay and appointed Deas-Thomson, his own son-in-law, in Macleay's place. The elder Macleay has an extraordinarily fine collection of insects. His son, Sir William

Macleay,[c] is the real naturalist and has attracted attention through one or two general papers on [ossifications?]. Dr Nicholson promised to introduce me to him, but had not done so even up to the time when I applied for the position at the Botanic Gardens. I went to Macleay myself, who promised at first to support me; but later on he turned out to be my greatest adversary just because the Governor, whose personal enemy he was, favoured my candidature. I've made the acquaintance of a Dr Welsh, who has married money out here, but I don't know if he's the one you mentioned. He's a young looking man of about 30, but I've seen very little of him.

I'm surprised that you've not yet received the letters I wrote to you in Sydney last February. I'm very glad that you've begun to correspond with my people. My brother Adolph has taken over my father's property, which keeps it in the Leichhardt family.

I can't tell you how deeply your friendly assurances have affected me, but I thank you over and over again. I will do everything I possibly can to requite you for what you have done for me.—And, should circumstances oblige you to follow me out here, my practical knowledge of the country, which is increasing day by day, will be at your entire disposition. I can't conceal the fact that I'm at present in a rather risky situation, somewhat to the confusion of my original plans; but even this may serve me better than I can foresee.

Good-bye, my dear William, and remember me kindly to your family.

Your affectionate friend,
L.L.

71 [TO JOHN MURPHY, Sydney, N.S.W.]

Newcastle[, N.S.W.,] 3 Debr. 1842

My dear John

I received your letter of the 28th of November, which is the 3d letter I received from you. I thought I had answered the first. Yesterday I returned from a very fatiguing walking expedition to Point Stephens, which forms the South head of Port Stephens, opposite to the grant of the Australian Agricultural Company. I was four days in the bush and found only once (at Telligerry) a hut where I could restore my exhausted strength. The heat was intolerable, fresh water was scarce, the provisions were soon consumed. It is difficult to find the way in the bush, as the people in this part of the country are not careful in marking the trees, to show that the footpath leads to an inhabited place, and as many great swamps exist here, which are the favourite places for cattle, which form in filing one behind the other tracks exactly resembling to the footpaths made by man. Thus I lost my way hundred times and I had only my compass to guide me. My dear boy, you have no idea how sweet a cup of tea with brown sugar and a piece of damper tastes after such a days journey in the bush. I prefer the tea in the bush to the most excellent wine and I am convinced that no beverage agrees so well with the constitution in this climate, as the tea. You know that the tea contains 2 principles, an exciting one and an adstringent one. The people in town make only an infusion of boiling water on the tea, in order to obtain only the exciting principle, which for itself alone is injurious to health. But in the bush they leave the old leaves and boil them again and again, adding allways a small quantity of fresh tea. They obtain the adstringent part of the tea united with a certain degree of the exciting one and this I imagine, is the cause of the wholesome effect on the stomach. The brown sugar and the milk weaken also its influence. I have found that even cold tea without sugar or milk quenches the thirst most effectively.— The bush was everywhere in fire and I was exposed to great danger in crossing one part, the trees changed into columns of fire and falling everywhere around me with a deafening noise. At night I lighted a fire and stretched myself under a forest oak (casuarina) and slept soundly, till the cold of the early morning made me uncomfortable and compelled me to stir my fire again and to dry the dew from my clothes. The deep silence was only interrupted by the browsing wallabis rustling through dry leaves or by the cry of the opossum. The first night I was however at the seaside and the roar of the breakers lulled me into sleep. Such a loneliness makes you feel your weakness and

Reconnaissance in New South Wales

approaches you more to the heavenly father and to his providence than the finest sermon in a full church: it is as if he himself was before you in this awful silence, which surrounds you, which makes you almost hear the beating of your own heart.— Yesterday I was travelling along a sandy beach 20 miles long. There was not a drop of fresh water and I had eaten nothing for the last 24 hours. I found a bed of shells (Donax) of which I devoured a dozen; but they were so briny, that my thirst was considerably increased. When my strength was almost entirely exhausted, I found a cask on shore, which was perhaps thrown over board by a vessel in danger; I went up to it and to my greatest joy and wonder I found it full of fresh water; I need not tell you how thankful I was and how I did drink. I am glad to hear that you are getting on well and that Mr Proudt [sic]ᵃ takes some pains with you. Make my compliments to your father and mother and to Proudts and Marshes if you see them.

Your affectionate friend,
L. Leichhardt.

[P.S.] Next week I hope to go up to Glendon and to Patricks plains; I want sadly a good companion, who had the same interest as I have. Perhaps I may still get one. Should you have time to walk out at a Sunday gather as many seeds as you can from all the plants, which are now in seed. You know that the seed is quite as characteristic as the flower (seed and seed [vessel]).

72 [TO W. KIRCHNER, Consul for Hamburg, Sydney, N.S.W.]

Newcastle, 4 December 1842

[Mein verehrter Freund,]

Seit Ihrem letzten, mir so angenehmen, Schreiben habe ich fast beständig meine Flügel geregt, und es ist mir deßhalb keine Zeit übrig geblieben, Ihnen einen zusammenhangenden Bericht über mein Thun und Treiben ab zu statten. Meine erste längere Ausflucht durch den Australischen Busch führte mich zum See Macquary und nach Brisbane water. Ich begleitete den Postmeister von Newcastle, welchem die Regierung den Auftrag gegeben, in jenen Gegenden die Quittrents zu sammeln. Es ist ein Mann, der viel gereiset ist, viel gesehen hat, und der mir besonders über Ostindien manches Neue zu erzählen wußte. Auf die Dauer wird das Wandern im Busche außerordentlich ermüdend, es fehlt an Abwechselung, die Vegetation ist auf hunderten von Meilen dieselbe — überall die zerstreuten, weißrindigen Eucaluptus oder Casuarinas, zwischen denen ein mittleres Strauchwerk sich findet. Wie die Bäume, Gesträuche und Pflanzen überall denselben Charakter tragen und weder durch die Frische und Grüne ihrer Blätter, noch durch die malerische Form ihres Wuchses und ihrer Combinationen uns erquicken, so ist auch der Boden, über den wir hinschritten, von denselben Gesteinen zusammengesetzt und das Terrain mäßig wellig und hüglig, ohne vorragende freie An und Aussichten. Das Gestein, welches fast ohne Unterbrechung den Boden von Newcastle bis Brisbane water oder wenigstens nach [Tukkerah Beach Creek] bilden, ist ein zersetzter Puddingstein, so genannt von seiner aus einer Menge kleiner Gerölle bestehenden Masse, welche einem Pudding mit reichlichen Rosinen nicht unähnlich sieht. Um Brisbane water tritt Sandstein zu Tage. Der See Macquary ist eine sehr ausgedehnte Wassermasse, welche wegen der Menge ihrer Büchten und der einspringenden, bewaldeten Hügel viel Mannichfaltigkeit zeigt und den Züricher See weit übertreffen würde, wenn eine frischere Vegetation das matte Grün des Australischen Busches ersetzte, oder reiche Ortschaften diese noch einsame Oede beseelten.

    Hier liegt die Besitzung Herren [Threlkelds], welcher die Eingebornen zu civilisiren trachtete. Ich besuchte seine Kohlenwerke, welche sich durch die Natur der Kohle wesentlich von denen in Newcastle unterscheiden. Ich sagte Ihnen früher, daß man in Newcastle vier Kohlenlager unterscheiden könne. Threlkelds Kohle entspricht dem obersten Lager, welches in

Reconnaissance in New South Wales

Newcastle außerordentlich schwach ist und nur an einigen Stellen brauchbare Kohlen zeigt. Threlkelds Kohle wird in großen Stücken herausgearbeitet, brennt sehr gut; doch bildet sie keinen Kuchen, sondern brennt, gleich Holz und Torf, indem sie eine feine, weiße Asche zurückläßt. Herr Threlkeld hat überdies den Nachtheil einer schwierigen Communication nach Sydney, indem der Eingang zum See Macquary fast ganz versandet ist und nur während hohen Wasserstandes für kleine Fahrzeuge schiffbar wird. In Newport, welches Herren Holden gehört, und das er als Township an den Mann bringen sucht,[a] leben jetzt nur einige kleine Pächter, welche der jungfräulichen Erde kaum eine Erndte abzudringen im Stande sind. Hier gab uns der Prediger von Brisbane water, Herr Rogers, eine Abendpredigt, bei welcher nur sieben von seiner eigenen Heerde gegenwärtig waren, obwohl er nur alle 4–6 Wochen nach Newport kommt. Sie sehen, daß die Leute eben nicht überreligiös sind. [In Tukkerah Beach Creek],[b] welches ungefähr 16 Meilen von Newport entfernt ist, wird der Busch frischer, der Boden besser, die Vögel lauter. Niedrige Gründe mit guter Viehweide breiten sich zwischen hier und Brisbane water aus. Es leben hier mehrere Ackerbauer und Viehzüchter, und in der Nachbarschaft von Brisbane water wird der Wanderer durch den Anblick eines schönen Gartens erfreut, in welchem an 900 Apfelsinen bäume, Aepfel bäume, Birn bäume, Feigen und Reben mehr oder weniger gut gedeihen. Dieser Garten wurde von Healy angelegt. Die Apfelsinen bäume sind nicht so kräftig wie die auf Ash island, doch sind sie durchaus frei von Blight und Scab, von einem schwarzen Fungus, welcher die Blätter bedeckt, und von Coccus, welcher Stamm, Zweige und Blätter heimsucht und auf Ash island fast über die ganze Pflanzung sich ausbreitet. Es ist indessen nicht zu leugnen, daß der sandige Boden im Garten von Madame Healey sich bei weitem weniger für Apfelsinen bäume eignet als der reiche Boden von Ash island. Brisbane water, an dessen Ufern zwei Townships (West und East Gosford) angelegt sind, ist außerordentlich malerisch, indem höhere

Bergmassen einen bessern Hintergrund bilden. Es ist bis jetzt die schönste Landschaft, die ich in dieser Colonie gesehen, und es ist wohl möglich, daß Gosford einst der Lieblingsplatz reicher Kaufleute von Sydney werden möchte, wenn ihnen Illawarra nicht besser gefällt, von dem Sie und mehrere andre gute Freunde mir erzählt haben. Ueberall wurden wir von den Colonisten gastfreundlich aufgenommen und mit Thee, Milch und Damper bewirthet, welche uns bald besser schmeckten als die wohlbesetzte Tafel in Newcastle. Wir kehrten auf einem andern Wege nach Newcastle zurück, indem wir zu einer Besitzung H. Foster's im Eingange des [Tukkerah Beach Sees] gingen und dann auf dem Meeresufer ungefähr 20 Meilen weiter schritten. Die Nacht überkam uns, und wir waren gezwungen, auf dem Strande zu übernachten. Das Brausen des Meeres, der dunkelblaue, gestirnte Himmel, unsere einsame Lage waren wohl geeignet, auf das empfängliche Gemüth einen tiefen, ernsten, feierlichen Eindruck zurück zu lassen. Am folgenden Tage hatten wir einen langen, ermüdenden Marsch zum Eingang des Maquary Sees. Unsere Nahrungsmittel waren erschöpft, Wasser war selten, und nur die honigreichen Blüthen der Lambertia formosa hielten unsere sinkenden Kräfte aufrecht. Wie wohl that es uns, als wir am Ende unseres Marsches in einer Fischerhütte gastfreundlich aufgenommen und mit Thee, Brod und Kalten Fischen bewirthet wurden. Am nächsten Morgen kehrten wir auf bekanntem Wege nach Newcastle zurück, und mein verehrter Gastfreund, Hr. S[cott] sorgte dafür, mich durch gute Pflege zu neuen Unternehmungen wieder tüchtig zu machen.

Wir gingen vergangene Woche zu Fuße nach Ash island, um dem wohlbesprochenen und übersonnenen Wassermangel auf dieser sonst so reichlich ausgestatteten Insel ab zu helfen. Die Kühe litten außerordentlich und der Garten noch mehr. Zu unserer großen Verwunderung fanden wir, daß die artesische Röhre[c], welche früher mit frischem Wasser gefüllt war, jetzt salziges Wasser enthielt, während der äußere Brunnen mit maßig salzigem Wasser gefüllt war. Wir versuchten, die Kuhe mit diesem letzteren zu tränken. Sie tranken; doch sagte man uns am andern Morgen, daß ihnen das Wasser nicht bekommen wäre, und daß sie purgirten. Da die Leute mir nicht in das Interesse des Herren S[cott] ein zu gehen schienen, und ihnen gewöhnlich jede Veränderung mißfällt, so glaube ich, daß die Behauptungen nicht begründet sind, da das Vieh im Busche sehr allgemein brackiges Wasser trinkt. Ich hoffe von einer Vertiefung des artesischen Brunnens wenig, ja glaube ich, die geologische Betrachtung und der eigenthümliche Charakter der Gesteine um Newcastle lassen auch kein Gelingen hoffen. Vergangenen Dienstag machte ich allein eine Fußreise nach Point Stephens, der südlichen Spitze von P. St. und den Besitzungen der Australian Agricultural Company gegenüber. Meine Absicht war, die eigenthümlich gebildeten Felsengruppen zu untersuchen, welche schon längst meine Neugierde erregt hatten, und von denen ich Proben in Sydney und Newcastle als Ballast auf den Werften gesehen. Ich glaubte in zwei Tagen zurück zu kehren; doch konnte ich nur nach viertägigem mühseligen Marsche, auf welchem ich zwei Nächte im Busche übernachten mußte und an 36 Stunden ohne Nahrungsmittel blieb, diese für mich außerordentlich erschöpfenden Reise vollenden. — Ich sah die Porphyrfelsen[d] in aller ihrer Macht und füllte meine botanische Büchse mit Handstücken. Zwischen ihnen und Newcastle dehnt sich in langem Bogen eine sandige, 20 Milen lange Küste, auf welcher ich keinen Quell, aber glücklicher Weise ein Faß frischen Wassers fand,[e] das wahrscheinlich ein in Gefahr schwebendes Schiff über Bord geworfen. Ich fand mehrere interessante Käfer und Larven, eine große Menge Pflanzen und mehrere gute Muscheln. Das war der Lohn meiner Erschöpfung, von welcher ich mich gemach erholen werde. Morgen haben wir den Bischoff[f] hier. Donnerstag gehe ich vielleicht nach Glendon.

Leichhardt's Letters

Newcastle, [N.S.W.,] December 1842

[My esteemed friend,]

I've been flapping my wings nearly all the time since I received your last most welcome letter, so I've had no time to spare in which to render you a coherent report on the things I've been doing and pursuing. My first long flutter through the Australian bush carried me as far as Lake Macquarie and Brisbane Water. I accompanied the Newcastle postmaster, who is under Government orders to collect the quit rents in these districts. He's a man who has travelled widely and seen much, and he was able to tell me a good deal that I did not know, particularly about the East Indies. It's very tiring to travel through the bush for any length of time, on account of the sheer monotony—a scatter of white-barked eucalypts or of casuarinas everywhere, with medium undergrowth between them. And, just as the trees and bushes and lesser plants are the same in character wherever you go, and fail to hold your interest, either through lively freshness of foliage or picturesque form and assemblage, the very ground across which we strode was of the same rock all the way. The terrain was heavily undulating and hilly, but without a single striking view or open, dominating prospect. The formation which extends unbroken from Newcastle to Brisbane Water, or at least to Tuggerah Beach Creek, is a weathered pudding stone, so called because of the numerous pebbles set in the mass of it, so that it looks not unlike a pudding thick with raisins. Around Brisbane Water sandstone makes its appearance. Lake Macquarie is a very extensive body of water which, on account of its numerous bays and wooded, plunging hills, offers great variety [of scenery] and would far surpass the Lake of Zurich in beauty if only a fresher green could replace the dull green of the Australian bush, or if there were a few settlements to bring life into this as yet uninhabited wilderness.

The estate of Mr Threlkeld, who tried hard to civilise the local blacks, is situated here. I went to see his colliery, which, from the nature of the coal, is mined in a different way from that at Newcastle. I've told you before that there are four distinct layers of coal at Newcastle. Threlkeld's coal corresponds to the top layer, which is extraordinarily slack at Newcastle and yields only here and there a coal fit for use. Threlkeld's coal is hewn out in great lumps. It burns very well, does not cake, and, like wood and *peat, leaves a fine white ash.* Mr Threlkeld, however, is under the disadvantage of poor communications with Sydney, because the entrance to Lake Macquarie is almost completely sanded up and is navigable only at high water and for small craft. At Newport, a property belonging to Mr Holden which he hopes to develop as a township,[a] there are at present only a few small tenants who are hardly in a position to force the virgin earth to yield them returns. Mr Rogers, the clergyman from Brisbane Water, preached to us here one evening, but only seven of his own flock attended the service, notwithstanding his inability to visit Newport more than once in 4 to 6 weeks. So you see that the people here can't be called over-devout. At Tuggerah Beach Creek,[b] which is about 16 miles from Newport, the bush is greener, the soil better, and the birds sing louder. Low-lying ground with good cattle pasture spreads out between here and Brisbane Water. Several farmers and graziers are established here, and not far from Brisbane Water the traveller can enjoy the sight of a fine orchard in which there are perhaps 900 fruit trees—sweet orange, apple, pear and fig, besides vines, all doing fairly well. It was laid out by Mr Healey [sic]. The sweet orange trees are not as vigorous as those on Ash Island, but they're quite free from blight and scab, from a black fungus that covers the leaves, and from Coccus, which attacks the stem, branches and foliage and has spread right through the orchard at Ash Island. All the same, the sandy soil at Mrs Healey's is not so good for oranges as the rich soil of Ash Island. Brisbane Water, beside which two townships, West and East Gosford, have been founded is very picturesque indeed, because bolder hills form a better background [than at Lake Macquarie]. It's the most beautiful scenery I've beheld in the colony up to now, and Gosford may very well become the favourite resort of Sydney's wealthier merchants, unless they prefer Illawarra, of which I've heard from you and other good friends. We were very kindly treated by the colonists everywhere, and we soon came to think that their tea, milk and damper tasted better than the good food of Newcastle.

We returned to Newcastle by a different route, which took us first to one of Mr Foster's properties, at the entrance to Tuggerah Beach Lake, and then along the sea shore for another twenty miles or so. Darkness overtook us and we were obliged to spend the night on the beach. The *foaming of the sea, the deep blue, starry sky, and our own isolation were circumstances of the very kind to leave a deeply serious, awe-inspiring impression on anybody of sensitive disposition. The next day we had to walk a long way, to the entrance to Lake Macquarie, and very fatiguing it was. We had eaten all our provisions, we found very little fresh water, and it was only the honey from the flowers of* Lambertia formosa *that kept us going. And how it did us good when, at the end of the day's march, a friendly fisherman asked us into his hut and offered us tea, bread, and cold fish! The next morning we took the beaten track to Newcastle, where my esteemed host Mr S[cott], through his kindly attention, ensured that I was soon fit for further ventures.*

Last week we went to Ash Island to do what we could to overcome the much discussed and over-publicised shortage of water on this otherwise richly endowed island. The cows were in a bad way and the orchard was even worse. To our great astonishment we found that the artesian bore,[c] *which had been sending up fresh water, was now running salt; and the well nearby was full of highly saline water. We tried to get the cows to drink from the well, and they did so; but we were told the next morning that it disagreed with them and purged them. As it seemed to me that Mr S[cott]'s men were not working in his interests and disliked having to do anything out of the ordinary, I doubt the truth of what they told us; for cattle often drink brackish water in the bush. I count little on the deepening of the artesian well—in fact, the geological considerations and the nature of the rocks around Newcastle give no promise of any success.*

Last Thursday I went alone and on foot to the southern headland of Port Stephens, Point Stephens, which faces the possessions of the Australian Agricultural Company. My purpose was to investigate some peculiarly disposed groups of rocks which had aroused my curiosity long ago. I had seen samples of the stone of which they are formed in the ballast on the wharves at Sydney and Newcastle. I meant to be back in two days, but it took me four weary days of walking, including two nights in the bush and about 36 hours without anything to eat, before I finished this extraordinarily exhausting journey.—I saw the porphyries[d] *in their full development and filled my vasculum with hand-specimens. Between these rocks and Newcastle there is a long, sandy,*

Reconnaissance in New South Wales

sweeping curve of 20 miles of coast. Nowhere could I find a spring, but by sheer good luck I did find a breaker of water that had probably been thrown overboard from a ship in distress.[e] *I found several interesting beetles and larvae, a great number of plants and a few good shells. That was the reward of my exhaustion. But I'll soon recover from the ordeal. To-morrow the bishop*[f] *is coming, and on Thursday I shall very likely set out for Glendon.*

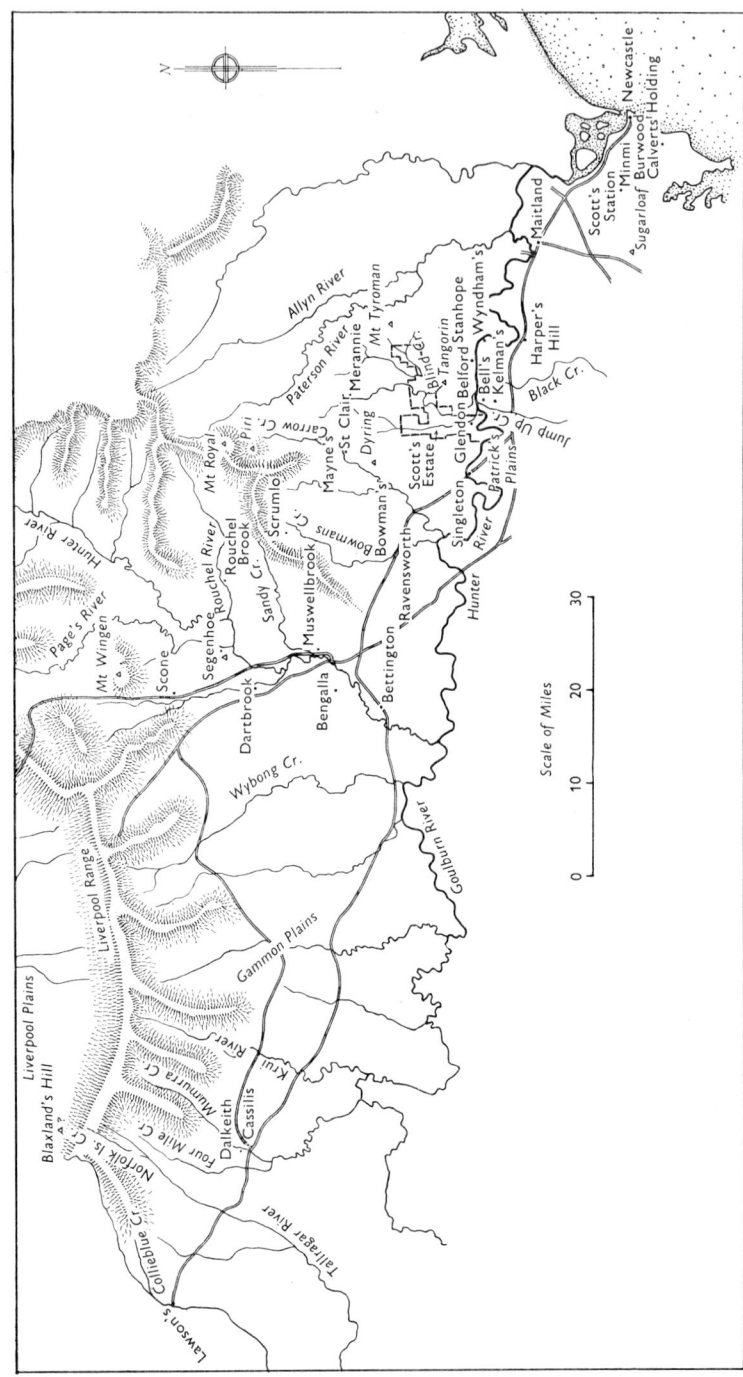

Fig. 8. Journeys on horseback, Hunter-Goulburn valley, N.S.W. December 1842–March 1843.

C. IN THE HUNTER-GOULBURN VALLEY ON HORSEBACK

73 HERREN C. SCHMALFUSS / Wohlgeb. / zu / Cottbus in der / Nieder Lausitz

Glendon den 16ten January 1843

Mein theuerster Schwager

Deinen Trauerbrief erhielt ich zu Anfange December 1842. Ich kann dir nicht ausdrücken, wie sehr mich der Tod unserer lieben Schwester Mathilde schmerzte. Doch ich danke dir für die sanfte nd kluge Weise mit welcher du verstehest, mir den bittern Kelch zu reichen. — Du fülltest meinen Geist mit einer so dichten Schaar von Erinnerungen aus der Heimath; ich begleitete dich mit Thränen im Auge Schritt vor Schritt den oft gewanderten Weg von Cottbus nach Trebatsch; ich war so voll von Wehmuth über die Veränderung aller Verhältnisse, daß die letzte schwere Nachricht nur den Eimer zum überlaufen brachte und ich mich meiner Schmerzen frei überließ. ⌐Doch auf der andern Seite war es mir angenehm, Adolph im Besitze unseres Vatergutes zu wissen. Ich freute mich, daß Barth ihm hierzu gleichfalls hülfreiche Hand geboten und daß unsere liebe Mutter nun nach allem Harren doch endlich in die gleichsam für sie bestimmte Wohnung eingehen kann. Mag Adolph nur in der Wahl einer Frau glücklich sein, welche Mütterchen liebevoll behandelt, daß ihre letzten Tage nicht durch eine unfreundliche Schwiegertochter verbittert werden.⌐

Nachdem ich ungefähr 8 Wochen in Newcastle gelebt und von dort nach allen Richtungen Excursionen gemacht hatte, kaufte ich ein Pferd und ritt nun gar sonderlich ausstaffirt den Hunters fluß hinauf. Wo immer sich höhere Berge zeigten, ließ ich mein Pferd auf guter Weide und stieg die Berge hinan, um ihre Bildung und die Gesteinarten kennen zu lernen, aus welchen sie bestanden. Einige Hemden, ein Theetopf, 1£ [sic] Thee und 2£ brauner Zucker waren alles, was ich zu meinem

Unterhalte mit nahm. Damper, ⌈eine Art Brod von unge⸺
säuerten Teige, welches in der heißen Asche gebacken wird⌉
und Fleisch, bisweilen Milch un Eier wurden mir von den
Ansiedlern gereicht, die mich gewöhnlich recht gastfreundlich
aufnahmen. Für den Fall, daß ich im Busche .tr freiem Himmel
zu übernachten hatte, führte ich eine wollene Decke mit mir,
in welche ich mich einhüllte und dann unter einem Baume
[mich] niederstreckte. Auf diese Weise habe ich auf dem
Meeres Ufer campirt, wo mich das Brausen der Wellen ein⸺
wiegte und im einsamen Walde, in welchem mich Kangooroos
umgraßten. — Der Wanderer im Walde von New Holland
hat nichts von wilden Thieren zu fürchten; der größter
fleischfressende Thier ist der wilde Hund von der Größe eines
Jackals, welcher zwar den Schaafen großen Schaden zu fügt,
doch sich nie getraut, einen Menschen an zu greifen. Die
Waldeinsamkeit, in der stillen Nacht, mit einem glänzend
Gestirnten Himmels gewölbe macht einen ausserordentl. tiefen
Eindruck auf einen einsamen Wanderer — und ich werde
einige Nächte dieser Art nie vergessen. Doch wechselt der
Wald nicht mit weiten bestellten Gefilden, wie wir sie in
unserer Heimath sehen. Mitunter findet man um die Busch
hütten des Ansiedlers einen beschränkten Raum von Bäumen
befreit, um ein wenig Weizen oder Türkischen Weizen zu
bauen. Doch größtentheils sind die Ansiedler mit Viehzucht
un Schaafzucht beschäftigt und kaufen sich ihr Mehl in den
Städten. Die Dürre war ausserordentlich anhaltend; die so hoch
gepriesenen Ufer des Hunters flußes hatten kaum Gras genug,
um mein Pferd spärlich zu nähren. Die ganze Gegend schien
wie versengt. Ungefähr 5 Meilen von Newcastle liegt Maitland,
eine viel bedeutendere Stadt, mit ungefähr 10000 Einwohnern.
Höher hinauf findet man zwar beginnende Städchen, doch sie
sind oft kaum Dörfer, obwohl mit bessern Wohnungen. Man
hat an mehreren Stellen bedeutende Weinberge angelegt. Die
Rebe befindet sich in der drückensten Hitze ausserordentlich
wohl. Das Laub ist stets frisch und grün, die Trauben sind
zahlreich nd groß, doch die Körner gewöhnlich kleiner nd

mehr getrennt, als ich sie in den bessern Weingegenden Europas gesehen habe. Man hat noch nicht guten Wein gekeltert, obwohl viele versucht haben. — Doch ist der Wein recht wohl trinkbar. Während mich in Newcastle ein Herr Walker Scott in seine Wohnung aufnahm, bewirthet mich hier in Glendon,[a] ungefähr 50 Englische Meilen weiter, sein Bruder Hr. Helenus Scott, der mit einem ältren Bruder sehr ausgedehnte Besitzungen hat. Ungefähr 80000 Morgen Land gehören ihm, auf welchen an 9000 Stück Hornvieh fast wild herum schweifen und eine Menge von Schaafheerden von Hirten herumgetrieben werden. ⌈Viele Hirten waren früher Verbrecher; jetzt werden keine Verbrecher mehr in die Kolonie eingeführt und freie Einwanderer verrichten ihre Dienste. Schäfer ist gewöhnlich der niedrigste Grad, womit ein Einwanderer anfängt.⌉ Er erhält ungefähr 175 Rtl. Lohn (25£) nd Wohnung nd Nahrung. Wären die Menschen hiermit zufrieden nd suchten sie sich ihre Wohnungen behaglich zu machen, indem sie kleine Gärten anlegten und einiges Land urbar machten, so könnten sie ohne Sorgen mit zahlreicher Familie recht glücklich leben. Doch die Einwanderer haben ihren Kopf von .sserordentl. Reichthümern so voll und sie wollen diese so schnell erwerben, daß sie stets nur denken andere Stellen nd höhere Lohn zu gewinnen. Die Folge davon ist, daß sie sich scheuen die Spade [sic] in die Hand zu nehmen, da sie ja eben nicht bleiben wollen, daß sie ihre müßige Zeit mit Nichtsthun verbringen, daß sie unstät von Ort zu Ort, von Herrsch[aft] zu Herrschaft wandern und so sich selbst und diejenigen schaden, in deren Lohn sie stehen. Denn der beständige Wechsel der Dienstboten, wo jeder Neue erst seine eigenen Erfahrungen zu machen hat, ist .sserordentlich unangenehm. — Man hat nun geglaubt, daß dieses unruhige Vorwärts streben, dieser Speculations geist besonders den Engländern eigen sei und daß es sehr vortheilhaft sein würde, Deutsche Familien zur Einwanderung zu bewegen, welche an Häuslichkeit gewöhnt, sich freuen würden auf dieser fruchtbaren Erde ihren Lebensgenuß ausserordentlich zu erhöhen.

Während der Mann dem Geschäfte des Herrn nachgeht, würde die Frau nd die Kinder ihre eigene Wirthschaft besorgen nd so allmählig für sich selbst und für die Landeigenthümer von großen Nutzen werden. Dieß scheint mir auch begründet; nur müßten Familen und nicht junge Männer auswandern, die sich in der Öde des Waldes nur nach der Heimath zurück sehen, die stets nur daran denken würden, so viel zu verdienen, daß sie nach Hause zurück gekommen un ab hängig leben könnten. Ich habe über diesen Gegenstand häufig mit den Ansiedlern gesprochen. Sie wollen besonders Schäfer: doch in den Gegenden, wo unsere größten Schaafzüchter leben, hat man auch nicht die leiseste Neigung zur Auswanderung. Die meisten Auswanderer kommen vom Rhein, von Hessen, Würtenberg [sic] nd Baden; doch dort ist Schaafzucht weniger bedeutend ⌐— und Weinbauer sind zwar sehr gesucht doch verhältnißmäßig weniger als Schäfer.⌐ — Dieser Theil der Kolonie ist ausserordentlich heiß und langer Dürre ausgesetzt, welche häufig die Waitzen Erndte zerstört welche gewöhnlich nur alle 3 Jahr gut .sfällt. Doch alle Frucht bäume gedeihen ausserordentlich gut. Pfirsichen, Apricosen, Feigen, Nectorinen, die Traube, Birnen, Aepfel ⌐sind in *ungeheuerer* Fruchtbarkeit — ich muß von dieser ungeheuerer Frucht barkeit indessen Birnen un Äpfel. wenigstens Theilweise ausschließen.⌐ Ein großer Nachtheil ist der Mangel an Wasser. Es giebt sehr wenige Quellen, sehr wenige fließende Wasser. ⌐Man hat tiefe Löcher (Wasserhöhlen) in welchen das Wasser das ganze Jahr über stehen bleibt und welche oft unrein sind. Künstliche ausgemauerte Behälter würden indessen diesem Uebel abhelfen. Dieß ist auch die Ursache, warum in diese Kolonie so erstaunlich viel Thee getrunken wird. Man kann das bloße Wasser nicht ohne Nachtheil genießen. Man kocht es deßhalb mit Thee und giebt ihm eine adstringirenden Geschmack. Thee und Zucker wird deßhalb als wöchentliche Ration jedem Diener nd selbst den Schäfern gegeben (Wöch entl. 1/4 Pfd Thee und 2 Pfd Zucker un 10 Pfd Fleisch nd 12 Pfd Mehl.) — Die Schaafe, welche jetzt schon bis zu 500

Reconnaissance in New South Wales

Englische Meilen landeinwärts in Heerden von 500-1500 herum wandern, sind klein; ihr Vließ ist weniger dicht, ihre Wolle weniger gekräusel[t]; die Kuhe geben weniger Milch und die Milch weniger Butter; gewöhnlich giebt eine Kuh in guter Zeit 1½ Pfd Butter in der Woche. Hierüber muß ich indessen erinnern, daß die Kälber nicht entwöhnt werden, sondern die Milch mit dem Herren theilen. Man strebt stets danach die Heerde zu vergrößern nd tödtet weder Bullkalb noch Kuhkalb. Nachdem es einige Monate alt ist, wird es ihm das Zeichen seines Besitzers eingebrannt nd nun kann es in den weiten Umzäunungen herum laufen wie es will. Diese Umzäunungen, welche oft viele 1000 Morgen umfassen sind sehr kostspielig; doch ohne sie würde man Kuh hirten nöthig haben, was bei der Höhe des Arbeits lohnes wenig abwerfen würde.⌐

Der Hunters fluß, welcher an seinem Ausfluße so groß erscheint, ist ungefähr 19 Engl. Meilen höher ein unbedeutende Bach. Doch seine Ufer sind sehr hoch und die heftigen Regengüße füllen sie rasch an; so daß man bei dem Mangel von Brücken im Winter oft lange Zeit zu warten hat, ehe man im Stande ist, den Fluß zu passiren. Die Bäche sind größten, theils nur Ketten von Pfützen. welche ein wenig Wasser den Sommer über behalten. Dieser Mangel an Wasser, welcher dem ganzen Erdtheil von Neuholland wenigstens in gewissem Grade eigen zu sein scheint, veranlaßt, daß wenn der Wind von NWesten oder Westen über das Innere hinueber fegt, jede Spur von Feuchtigkeit absorbirt wird und nun die volle Gluth des dem heißen Sonnenstrahl ausgesetzten Bodens uns an der Ostküste ins Gesicht weht. ⌐Die Hitze ist in diesem Falle ausserordentlich; der Schweiß läuft in Strömen am Körper hinunter; man wirft alle Kleidungsstücke bis auf Hemd un Hosen ab und man würde sich selbst diese gern entledigen, wenn es die gute Sitte erlaubte.⌐ Die Blätter der Bäume verdörren, die junge Saat ist oft verloren, die Früchte backen fast auf den Bäumen. Doch da die Verdunstung des Schweißes wiederum verhältnißmäßige Kälte erzeugt, ist das Gefühl der

Hitze nicht so groß, als es in einer nassen Atmosphäre, wie
z. B. in der Ostindischen sein würde. Dieß ist auch die
Ursache, warum die Hitze dem Körper weniger schadet nd
der Mensch hier weniger Krankheiten ausgesetzt ist, als in
Ostindien. Indessen haben mich die letzten heißen Tage doch
ein wenig rheumatisch gemacht. Wir hatten eine Hitze von
39°-40° R (120° F), welche um 9 Grad die Blutwärme
übertrifft.

Auf meiner letzten Fußreise fing ich ein junges Kanguruh,
welches mir jetzt viel Spaß macht. Ihr wißt, daß das Mutter,
thier die jungen bis zu einem sehr vorgerückten Alter in einer
Tasche mit sich herumträgt. Wird nun das Thier verfolgt nd
sieht es sich in großer Gefahr, so hebt es mit seinen Vorder,
füssen das Junge aus der Tasche nd wirft es so zu sagen über
Bord. Dieß war der Fall mit dem Meinigen. Es war so
unschuldig, daß es, als ich ihm ein Säckchen vorhielt, sogleich
Kopf über in dasselbe hineinsprang, wahrscheinl. glaubend,
daß es zur Tasche der Mutter zurück kehre. — Die Ameisen
sind in diesem Lande .sserordentlich zahlreich; man findet sie
von der Große einer Wespe und gleich dieser mit einer
Stachel versehen, bis zu der eines Sandkorns von fast micro,
scopischer Kleinheit. — Die Termite baut sich 5 Fuß hohe
kegelförmige Wohnungen aus gelbem Thon, oder erfüllt das
Innere hohler Bäume mit ihren Zellen. — Schöngefärbte
Papageien sind die Hauptfeinde der Trauben und Früchte; sie
fliegen in Volkern von 7-12 nd mehreren. Der Kakadu
zerstört die Türkischen Waizen felder, auf die er sich zu
Hunderten niederläßt.

Lebe wohl ⌜mein lieber Schwager und meine liebe
Schwester; küßt Euere Kinder, grüßt Mütterchen und Barths,
Adolph u Hilgenfelds von
                    Euerem herzlich Euch liebenden⌝
                                    Ludwig

Reconnaissance in New South Wales

C. *Schmalfuss, Esq., Cottbus, Nieder Lausitz*

Glendon[, N.S.W.,] 16th January 1843

My dear brother-in-law,

I received your sad news at the beginning of December 1842. I can hardly tell you how grieved I was to hear that our dear sister Mathilde was dead. But I thank you for the quiet understanding with which you passed me the cup of sorrow.—Vivid memories of Home came crowding into my mind; I could not hold back the tears as I followed you step by step the whole way along the familiar road from Cottbus to Trebatsch; and I was so acutely aware that everything in the world would now be different that the last news just filled the vessel to overflowing and I gave way to my grief. ⌜On the other hand, I rejoiced to hear that Adolph had come into possession of father's estate. I was glad to know that Barth too had offered Adolph a helping hand, and that mother, after having to wait so long, can now go to live in the house that was likewise meant for her, at last. All that I hope is that Adolph will be fortunate in the choice of a wife, and that mother will have a daughter-in-law who will cherish her, not an unkind one who might burden and embitter her declining years.⌝

After spending about 8 weeks at Newcastle and making excursions in all directions, I bought a horse and then rode up the valley of Hunter's River in fine style. Wherever a hill higher than the rest showed up, I left my horse where he could browse contentedly, and climbed up to the top, to study the nature and structure of the rocks of which the hills were formed. A few shirts, a billy can, 1 lb of tea, 2 lbs of brown sugar, were all that I took to keep me going. Damper⌜, a kind of bread made from unleavened dough, baked in the ashes,⌝ and meat, and now and then milk and eggs, were offered to me by the settlers, who received me as a rule very hospitably. In case I might have to sleep under the stars while in the bush, I took a blanket with me, and used to wrap it around me and then lie down under a tree. I've camped like this beside the sea, lulled by the sound of the waves; and alone in the forest, with kangaroos browsing around me. Roaming in the forests of New Holland there's nothing to fear from wild beasts. The biggest carnivorous animal is the wild dog, which is about the size of a jackal. It certainly does great damage amongst the sheep but it never makes bold to attack a man. The *sense of isolation one has in the bush, on quiet nights under a clear and starry sky, is profoundly felt by anybody travelling alone, and there have been a few such nights that I shall never forget. But there's no alternation of forest with stretches of orderly fields, like there is at home. At some places in the bush you'll find that a small area around a settler's hut has been cleared so that he can grow a little wheat or Indian corn, but most settlers are so busy rearing cattle and sheep that they buy their flour in the towns. The drought has persisted for an extraordinary time, and there was scarcely enough grass on the vaunted banks of Hunter's River to provide my horse with scanty meals. The whole region had a scorched appearance. About 5 [German] miles from Newcastle lies Maitland, a much more important place with about 10,000 inhabitants. Farther up the valley little towns are springing up, but they're hardly the size of villages, though most of them have better houses [than you find in villages]. Vineyards of some size have been laid out at several places. The vines withstand even the most oppressive heat, the leaves keep fresh and green, the bunches are big and the grapes large, yet the stones are smaller and farther apart than I've seen in the better wine-growing districts in Europe. No good wine has yet been pressed though many have tried [to produce it].—The wine is nevertheless quite palatable. Whilst I was at Newcastle I was entertained by a Mr Walker Scott, and here at Glendon*[a] *about 50 English miles farther on, I am the guest of his brother, Mr Helenus Scott. The latter, and an elder brother, own very extensive properties here. They have about 80,000 acres of land, on which about 9,000 head of horned cattle are running almost wild, and a number of herds of sheep are being moved about by shepherds.* ⌜*Formerly many of the station hands were convicts; but convicts are no longer being transported to the colony, and the work they used to do is now done by free immigrants. Shepherds come lowest in the scale, but that's where the immigrant begins.*⌝ *His wages are about 175 Rtl. (£25) with living quarters and keep. Were the men satisfied with this, and willing both to make their quarters comfortable and to bring a plot of land into use as a small garden, they could enjoy a carefree and happy existence here, even with large families. The immigrants, however, are so bent on making fortunes, and on making them quickly, that they think of nothing but getting better positions and higher wages.*

Reconnaissance in New South Wales

The consequences are that they fight shy of handling a spade as they're not even willing to stay in one place; they squander their spare time in idleness; and they move around restlessly from place to place, from employer to employer, to the detriment both of themselves and of those who employ them—for constant change of labour is anything but satisfactory when every new comer has first of all to learn how to do the work.—They've come to think that this restless urge to get on, this spirit of speculation, is particularly strong amongst the English, and that it would pay to encourage German families to come out here. Germans[, they think,] with their understanding of family life, would eagerly embrace the opportunity of doing so much better, in a fertile country like this, than they can at home. Whilst the husband [of a German family] was attending to his employer's affairs, the wife and children could be attending to their own, to the gradually increasing advantage both of themselves and of the owner of the property. And there is reason in this, I think; only, however, if whole families migrate, and not merely young men. For young men, alone in the bush, think of nothing but home; and young Germans would work only for the sake of making enough to enable them to return home and live independent lives there. I've often discussed the subject with settlers. The great need is for shepherds; but hardly anybody wants to emigrate from the parts of Germany where most of our sheep are reared. Most emigrants are from the Rhine, Hesse, Württemberg and Baden, where sheep-rearing is not so important ⌜*—wine-growers are badly needed too, though not so much as shepherds.*⌝ *—This part of the colony is very hot, and subject to long spells of drought. They can ruin the wheat harvest, which turns out well only one year in 3 as a rule. Yet fruit trees all do extraordinarily well. Peaches, apricots, figs, nectarines, grapes, pears and apples are bearing* prodigious quantities of fruit⌜—*though I must qualify 'prodigious' at least as regards some of the pear and apple trees.*⌝ *One great disadvantage is the shortage of water. There are very few springs and there's very little running water.* ⌜*But there are deep depressions (waterholes) which have water in them throughout the year, though they're often muddy. The excavation of lined tanks would overcome the difficulty. It* [the shortage of water] *also explains why they drink such gallons of tea in the colony. You can't enjoy the plain water without* risk of harm, so they boil it with tea, which gives it an astringent taste. Tea and sugar are therefore issued as a weekly ration to all employed hands, including shepherds (1/4 lb tea, 2 lbs sugar, 10 lbs meat and 12 lbs flour each, every week).—The sheep, which are now running in flocks of 500 to 1,500 as far as 500 English miles inland, are small, with a lighter fleece and less crinkly wool [than ours]; and the cows give less milk and the milk less butter. In good seasons one cow generally gives 1½ lbs of butter a week. Don't forget, however, that the calves are not weaned, so they share the milk with the owner. Everybody is trying to increase the size of his herds, and neither bull calves nor heifers are slaughtered. The calves are branded with the owner's mark when they're a few months old, after which they're free to roam about in the huge, fenced runs. The enclosure of these runs, which may be many thousands of acres in extent, is very expensive; but they save the expense of stockmen, whose high wages would cut deeply into the profits.⌝

Hunter's River, which is so broad where it flows into the sea, is little more than a brook 19 English miles farther up. Its banks, however, are very high, and the river is quickly bank-full after heavy rains. As there are few bridges, this means that you may have a long time to wait before you can cross the river in the Winter. The brooks, for the most part, are mere chains of ponds in which a little water still remains over the Summer. This dearth of water, from which the whole continent of New Holland seems to suffer at least in some measure, accounts for the drying up of every drop of moisture when westerly or north-westerly winds sweep across from the interior. When that happens, the full heat of ground exposed to the Sun strikes us in the face, like radiation from a furnace, on the East coast. ⌜The heat, in these circumstances, is amazing. Sweat pours down off your body; you throw off everything but your shirt and trousers, and if decency allowed it you'd get rid of these too.⌝ The leaves of the trees wither, the springing crops are often lost, fruit is almost baked on the trees. Nevertheless, because the evaporation of sweat leads in turn to relative coolness, you feel the heat less than you would in a humid atmosphere like that of the East Indies, for instance. And this is also why the heat does less harm to the system, and why people are less subject to illness here than in the East Indies. All the same, the recent hot weather has made me slightly rheumatic.

Reconnaissance in New South Wales

We had a temperature of 39°–40° R. (120° F.), which is about 9° above blood-heat.

The last time I was out on foot I caught a young kangaroo which is now giving me a lot of amusement. You know that the mother animal carries her young about with her in her pouch until they're quite well grown. If she is pursued and feels dangerously hard-pressed, she lifts the young one out of her pouch with her fore paws and, so to speak, hurls it overboard. This is what happened to mine. It was so innocent that when I held a bag out in front of it, it jumped straight in, head over heels, thinking, very likely, that it was hopping back into its mother's pouch.—Ants, in this country, are extraordinarily numerous. You find them from the size of a wasp—and with a sting like a wasp—down to that of an almost microscopical grain of sand.—The termite builds a conical anthill 5' high out of yellow clay, or fills the insides of hollow trees with its cells.—Brightly coloured parrots are the worst enemies of grapes and fruit. They fly in flocks of 7–12. The cockatoos do a lot of damage to the fields to Indian corn, on which they settle in hundreds.

Good bye⌐, my dear brother-in-law and my dear sister. Kiss your children for me and give my love to mother, and the Barths, Adolph and the Hilgenfelds, from

Your affectionate⌐
Ludwig.

74 [TO LT. ROBERT LYND, Barrack Master, Sydney, N.S.W.

Glendon, N.S.W., 16th January 1843.]

My dear friend

It is now more than four weeks that I left Newcastle and I suppose you are curious to know where I am and what I am about. My first resting place was at the foot of the sugar loaf to the top of which I ascended in order to ascertain whether Dr Nicholson's map[a] was correct, in which it was marked as being composed of Trapp. I found that it was not the case, but that it was the same pudding stone which extends over an immense country from Newcastle round Lake Macquary almost to Brisbane water and perhaps to the Liverpool Range, as it is here in Glendon even more powerful than down in Newcastle.—But this pudding stone appears to me the museum in which we have to study the former state of this country, as it contains pebbles of a great variety of rocks, probably all from the upper country, though New Zealand, Van Diemensland and the bottom of the sea might have furnished a great part of them.—From the Sugar loaf, where Doryanthes excelsa was in every state of developpement, I went to Harper's Hill, Mrs Harper the wife of the late Surveyor General[b] received me very hospitably and I lived 3 days in her family. At Harpers Hill I found a great number of fossils particularly Trochus and Spirifer and several species of bivalves and Zoophytes. One single impression of Equisetum seemed to me to indicate the coeval formation of this sandstone of great power and hardness with the clayy sandstone at Newcastle, perhaps between the first and second seam of coal.

The country looked exceedingly barren. My poor horse did scarcely find enough to sustain itself. But it was interesting to see how well the European fruit trees bore the dryness, looking all fresh and green, whilst the native trees looked [pale] and the turf was all parched up. The vine struck me particularly by the deep rich green of its foliage, though the grapes were rather few. I found here several little plants new to me—one a regular weed spreading its prostrate stem[s] in every direction over the ground, belonging to the Rubiaceae and perhaps nearest to Asperula.—Another weed which I found at first at black Creek is a little Sida with pretty looking yellow flowers. At Glendon a sensitive plant—whether Mimosa or Hedysarum I know not—grows very frequently in the paddocks; two other leguminous plants (one very like Orobus) are also frequent at Glendon. Arum with a large purple Spathe is just out of blossom. Sterculia covered with white purplish flowers, which resemble at the first aspect those of Tecoma australis grows as a powerful tree on the mountains; there also Hibiscus heterophylla was found with its large magnificent flowers.—A new Clianthes made its first appearance at Mrs Harpers, the 2 kinds

Reconnaissance in New South Wales

of Notolaenas, which you find in my collection of ferns were frequent in some limited shady places in the mountains. There grows a little tree, with rich green oblong lanceolate serrate leaves, with racemes of white monopetalous flowers and yellow fruits at the banks of the Hunter and of Glendon brook generally forming little groves with the most hospitable shade and at present with an inviting turf beneath it. Unfortunately I cannot determine a single plant and I have to give them my own names to keep them separate in my memory.—

From Harpers I went to Mr Windham [sic], whose vineyard I wished to see and who was spoken of to me as a man of great experience.[c] The present state of the colony has made him leave his vineyard to the cattle, though he is far from discouraging any other from the pursuit of cultivating the vine. His wife was a very interesting woman, she must have been extraordinarily beautiful, for she was still a fine woman though perhaps 30 years of age.—I communicated to him my plan of a squatting expedition to the interior and he believed that it would not only be the least expensive but at the same time the most profitable for science.—I went from his house to Mr Mitchells ground at black Creek to examine the position of the limestone. It occurs to me that the place where the limestone is found belonged formerly to the basin of a limited lake, in which carbonate of lime was deposited in a tufaceous form, perhaps from the water[s] of a spring.[d] It is very insignificant, in detached pieces surrounded by a white powdery stuff—but becomes thicker at the government ground. But even if there was a sufficient quantity, the question would still be, whether the quality of the lime would be good enough for working the stone.

A fortnight ago I arrived at Glendon, but before this I saw the vineyard of Mr Kilman [sic], who treated me with a very good glass of wine of his own making. His principal object however is to produce a good table grape. To obtain this he does not prune during the Summer, but leaves rich shade, which decreases the quantity but increases the quality of the grapes, they become closer, the berries larger and their skins more tender.

Glendon is like a little village; the cottage of Mr Scott being surrounded by a number of buildings, stables, stores, workshops, and lower down a long street of little cottages for the people which Mr Scott employs on his farm. A fine garden and a nursery and a vineyard are connected with the farm. After having examined the geological neighbourhood of Glendon, I went up Glendon brook, to see the coal and to find limestone if possible. It is very difficult to identify the coal with any of the seams of Newcastle. But I think that it is the second and that the fine sandstone, which is quarried near Glendon for building purposes is identical with that of Harpers, so that the formations from Maitland to Glendon and perhaps higher up range between the first and second seam of the Newcastle coal. If we suppose that these rocks were formed in a basin, it would agree very well, to find the greatest number of coal seams in the midst of the basin, which I suppose was near Newcastle, the fewest in the contrary at the outlines, as at Glendon and between lake Macquary and Tukkerah Beach lake.—I for my own part, would not use the word basin, because the idea which I have in my mind of the formation of the country does not agree exactly with the term.—All the country east of the blue Mountains was probably forming at the same time; in one place a large river came from an existing continent and carried during high flood the vegetable matter, trees and leaves down to its estuary [where] in this manner beds of coal were formed— at the same time sandstone and pudding formed higher up to the South or lower down to Moreton Bay. At Glendon Brook I met the first time with a more extensive igneous formation. A beautiful arc of blue mountains stretched from South east to east, its two terminal points being the highest; at N West and West several high mountain masses formed a corresponding though interrupted chain.—Long stretched hillocks extended from these mountains towards the valley: numbers of gulleys, of creeks, of brooks, partly running in consequence of the last rains gave to the surface a great variety. These singularly formed hillocks were generally composed of a trachytic rock, including in a feldspathic paste fine cristals of feldspar, more or less conspicuous, often reminding me almost of the porphyr of the ancients. Cristals of mica were frequent, and they were often of the most beautiful gold colour, particularly in more decomposed rocks. Quarz was often present, but sometimes those cristals with vitreous cassure, looked more like feldspar and Topaze, than like quarz. Chemical analysis and well defined cristals will soon determine their nature. [*From margin:* Besides these substances either Hornblende or Pyroxene are very generally present.] A granitic rock, or at least belonging to this group was found on the highest mountain tops, whilst those long stretched [ditch] like hillocs were composed of Trachytic roc. Often these hillocs had a direction from NW–SE or perpendicularly to it from NE to SW.—If this was general it would certainly be extremely curious, the former direction being that of the basaltic dykes at Newcastle.—The pudding and sandstone and a kind of red conglomerate below it were found at the flancs of the mountains frequently altered by the influence of their igneous neighbours. Of mineral substances only veins of carbonate of lime were found in the red conglomerate and in the Trachytic roc. I never did see them in the upper pudding or in the sandstone, or in the granitic roc. Once I saw a trace of gypsum. All these substances were not frequent enough to become useful.

Mr Scott has preserved partly the native names to the different mountains.[e] The S East corner is called Taingering; the east end of this arc and of the continuation of Taing, is Tyroman, at the north flanc of which passes the road to the Paterson. From Tyroman to the north the dividing range extends. More detached ranges are Meranni, Jack Shea and Wargungunnie. I believe that the mountains range between 1500–2000′, but I have no exact observation to go upon. Tyroman is the highest. There are several fine localities, which at a favourable season of the year would give a large harvest to the botanist. At present very few plants, perhaps in consequence of the late rain have lifted their little heads. I have mentioned allready the principal ones. After a very tedious journey of about 17 days, at every one of which I was either ascending a mountain or visiting and exploring valleys [*From margin*: the 6th to Tyroman the 8th to Jack Shea the 7th to Bukembelong or Blackguard hill the 9th to Wargungunnie] I returned very much exhausted to Glendon, where the rich fruit and the good table soon restored my lost powers. I caught a number of opossums, of native cats, one kangooroo rat and a young kangooroo, a beautiful little creature which the mother had jerked out of her pouch and which jumped head foremost into the bag in which I carried my specimens. Should you see Marlows, tell Miss Maryanne that I was very sorry of not being able to send it to her, for a better and more interesting little pet might scarcely be found.—

⟨Not. When I was passing Maitland Mr Rusden,[f] the clergyman and the father in law of Mr Helenus Scott did speak to me of the establishment of a school and of my taking part in it. But I am again as I was from the first moment I saw the blue distant hills, continually strifing to get to them and over them, and new ranges appeared then which excited only new desire! Poor miserable creature!⟩

At present these new mountains are before me, the Wollombi range, Dyrinne and Mount Royal. I visited the Paterson and Allan [*sic*] River and they carried such a quantity of basaltic pebbles, that I hope to find at Mt Royal a good lump of basalt. The Wollombi is said to be a granitic range and Dyrinne has decidedly trachytic roc in its composition. I shall soon become acquainted with the fathers of the land.

75 [TO WALKER SCOTT, ESQ., Newcastle, N.S.W.

Glendon, N.S.W., *c.* 15th February 1843]

My dear Sir

You will certainly believe Mr Bolton,[a] that I creep like a snail through the country, for instead of finishing my journey in 6–8 weeks, I am after two months scarcely half way. I made

Glendon the centre of my excursions to all the neighbouring mountains. I visited on a journey of about 17 days all the stations and mountains of the Glendon Estate and lived a hermits life for about 3 weeks at Mt Royal, the geology of which as well as the brush of its easterly flancs yielded me a rich harvest of interesting objects. The letter of Major Crummer made Mrs Harper receive me very kindly and your letter has only transferred your own kind and hospitable feelings towards me to Mr Helenus Scott, who has done everything, to render my rambles through the bush less fatiguing. During my journey to Glendon I omitted no opportunity of gathering information on the cultivation of the vine and the wine making. I saw Mr Windham, who has however unfortunately turned his cattle into his vineyard, and Mr Kilman, who has perhaps bestowed the greatest attention on the subject. His wine is good. His principal object however is not wine making but to obtain fine table grapes. He does not like Summer pruning, but leaves much wood and a very rich shade, which seems indeed favourable for the forming of large full grapes. Here in Glendon they have a very considerable vineyard, though Mons de Vonge[b] might have found by far superior soil for the vine at the chief station. My wanderings left me little hope to partake in the pleasures of the vintage, which generally brings back to my memory so many agreeable recollections of bygone days. Mr Helenus Scott however waited so very long and was so little decided, whether to make wine, that I found the over ripe grapes still on the vines when I came back from Mount Royal. Threatening rain prevented me from going to the Paterson where I wished to examine the position of the limestone, analogous layers of which I have seen between Dyrinne and Mount Royal in Fall brook and Carro [sic] creek. It seems that springs containing carbonate of lime deposited this limestone in various thickness at the bottom of a lake. At Dr Mitchells ground, where Mr Hel. Scott has made a kind of quarry it is rather a powder with some bigger detached pieces of limestone containing and surrounding pebbles. These pebbles become lower down more frequent and form a regular pudding stone of great hardness, small veins of carbonate of lime enter into the pudding. There is no hope of working this at the present place. At the neighbouring government land pieces of limestone often one foot long and about 3 inches thick lie in a rich red loam; the limestone is here more solid, but contains equally gravel. If it was really good, it might be worked, but I doubt much the good quality of it and experiments in Glendon do not speak favourably. There is everywhere about Glendon indications of limestone. It is found in small round concretions, in the clayy alluvial soil, it is found in pseudocrystals[c] in the clay beds of the banks of the river, it is also met with in more or less considerable veins in a red conglomerate and in some igneous rocks at the different stations. Detached blocks of marble with very coarse crystallisation are found in an analagous red loam near Stanhope opposite Mr Gordons place and at the ranges beyond the yellow rock towards Mr White's station, if you know it. But at no place it would pay the labour to collect or to work it. The Paterson limestone lies below all the formations of Glendon.[d]—At Meranni stalacktites are found in the caves of the pudding stone, but it is more object of curiosity, than of practical use.—The geology of Glendon estate presents some very interesting circumstances. But I'll not fatigue you with a long explication. I'll sollicit your hearing when I see you again. I have to relate many an accident. I was between the horns of a wild bull, I was almost burnt to death in a hollow tree—so I hope to become seasoned for the interior.—

Should you write to Mr Kirchner, I beg you to mention me kindly to him and his little family. I shall write as soon as I can. I scarcely venture to ask another favour. I lent the Zoophytes of Blainville[e] to Mr Wilton. When I saw them last at his house, they seemed to me rather roughly handled, but I forgot to take them with me. Would not you take back this valuable book by pretending to see some of the drawings? I know my dearest Sir, that it is a great liberty to take, to ask you such a thing, occupied as your attention is by your political dealings

etc; but have pity with the poor solitary traveller. I forgot to ask you, whether my money was paid in the Commercial or Australasian(?) bank in Maitland. Should any letters be at Newcastle for me, would you send them up to Glendon and write just at the back of one of them: Com. B. or Austr. Bk.
I wish you full victory in your political fight, rain at Ash island and dry winds at the North shore. Many compliments to Major and Mrs Crummer and to Mr Bolton.
I beg you to believe me ever faithfully yours,

L. Leichhardt

76 [TO LT. ROBERT LYND, Barrack Master, Sydney, N.S.W.

Glendon, N.S.W.,] den 19th Februar[, 1843.]

My dear friend.

I received your letter of the 28th Jan. at Capt. Mayne's[a] who lives between Dyrinne and Mt Royal, when I was returning from Mount Royal. Listen quietly, I have much to relate. But before I commence, I shall tell you, that I never forget to write to you, but as I wish at the same time to give you something with a head and a tail, I wait till I have finished an excursion or several ones, to give you afterwards a clearer account of them. I cannot express to you how much pleasure I derive from your letters, they are so well written and they contain so much interesting matter; that I read them generally a dozen times.—

The 23rd January I went to Meranni, about 10 miles from Glendon in order to examine caves, which were said to be covered with stalactites. After having searched long time, without finding them, I returned to the entrance of a creek, where I had left my horse grazing. But before I came to my horse I met a wild bullock at one of the little flats near the creek. I had often crossed herds of cattle without being molested by them and therefore, when I saw this bullock lowering his horns, I thought I would frighten it away and approached him therefore with my hammer uplifted making as big a noise as I possibly could. But the bullock far from being frightened ran at me and would have most certainly crushed me, if I had not jumped behind a tree. The animal ran against the tree and bored round it, attacking me a second and third time. The tree was big and I was unable to observe the movements, so it was that he caught me fairly the third time between his horns. I had my hammer in my hand and gave him a strong blow with it—be it that this stunned him or was he tired—I managed—how I know not, to get from him and again behind the tree, trembling in consequence of the violent exertion over the whole body. Seeing a second tree behind me I retired to this, the first tree hiding my movements, afterwards to a third, which brought me about 15 yards from the animal and to the banks of the creek. By a great circuit I came to my horse and returned to Glendon. But I left my hammer at the place, as it fell out of my hand, when I knocked the beast. Unfortunately I had no dog with me which would have distracted the attention of the animal from me.[b]

The next day I commenced my journey to Mt Royal with an old sawyer, who had cut timber for more than 9 years in the brushes of Mt Royal. He pointed out to me several forest trees which I had not seen before, for instance the *box*, a species of Eucalyptus, which has not a scaly bark, but more like an old elm at home, the fissures forming a network drawn out in a vertical direction.—the *black butt*, a fine tree, resembling in its bark the *stringy bark tree*. Its young branches are however smooth. Besides these two the following are easily distinguished: the *spotted gum* by the spotted appearance of its scaly bark, the *iron bark* by the deep vertical fissures in a rough darkish bark; the *stringy bark*, by the bark pealing off in long stripes; the *forest gum* resembling much the spott. gum, but the scales are longer and broader the blue gum resembling the forest gum, but the bark is of a peculiar colour. These trees have different leaves, but the leaves of young and old are very different and the trees are generally so very high, it is rather difficult to get them. The best caracters seem to me to be the marginal and secondary veins of the leaves, leaving the form and length of them altogether, and the colour and form of the young branches, which are sometimes angular, sometimes round—red—or green. Besides this the smell is distinctly different and a good nose, with a good nasal memory would distinguish them easily.

The family of Capt Mayne, to whom I had a letter from Mr Scott received me very kindly and the garden was full of the finest peaches, which I tasted in the colony. His house is situated at the North westerly or Northern slope of Dyrinne, a considerable mountain of perhaps 2500′ above the sea which I ascended the following day. The base of the mountain is formed of red conglomerate, the top of sandstone and pudding stone, which composes the greater part of the ranges between Dyrinne and Mt Royal. A kind of granitic rock[c] makes its appearance from time to time either at the flancs of the mountains, or in low ranges.—As soon as we come to the highest terrace of Piri,[d] which lies before Mt Royal, the sharp edged stones which cover the ground tell us that we are on another formation and the hammer shows us a bluish dark rock with Olivin of a yellow transparent colour, and with white cristals, which are probably Arragonite (carbonate of lime with prismatic cristallisation). Another substance I think to be Titanite of Iron. These are accidental substances caracteristic of basalt and I shall call therefore the rock of Piri of Mt Royal and probably of the whole Liverpool Range, of which Mt R. is a spur—not Trapp but basalt Piri rises very rapidly from its Sandstone base It is a long sharp range with an undulating outline, with a direction from South to North: a low connecting *wall* runs in the same direction from its northerly slope to the foot of Mt Royal, which rises still more rapidly and by far higher, its top being perhaps 3000′ above the sea. Though Mt Royal appears as a cone from the South it is like Piri a sharp range. Near the top it becomes so sharp, that you have to clime along a narrow wall to get to the top, which seems to be a heap of broken prismatic pillars 2–3 feet long. At several places at Piri as well as at Mt Royal I observed the inclination of the roc to prismatic division—another very general caracter of basaltic roc.—The base of both mountains is covered with forest, the principal tree being the black butt. The westerly flancs are covered with a rich grass, devoid of trees at Piri and with very few ones at Mt Royal. The easterly flancs are covered with a rich brush to the even edge of both mountains; this brush covers even the more expanded top of Mt Royal, though it seems only a loose heap of Ruines. During my stay at the mountain I observed with few exceptions a S Easterly wind. This wind carries the moisture of the sea, which is scarcely more than 50 miles distant to the mountain, and is here the cause of frequent morning mists and of rain. This rain supplies the East side of the mountains with abundant moisture and is so the cause of the rich vegetation, which covers them. At Mt Royal the moisture is perhaps equally attracted and better kept by the loose stones of the top and I explain so the dense brush extending over it.—And oh that you could have been with me in these brushes! A great variety of trees of great height tied together by vines form the body of the brush. Here grows the nettle tree about 80 and 90′ high with its large leaves, and the noble red *cedar* (what is its scientific name?) the red Sterculia, the Sassafras, the Ricinus, the Rosewood, the cohiti wood. It will take some time before we find even their real names. It was impossible to get branches from many of them, or to identify bark and leaves. In little gullies, where the waters went down the fern tree grew luxuriantly about 12–15 feet high with long leaves of 8–9′ long. [*From margin:* Underneath it a peculiar plant, if I am not mistaken a kind of Dorstenia, and Greliola and the common nettle covered the ground.] Polypodium, Asplenium, Aerostichum grew everywhere, mosses hung down in festoons and a species of bird had knowingly made use of them to hide its nest.—Lichens of various colour covered the rotten and the living trees. The lyre bird, the native Turkey with its peculair nest of leaves, the fermentation of which hatches the eggs, a kind of rat, the Echidna and many curious animals live here, though I had no means of getting them, I myself being no shot, and having no terrier to find the small quadrupeds out for me. And now let me tell you all my disasters. The day when I went to Capt Maynes the horse fell with me, I was violently shaken but broke fortunately no bone. When I came up to Piri, my horse broke the bridle and went back about 14 miles to Capt. Mayne. Next day I went to search for it, but could not find it. The man returned to Glendon to fetch provisions, so I lodged myself in a hollow black butt tree in which I could just extend the full length of my body.—I took the long leaves of the fern tree and spread them over the floor and over the walls of my lodgings.—A cheerful fire was burning before it. So I lived alone, accompanied only by my dog for four days, making excursions to the different parts of the mountain. For Breakfast, Dinner and Supper I had my tea and suggar, damper and bacon. Before sunset I ascended frequently Piri, from the top of which I enjoyed the most beautiful view over the distant mountain Ranges, which have very generally a direction from North to South. This view is not so grand as that from Mt Royal, which is above all description magnificent, laying this mountainous country like a map before your feet, with all the valleys and creeks or brooks which rise from its flancs. Fallbrook from the west, Carrocreek or Busby creek from the East, the Paterson and Allan River from the North or North East. At nightfall I descended to my hollow tree prepared my tea, wrapt myself in my blanket and watched Orion and Sirius gliding slowly through the foliage of the black butt. The flying squirrel commenced to call, the Wallabi came from the brush to browse on the rich grass of the forest and the mountains. I felt myself exceedingly happy, dreamt with open eyes, till the eyelids became heavy and the head sunk to the saddle which formed my pillow.—My provisions were however short. I had to live only on bacon notwithstanding my returning to Capt Mayne at Monday next. One may live on bacon without bread, provided one takes only small quantities at the time. After two days rain, during which I stayed at Capt Mayne's, the dray came up from Glendon with provisions for the sawyers and myself and I returned with my horse to my hollow tree at Piri. Mr Scott had sent me 5 quires of paper to collect plants for myself and for him and I wished to study the brush with leisure. But scarcely arrived at the mountain my horse went off a second time, I had to track it for 10 miles and to ride back without saddle and bridle (which latter I had forgotten in the hurry of the pursuit). As I was drying my paper the fire crept through a hollow branch up to it and destroyed not only about 2 quires but also my shirt which I had just washed. At night the wind blew rather hard and carried one of the fern leaves of my lodging, which were very dry, in the fire. The fire ran along it over to the tree to my blanket and at scarcely a moments notice I was all surrounded by the flame. As I was fortunately awake, I jumped up like lightning and threw blanket and everything from me. To fill the cup of my misfortunes, I lost my pencil next day and could not put any notes down.—Rain commenced, I was compelled to send my plants home half dry and in order not to loose them, I had to follow them about 4-5 days afterwards.

Is this not enough to discourage a poor soul, with no friend and comforter near him? I took it however very cheerfully, except loosing of the pencil, for I knew that I had kind friends who would say afterwards 'thank heaven that you are safe now' and so I thought with Eneas 'Et haec meminisse juvabit'.[e]

I have found a good number of fine plants, which will increase considerably our collection. I wished I could send you all I have now. A new asplenium, a new pteris some of the Norfolk island ferns. A number of little leguminous plants, some very pretty ones. leaves and branches and sections of wood and some fruits of the brush trees for instance the red Sterculia. —A fine Orchis with a pink flower was found at the hill before Piri. Some specimens are not well dried, but the greater number is tolerable. [*From margin:* A fine water lily, which belongs probably to the family of Burtomeas, a little Scrophula.]

If many of my Newcastle specimens were bad, you must

remember that I had to dry and to make excursions. I follow however now a quite different plan—whenever time permits, I get my plants dry in 2–3 days, being all the time occupied with it. The collection of small specimens which you make is just the thing; we want.—It is very strange, that I did never find any of the Ophioglossumes an[d] Botrychin[..] The Gleichenias are quite wanting here about. [*From margin:* The description of the Mimosa which you sent me, agrees exactly with the little plant here. I have it in fruit and blossom. It is very abundant. This district belongs perhaps to the flora of Pt Jackson.] I shall still remain for some time in Glendon. I shall therefore expect your answer here. I am going afterwards to Dalkeith at the Liverpool Range. Mr Boydel has invited me to come to the Paterson, but I shall probably delay till I return. We have bushrangers here who have robbed several gentlemen, one of which Mr Hentig^f escaped very closely. The long rains have set in Friday last and they are very much required, for the grass which was so rich when I left for Mt Royal was all gone when I returned.

I have one commission for you my dear friend. When I left Sydney Mrs Barney promised me to save geological specimens from Illawarra and New Zealand for me, which Cornel [*sic*] Barney had in his office. As Colonel B. is now preparing to go home, he might forget his promise and I wished to remind him of it. I propose to you several ways. I am convinced Mrs Barney would receive you very kindly, if you went with my compliments to her.—Quid rides?—I really think she did not deceive me; she was always exceedingly kind and I cannot imagine for what purpose.—Should you not like to do this, write a note to Capt. Marlow, and he will remind Colonel Barney, if you should not prefer to do it yourself. I wished only, you saw the specimens (particularly the fossils from Illawarra) yourself. I for my part think the last way would be the best.

Whenever you make excursions, recollect, that we are short of the seeds. When I come home I'll make little boxes to receive them to make a collection of them.

Adieu my dear friend Make my compliments to Marlows and to Dr Macdonald and do not forget writing
to your affectionate friend
Lud. Leichhardt

77 [TO HERR W. KIRCHNER, Consul for Hamburg, Sydney, N.S.W.]

Glendon, 21 Februar 1843

[Mein verehrter Freund,]

Ich weiß nicht, ob Sie meinen letzten Brief von Newcastle beantwortet haben. Thaten Sie es, so bin ich Ihnen lange eine Antwort schuldig geblieben. Ich hätte Ihnen vieles zu erzählen, besonders wenn Sie in meiner Wissenschaft folgen wollten. Lassen wir den besten Theil bis zur mündlichen Unterhaltung und hören Sie geduldig zuerst einen historischen Bericht meiner Fortschritte. Ich verließ Newcastle am 9. Dezember und ging zu Fuße bis zu dem Orte, wo mein Pferd mich erwartete. Mein erstes Reiseziel war 'the Sugar Loaf' (der Zuckerhut), jener blaue Berg, welchen man gerade im Westen von Newcastle erblickt. Man hatte mir gesagt, daß er aus Trapp, einem feurigen Gestein, geblidet wäre, und doch hatte mir ein früherer Besuch am Fuße nur Pudding bezeigt. Ich fand, daß dieser Puddingstein gleichfalls die höchsten baumlosen Klippen zusammensetzt und daß kein Trapp dort existire. Während meines Aufenthaltes in [Minmi], der Vieh⁄ Station des Herrn S[cott], hatte ich Gelegenheit, mehrere Tettigonias (locusts, wie sie die Leute nennen) zu untersuchen, deren Trommel ich schon längst zu sehen neugierig gewesen. Das wahrhaft Lärm machende Organ ist nicht so leicht gefunden, wie man glauben möchte. Es scheint mir durch die leisen, außerordentlich schnellen Contraction eines starken Muskels veranlaßt, der sich in eine starke Membran inserirt und diese in entsprechende Schwingungen versetzt. Es ist kein schwingendes Häutchen, durch einen darüber hinstreichenden Luftzug in Bewegung versetzt, wie das menschliche Stimm⁄

Reconnaissance in New South Wales

organ; es ist keine Pfeife, in welcher die Luft selbst schwingt und den Ton hervorbringt — sondern es ist eine Art Knarren. Von [Minmi] ritt ich nach Maitland, wo ich Herrn R[usden] sah, sodann nach Harper's Hill, wo ich eine große Menge Versteinerungen sammelte. Die ganze Gegend erscheint so versengt, so unerquicklich, daß meine Hoffnung, eine der reichsten Landschaften der Colonie zu sehen, bedeutend geschmälert wurde. Das Interesse, welches ich von Jugend auf an der Bereitung des Weins und der Cultur des Weinstocks nahm, führte mich zu Herrn W[yndham], der einen bedeutenden Weingarten besitzt und mehrere Jahre Wein gekeltert hat. Er hatte indessen sein Vieh in den Weingarten getrieben, und schien alle Lust verloren zu haben, sich ferner mit Weinbau zu beschäftigen. Nicht so Herr K[elman], ein Schotte, welcher mir ein recht gutes Glas Wein vorsetzte, und von dem ich Manches in der Ziehung der Stöcke gelernt.

So kam ich dann nach ungefähr 14 tägigen Querzügen nach Glendon, wo man mich außerordentlich freundschaftlich aufnahm. Ich wanderte zuerst in den nächsten Umgebungen umher, um mich heimisch zu machen, dann unternahm ich längere Wanderungen — und endlich machte ich eine geologische Reise über die ganzen Glendonschen Besitzungen der Gebr. S[cott]. Viele interessante Verhältnisse wurden gesehen, vieles gelernt und zwar in jeglicher Beziehung, von den Felsen, welche den Boden bilden, bis zu den Schaafheerden und Schäfern, welche über ihn hinziehen. Nach mehr denn 14 Tagen kehrte ich wieder nach Glendon zurück, und meine Fels- und Pflanzensammlungen begannen, sich zu mehren. Die Gewitterstürme des 16 Dezember hatten die verbrannten Flächen und den verwelkten Busch in lachende Anger und schattenreichen, das Auge stärkenden Wald umgewandelt und eine Menge von Pflanzen begannen, sich zu erheben und zu blühen, von deren Daseyn ich früher nicht die geringste Ahnung gehabt. Nie habe ich die Natur so wunderschön gesehen. Nachdem ich mich ein wenig von dem Buschleben erholt, riefen mich die blaue Berge wiederum in die Ferne. Doch ehe ich Glendon

verließ, sollte ich mit den Gefahren des Busches ein wenig
bekannter werden. Ich wurde auf einer Wanderung von einem
wilden Ochsen angegriffen und entkam ihm, ich weiß selbst
kaum wie — nachdem er mich schon zwischen seinen Hörnern
gefaßt. Ich hatte dabei das Unglück, meinen geologischen
Hammer, einen alten Reisegefährten, einzubüßen, mit welchem
ich dem Thiere einen Schlag zwischen den Hörnern versetzt
hatte. Ich sage, zitternd vor Ermattung entkam ich. Den
nächsten Morgen begann ich meine Reise zum Mount Royal,
ungefähr 34 Meilen von Glendon, begleitet von einem alten
Cedernholzsäger, der 9 Jahre lang in den Gebüschen jenes
Berges Cedern gefällt. Wie viel wäre von dem Reichthume
jener Gebüsche zu erzählen! Dem, was Sie am meisten interes-
sirt, konnte ich die wenigste Aufmerksamkeit schenken, ich
fand nur, wie durch Zufall, einige interessante Käfer. Ich sah
Raupen; doch wie konnte ich sie transportiren? Sicherlich
waren sie neu für Sie, denn sie lebten auf Pflanzen, welche
nicht in der Nähe von Sydney wachsen, z. B. dem Farrenbaum
(ferntree).

Mein Begleiter verließ mich, ich war nur von meinem
Hunde in dieser Wald- und Bergeinsamkeit begleitet. Mein
Pferd selbst war mir entlaufen. In einem hohlen Baum hatte
ich meine Wohnung aufgeschlagen. Vor demselben brannte
ein munteres Feuer. Ich lebte von Damper und Speck, Thee
und Zucker. Ungeachtet dieser Beschränktheit, ungeachtet
einer Menge von unangenehmen Umständen, indem ich
kochen, backen, waschen, Wasser holen mußte, fühlte ich
mich außerordentlich heiter. Einer einsamen, großartigen
Natur allein gegenüber zu stehen, ist außerordentlich erhebend.
Man fühlt sich in ihr, als ein ihr zugehöriger Theil. Wenn ich
Abends spät von meinen Wanderungen zur ausgebrannten
Baumhöhle zurückkehrte, welche ich indessen mit den 10 Fuß
langen Farrenbaumblättern grün geschmückt, bereitete ich
meinen Thee und genoß mein einfaches Abendbrod, aus
Damper und rohem Speck bestehend, dann hüllte ich mich in
mein Blanket (wollene Decke) und blickte in die dichter

Reconnaissance in New South Wales

werdende Finsterniß hinaus, oder in den gestirnten Himmel hinein, dessen leuchtende Sternbilder so groß und ruhig vor mir vorüber glitten. Allmählig wurde der Kopf schwerer und sank auf den Sattel zurück, welcher mein Kopfkissen bildete. Um ungefähr 2 Uhr des Morgens weckte mich die Kälte, ich mußte mein Feuer wieder schüren, um dann warm und ruhig dem hellen Morgen entgegenzuträumen; dann kamen das Frühstück und die Geschäfte des Tages und so fort. Tyra [sic] und Mount Royal, welcher von Glendon als ein hoher Hügel erscheint, sind langgestreckte enge Grate, nur die Kuppe von Mount Royal breitet sich etwas aus. Beide Berge sind aus einem dunkelblauen Gestein, dem Basalt, gebildet, welcher nichts, als eine vorweltliche Lava ist. Er scheint durch eine lange, von Süden nach Norden laufende Spalte, durch Puddingstein und Sandstein, hervorgedrungen zu sein. Die östlichen Gehänge beider Berge sind mit dichtem Gebüschen bedeckt, in welchen die rothe Ceder, ein edler Baum von 90 bis 100 Fuß und höher, der Nesselbaum mit seinen breiten brennenden Blättern, der Farrenbaum wachsen — alle von Schlingpflanzen und Lianen zusammen gebunden, von kriechenden Gewächsen und Schmarotzerpflanzen, von Moosen und Lichenen bedeckt. Der Leyervogel, der native Turkey und eine Menge interessanter Geschöpfe leben in diesen Gebüschen. Doch war es mir nicht möglich, mich ihrer zu bemächtigen, da ich kein Schütze bin und kein Schütze mich begleitete. Ich war drei Wochen von Glendon abwesend und hätte noch viel länger auf Mount Royal verweilen können, wenn mich eintretender Regen nicht gezwungen hätte, meine gesammelten Pflanzen mit dem gerade angekommenen Ochsengespann, welches Nahrungsmittel für die Cedern-Säger gebracht, nach Glendon zu senden. — Ich hatte übrigens noch manches Unglück zu erdulden. Während einer Nacht, in welcher der Wind heftig wehte, fiel ein Farrenblatt aus meiner Baumhöhle ins Feuer, das Feuer lief an ihm entlang in die Höhle und setzte diese in volle Flamme. Glücklicher Weise erwachte ich zeitig genug, um herauszuspringen,

Blanket und die Blätter weit hinauszuwerfen. An einem Tage verbrannten mir mehrere Bücher, Papiere und mein Hemd, während ich mit Pflanzentrocknen beschäftigt war. Dann verlor ich meinen Bleistift und konnte keine Bemerkungen mehr niederschreiben. Mein Pferd lief mir zweimal davon und ich hatte es ungefähr 10 Meilen weit wieder zu holen. Doch alles dies entmuthigte mich nicht, und nachdem ich es erduldet, gedenke ich lachend der bestandenen Gefahren. — Doch, mein lieber Freund, keine Raupen, keine Insekten, oder nur wenige für Sie. — Ich bin jetzt mit Weinmachen beschäftigt. Sehen Sie vielleicht Herrn Robert S[cott], sagen Sie ihm, daß ein guter 43er ihn erwartet, wenn er hier auf seine freundlichen Güter zurückkehrt. Doch ich muß nicht zu voreilig preisen. Ich hoffe nur...

Glendon, 21 February 1843
[*My esteemed friend,*]
I don't know whether or not you've answered my last letter from Newcastle, but if you have, a reply from me is long overdue. There's so much I'd like to tell you, particularly as you wanted to follow the course of my scientific pursuits; but let us leave the best of it to be told by word of mouth, and bear with me first of all by listening to a historical account of my progress. I left Newcastle on the 9th of December and went on foot to the place where my horse was awaiting me. My first destination was the Sugar Loaf, the blue mountain that shows up due West of Newcastle. I had been told that it was formed of an igneous rock, trap; yet, on an earlier visit, all I could find at its base was pudding stone. This time I found that even the rocks on its bare summit were pudding stone, and there's no trap there at all. Whilst I was at Mr S[cott]'s cattle station at [Minmi], I was able to investigate several Tettigonias or locusts, as the people call them. I had been curious about their drumming for a long time. The true sound-producing organ is not so easy to find as you might think. It seems to me that the sound is caused by rapid but barely perceptible contractions of a strong muscle inserted into a stiff membrane which it sets into corresponding vibration. It is not a tiny

Reconnaissance in New South Wales

tympanum set in vibration by an impinging current of air, like the human vocal cords; nor is it a pipe in which the air itself vibrates and so produces sound. It's a kind of creaking. From [Minmi] I rode on to Maitland where I saw Mr R[usden], and went thence to Harper's Hill where I made a big collection of petrifactions. The whole district looked so burnt up and unattractive that my hopes of seeing how fertile one of the richest parts of the colony can be, are waning. My abiding interest in wine-making and the cultivation of vines, led me to visit Mr [Wyndham], who has a considerable vineyard and has been pressing wine for several years. But it happens that he has turned his cattle into the vineyard, and he seems to have lost all interest in making wine. Not so Mr K[elman], a Scotsman, who offered me a glass of really good wine, and from whom I learned a great deal about the training of vines. Making frequent detours like this, it took me about 14 days to reach Glendon, where I received a warm welcome. The first thing I did was to get to know the immediate surroundings of the place. I then made longer excursions, and finally went on a geological tour over the whole of the Glendon estate of the brothers S[cott]. I saw and learned a great deal about how things are disposed—all sorts of things, from the rocks that form the soil to the sheep and shepherds that roam over it. After more than a fortnight I came back to Glendon, with many additions to my collections of rocks and plants. The thunder storms of the 16th of December had transformed the scorched surfaces and the withered bush into bright meadows and shady forest, a relief to the eye; and a number of plants began to come up, and to come into flower, of whose existence I had had not the least idea. Never before had I seen the world so fair. When I had got over some of the hardships of life in the bush, the faint, blue mountains beckoned from afar. Yet before leaving Glendon I was to learn a little more about the dangers of the bush. On one of my excursions I was attacked by a wild bullock, but I got away from him—though I hardly know how I managed it—after he had caught me between his horns. I had the bad luck to lose an old companion of my travels, my geological hammer, having struck the beast between the horns with it. I was quivering from exhaustion when I got away from him, I can tell you! The next morning I began my journey to Mount Royal, which is about 34 miles from Glendon. I went with an old sawyer who had been felling cedar in these mountain forests for 9 years, and you'd hardly believe how much he had to tell about the richness of the forests. I had to give the least attention to just what interests you the most, and the few interesting beetles I did find were found, as it were, by chance. I saw caterpillars enough, but how was I to carry them? They would undoubtedly have been new to you because they were feeding on plants like the tree fern, which don't grow anywhere near Sydney. My companion left me, and I was alone with my dog in the remoteness of these mountain forests. Even my horse had strayed away. I took up residence in a hollow tree and had a lively fire burning outside the door. I lived on damper and bacon, tea and sugar. I didn't mind the austerity or the thankless tasks of cooking, washing clothes and fetching water, and I felt extraordinarily happy. To confront the grandeur and the solitudes of Nature quite alone produces a strange feeling of exaltation. One becomes aware of his insignificance and his weakness; but he feels that he is part of Nature, with a proper place in the scheme of things. When, late in the evening, I returned from my wanderings to my hollow tree (it had been burnt out but I decorated it [inside] with tree-fern fronds 10 feet long), I made some tea and enjoyed a simple supper of damper and bacon. After that, I wrapped up in my blanket and peered out into the deepening night, or gazed up into the starry sky where the bright constellations wheeled so grandly and quietly overhead. I became drowsy and my head sank back on to the saddle that served me for a pillow. About 2 o'clock in the morning I would wake up feeling cold, and would have to put more fuel on the fire so that I could dream on, in warmth and comfort, until full daylight. Then there was breakfast, plans for the day, and so forth. Tyra [for Piri] and Mount Royal, which look like a single hill from Glendon, are long, narrow ridges, except that the top of Mount Royal is somewhat flattened. Both mountains are formed of a dark blue rock called basalt, which is nothing but a prehistoric lava. It seems to have been erupted through a long fissure which runs from South to North through pudding stone and sandstone. The easterly slopes of both mountains are covered with thick brush, in which the red cedar, a noble tree of 90 to 100 feet and more, the nettle tree with its broad, stinging leaves, and the tree fern are found— and they're all bound together by climbing plants and lianas, and covered with creeping and parasitic plants, and mosses and lichens. The

Reconnaissance in New South Wales

lyre bird, the brush turkey and a number of other interesting creatures live in these brushes. I was not able to secure specimens of them however, because I'm a poor shot and had nobody with me who could shoot them. I was three weeks away from Glendon, and could have remained at Mount Royal for much longer, had not the threat of rain constrained me to send my collection of plants back to Glendon by a bullock team that had just come up with provisions for the cedar cutters.—I had to put up with a good deal of bad luck too. One night when the wind was blowing hard a fern frond fell out of my tree on to the fire. The flame ran back along it to the tree and set everything on fire. I woke up soon enough to be able to jump clear and to hurl the blanket and the fronds outside. One day when I was busy drying plants, some of my books and paper, and my shirt got burnt. Then I lost my pencil and could no longer make notes. My horse gave me the slip twice and I had to go about 10 miles off to fetch it back. But I was not discouraged by all this, and, when they're over, I can laugh at dangers passed.—Well, my dear friend, no caterpillars, no insects—or just a few for you. At present I'm busy making wine. If you happen to see Mr Robert S[cott] tell him that there's a good 43er awaiting his return to his pleasant estate. But I mustn't sing its praises too long before they're due. I only hope...

78 [TO LT. ROBERT LYND, Barrack Master, Sydney, N.S.W.

Glendon, N.S.W.,] 2tn März[, 1843]

My dear friend.

Let me tell you how I was occupied these last 12 days. When I came back from Mt Royal I found that Mr Scott had not yet done vintage and that his grapes were spoiling fast. My wine mania rose and I set immediately to work to show my skill. The small Burgundy grapes were collected and crushed by two rollers, which I had seen used to crush the beetroot, and which Mr Scott had made according to my plan. I put the juice of an excellent quality, which I determined by a saccharometer I made myself, into a big wine cask and left it to ferment in a kind of cave dug in the neighbourhood of the vineyard. Two days afterwards the yng [young?] and sweetwater grapes were gathered and prest and put also to ferment on another plan, which will prove of great influence on colonial wine making. After having put the liquid in the cask one shuts it airtight and passes one leg of a bent lead or tin tube (a syphon) into the bung hole. The external leg is placed in [a] mug filled with water, preventing any communication between the interior of the cask and the external air. This method is followed in France and I saw it followed here in the fermentation of sugar beet. The carbonic acid produced during fermentation pushes its way through the water, but the atmospheric air never can get into the cask.—you see that I had two improvements (that of crushing the grapes by rollers instead of treading them with feet) and the Syphon, in my favour. The rainy weather com⁄ menced and kept the temperature from 68–71° (69° being the regular temperature of good fermentation. This proved also very favourable. My juice fermented in 10 days and when I racked it off it proved to be a strong spirituous wine. I have put it now in other casks and time must clear it. You shall assuredly taste it one day.—This success under other very unfavourable circum⁄ stances, will prove useful to me in case I have to turn my attention to the wine making.

Yesterday I found a great number of the little fern, you found at the Gaol grounds—in fact it formed a considerable part of the turf in the paddock. Tomorrow morning I start for Dalkeith hundred miles up the country at the Liverpool Range. From there I shall visit Mr Robertson, then I shall go to the Gwydir and at last to New England. After having done this I shall return to you my dear friend, who takes such an interest in my welfare and whose bill of Kindness would make any man bankrupt.   Vive Vale Memor esto mei.[a]

Reconnaissance in New South Wales

79 HELENUS SCOTT ESQ. / Glendon

Dalkeith the 24th of March 1843

My dear Sir

When I stayed at Bengalla,[a] I had long conversations and discussions about vine growing and wine making with my countryman Mr Luther. The soil which he has chosen seems rather too stiff in some places, but, as a layer of carbonate of lime lies below, a deep trenching or even ploughing will produce such a mixture that it may turn out a good soil. Though there is a considerable resemblance between this soil and that of Glendon, in which the concretions of limestone are found, there is this great difference, that the subsoil of that in Glendon is clay, whilst the subsoil in Bengalla is sandstone or puddingstone. In Germany we do not consider sandstone a good subsoil for a vineyard. Mr Luther knows well the fermentation with a syphon and he thinks it an excellent plan. I did never think wine making would become such an important branch of agriculture, as Mr Luther showed and proved it to be—and if my mind was more settled about my scientific plans, I should immediately turn vine grower.—I do not expect that any of the wine made this year will turn out good, except the one cask of Burgundy grape in the cellar under the house, but I know the causes which make me doubt about the other casks. [Mr Kelm]an has given me much information about cutting [and pr]uning the vines. I think I became useful to him in...of fermentation. I shall communicate to you, all I have...whenever I return to Glendon. I am very sorry that [I cann]ot give you a good account of your estate at Dalkeith. [The lo]ng continued drought has destroyed the grasses almost to the root and it seems to me that the turf, if you can call it so, is naturally very thin even for Australia. The layer of soil, which covers the ridges is very thin and the box, which is almost the only forest tree, is very stunted. There are however some Sterculias, particularly on the flat tops of the ranges. All the creeks are without water. Their bed is covered with depositions of limestone whilst the ridges are formed of a basaltic roc. The natural springs are very interesting and would give a rich supply of water (a far richer, than at present) if the[y] were regularly cleaned. I am convinced, that half the money spent in the wells would have secured a constant supply of fine water from these springs. One of them lies on government ground, but it is so situated, that nobody would ever disturb you in its possession. The plains about Bennigelliroy [have a] rich deep soil; but I fear, that they are subject to inundations during wet years, which will most certainly return. During dry seasons, at least at present scarcely a trace of vegetation exists. And yet the garden which Capt. Wilkinson[b] laid out produces after three years continual drought several kinds of beans, salads, cucumbers, water melons, rock-melons—in short any thing, you plant and which you take care of. But this is the deficiency. There is nobody who *will* take care; since Capt. Wilkinson is gone. Mr Barclay is a well informed person. He has travelled an[d]...seen much; but he is by no means the fit person to...such an establishment, particularly not being under your im[mediate] control. He is an Irishman, lazy, taking professedly no in[terest in] anything which concerns this colony. The store keeper is, as he...a perfectly supernumerary person, from whose services you der[ive no] benefit. The little which he has to do, could easily be done by...superintendent.—The servants are like the masters slow, sloven...After having seen the establishment, my impression was, that it was not your favourite child, that it was a kind of outpost, in which you took less interest. It is true, the present state of nature is by no means cheering, the country looks bleak and naked, the young grass springing up after slight showers of rain, being nipped off by a night frost. The cattle have to wander 6 miles and more to get water; they have to leave their calves for a long time, during which the native dogs pounce on them and devour them. This is the cause, why increase of stock is very trifling. It was interesting to see, how the poor creatures crowded round the well at Bennigelliroy. As soon as the man leaves the house to fill the troughs there is a general stirring

amongst them and they come bellowing from all sides to par‑
take in their share of fresh cool water.—Considering a great
number of leguminous and o[ther] herbs, which grow on the
ranges, I am inclined to think, that this country would be more
favourable for sheep than for cattle, and I guess 10–15000 sheep
would do admirably here. You will laugh at my judgment and
I know well that I have to learn more of it; I make only a
botanical induction.—

They showed me a piece of Massicot (Oxyd of lead)[c] found
near your station, washed out of the alluvium. I examined the
spot, [but I co]uld not find another trace of it. This lead ore is
rare...there would however be other ore with it, could we
discover...[wher]e it came from. They spoke with me about
rock salt; but it [turned] out to be salt scattered over the sheep
run by Mr Dennis...not angry with me, of writing you such
a long letter. [I trie]d to give you a true picture of the state of
Dalkeith, which...far out of your sight. I beg you to present
my compliments to [Mrs Sco]tt and to your little ones and to
believe me

Your sincere friend
Ludwig Leichhardt.

[P.S.] I beg you not to b[ring u]p those casks which are pro‑
vided with a syphon. Should the other ones throw out their
bungs, it would be best to provide them equally with syphons.
But take care of the water in the vessels, etc. that the opening
of the tube be continually shut.

80 [Possibly to MESSRS WALKER AND ROBERT SCOTT,
Newcastle, N.S.W.

Possibly from one of Lawson's stations,
late March or early April, 1843]

...in my hands. I have now travelled through [a great]...of
country and am preparing for an overland journey to Moreton
Bay. I wish you might have some benefit from the experience
I try to obtain amongst cattle and sheep rearers. If I cannot find

anything like profitable emploiment in Sydney I shall not only come down to you in scientific pursuits, but I shall bring what money is left to me and then you may give me the chance of becoming useful to yourselves and to me. According to my experience their are several pursuits which would give me an independence. The people at the Hunter commence to think seriously of wine making, which I have shown them to know as well as anybody. I have lived for a long time with sheep rearers who superintend their establishments themselves, and they assured me that sheep would pay at least 10 pr cent in... [think]...

IV
TO THE MORETON BAY DISTRICT

[to face page 651]

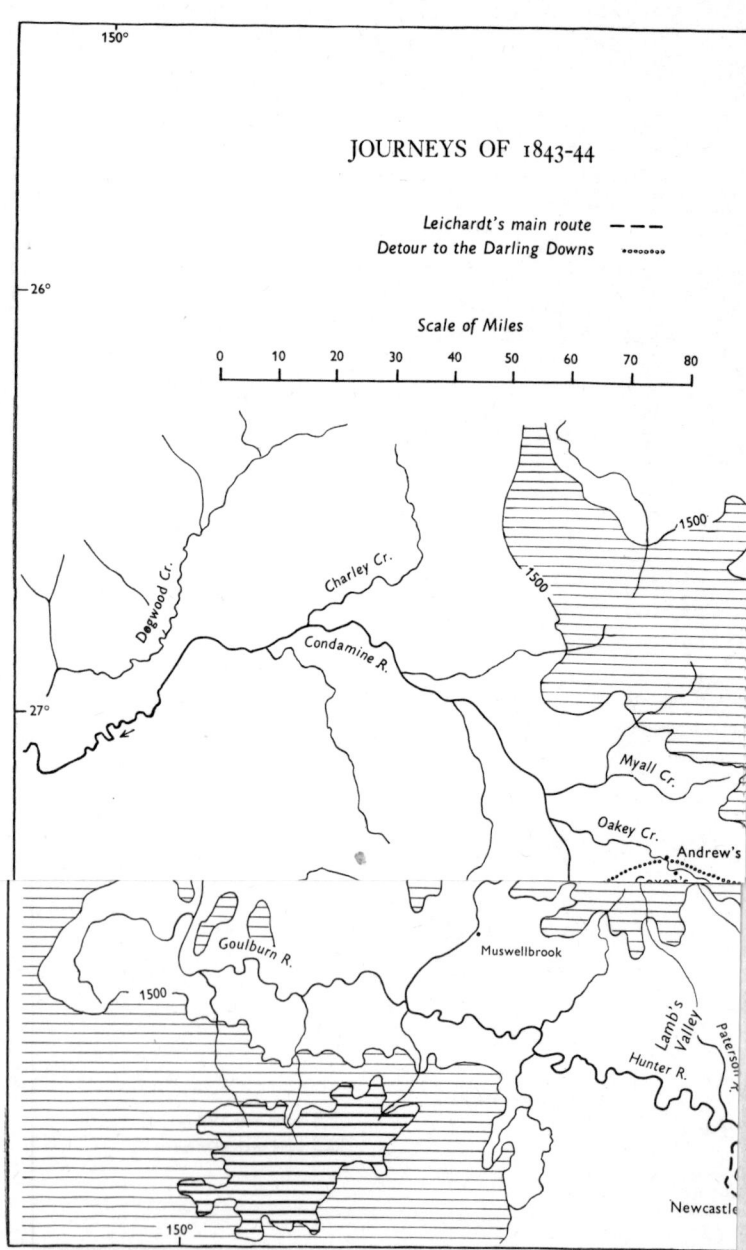

Fig. 9. Journey on horseback, Liverpool Range to Wide Bay ar
Positions of stations approxim

Moreton Bay

81 [TO LT. ROBERT LYND, Barrack Master, Sydney, N.S.W.

Pringle's Station, Rocky Creek,
Liverpool Plains district, N.S.W.,]
10th May, 1843

[My dear friend,]

... ... I went up the Liverpool Range (from Cassilis) with a gentleman[a] who was going to establish a sheep-station at the Range. All the spurs and secondary ranges, as well as the Liverpool Range itself, are entirely flat at the top. You climb up with great difficulty over loose, sharp-edged, hard rocks, and you find yourself, to your agreeable surprise, on a plain so smooth that you might drive in a carriage. The plain of the Liverpool Range is almost three miles broad. The rock becomes frequently cellular, and the cellules are filled with white crystalline substances—different species of zeolithe. The principal trees are the bastard box, the white gum, and a gum which the sawyers call blackbutt at Piri, but which they call forest mahogany here. It resembles much the stringy bark in its external aspect.

Next day I descended into the Liverpool Plains—an extensive level country, showing black soil covered with grass, with *Compositae* and *Leguminosae*, formerly the bottom of a large inland lake, with hills and ridges rising like islands. These hills are either sandstone or basalt. The sandstone is coarse and soft. The water and atmosphere have washed the sand off, and formed a layer of sand from one to three miles broad round the principal mountains. ... ...

82 [TO LT. ROBERT LYND, Barrack Master, Sydney, N.S.W.]

Mission Station, [Moreton Bay,] 23rd June, 1843.

[My dear friend,]

... ...From the Condamine river, the country rises very gently, almost imperceptibly, till the road passes between two hills or ranges, when the basaltic rock re-appears again. Very extensive shallow valleys or plains, generally with a creek, overgrown with reeds, covered with high rich grass, were spread before my eyes, when I passed these hills, the right of which goes under the name of Rubislaw, and the left under that of Sugarloaf. Here and there the grass-tree is seen either single, or in groups and groves. It is one foot and more in diameter, and eight to ten feet high. Till then I had never seen the grass-tree in rich soil; on the contrary, it was the sign of the poorest sandstone rock and sand. Here the case is reversed, the grass-tree grows in the finest soil, and generally in plains. The ranges which border the plains are covered with box, with a gum tree called the Moreton Bay ash, with a different species of angophora, and with another white gum. The trees are generally very scattered, and the forest becomes only denser, the vegetation more powerful, as we approach the range of its eastern slopes. All this country, from the Condamine to the range, is called the *Darling Downs*. There is no equal to them over all the colony for sheep rearing, for the fatness and tenderness of the mutton, for the excellent qualities of the wool (which, however, is not generally admitted,) and for the cheap rate for which flocks can be managed. One shepherd can look after 2000 to 3000 sheep, which would require four shepherds in other parts.

As we approached the settlement, (Brisbane Town,) unknown trees and shrubs increased in number, and the settlement itself, well situated on the banks of a noble river, was surrounded by acacias in the full pride of golden blossoms. ...
... ...

The philanthropist could never find a purer and better nucleus for the commencement of a colony than these seven families of Missionaries are: they themselves excellent, tolerably well educated men, industrious, with industrious wives. They have twenty two children, though very young, yet educated with the greatest care—the most obedient children I have seen in this Colony or elsewhere. If the Governor was in any way a man of more comprehensive views, and if he considered the moral

Moreton Bay

influence of such a little colony on the surrounding Settlers, he would not grudge them the few acres of land which they are at present in possession of—he would grant it to them for the five years of suffering they had to pass. The Missionaries have converted no black-fellows to Christianity; but they have commenced a friendly intercourse with these savage children of the bush, and have shown to them the white-fellow in his best colour. They did not take their wives; they did not take bloody revenge when the black-fellow came to rob their garden. They were always kind, and perhaps too kind; for they threatened without executing their threatenings, and the black-fellows knew well that it was only *gammon* ... ...

83 [ADDRESS NOT PRESERVED.]

Moretonbay den 27tn Juny 1843

Meine theuerste Mutter

Ich habe nun schon große Reisen in dieser Kolonie vollendet — ich habe vieles gesehen und manches gelitten. Oft bin ich in dem einsamen Walde, auf hohen Bergen, allein gewesen und bei meinem nächtlichen Feuer in meiner Wollendecke gehüllt habe ich zu dir und allen meinen lieben daheim hinüber gedacht. Dein Sohn führt ein wunderliches Leben. Arm, wie er ist, kämpft er mit den Umständen so wacker, wie er kann und stets kommt ein helfender Arm, wenn sich die Aussicht trübt. Denke dir einen jungen Bauer mit Jacke und Hose auf einem kleinen schwarzen Pferde mit einer Wollendecke und mit Quersack über den Sattel geschnallt, mit einem schweren Hammer am Sattelknopfe und mit einem kleinern in der Tasche und du hast eine ziemlich gute Vorstellung von deinem lieben Sohne, wie er durch den Wald von Australien reitet. Ein weißer Hühnerhund ist ausser dem Pferde oft meine einzige Begleitung. — Kommt es darauf an vorwärts zu kommen und zeigt die Gegend nichts merkwürdiges, so reite ich ungefähr 5–6 deutsche Meilen. Ist die Gegend merkwürdig, so mache ich halt, gebe meinem Pferde Rast und wandere durch die Nachbarschaft, um zu sehen un zu sammeln, was mir interessant scheint. ⌐Die Ansiedler sind im Allgemeinen sehr gastfreundlich. Es sind gewöhnlich junge unverheirathete Männer, welche 500-1500 Stück Rindvieh haben, welches sie so viel als möglich zu vermehren trachten. Die Verschiedenen Ansiedlungen sind ungefähr 4 Meilen von einander entfernt nd bestehen nur aus einigen Bretterhütten, welche in diesem milden Klima zum Schutze gegen Wind nd Wetter hinreichen. — Das Rindvieh ist mitunter sehr schön und nie habe ich auf meinen Reisen in Europa schöneres Rindfleisch gegessen oder fettere Thiere gesehen. — Andere Gegenden sind mit Schaafheerden bedeckt. Viele Tausende von Schaafen wandern über die Hügel und Ebenen von New England nd Darling downs. Diese Gegenden sind besonders für die Schaafzucht geeignet, indem oft 2-3000 Schaafe in einer Heerde gehalten werden können. Dieß macht die Schaafzucht viel vortheilhafter da der Lohn der Schäfer so hoch ist. Diese Leute erhalten 140-190 Rtl. Du mußt indessen wohl bedenken, daß ihre Bedürfnissen im Verhältnisse sehr theuer sind. Man lebt im Innern der Kolonie auf diesen Rindvieh und Schaafstationen entweder ausschließlich von Rindfleisch oder von Schaaffleisch; hierzu ißt man eine Art ungegohren[es] Brod, welches man Damper nennt, nd welches man, nachdem der Teig wohl geknetet ist, in frischer Asche bäckt. Thee wird in unglaubliche Menge getrunken, denn zum Frühstück, zum Mittag⸗ nd Abendbrod wird immer eine Kanne Thee mit getrunken. Es scheint die Suppe zu sein. Selten haben sie Kürbisse, welche indessen schön wachsen und hier sehr mehlig sind. Wenn die Leute nun auch Fleisch haben, so viel sie immer nur essen mögen und jeder Mann ißt gewöhnlich fast 2 Pfd täglich (sie erhalten 12 Pfd für der Woche), so haben sie doch auf der andern Seite nicht die Gemüse, deren Ihr Euch erfreut. Selten findet man Kartoffeln un Kohl, da die Hitze des Sommers gewöhnlich alle Pflanzen zerstört.⌐ Ich habe einige Deutsche gesehen und schätze mich glücklich, sagen zu können, daß der Deutsche überall durch seinen Fleiß und

Moreton Bay

durch seine bescheidene Genügsamkeit sich die Liebe seiner Nachbarn erwirbt und sich unabhängig zu machen versteht. — Einige Deutsche sind als Schäfer sehr geschätzt, ein anderer war Weinbauer. Hier in Moreton bay lebe ich in der deutschen Mission, welche aus lieben, wackern Leuten besteht, die manches erduldet haben, um die Schwarzen zu bekehren, doch leider wenige Fortschritte machen. Sie sind alle verheirathet nd zusammen 7 Familien mit ⌜gegenwärtig⌝ 22 wohl erzogenen Kindern. Ich fühle mich wohl zwischen ihnen, da es gar zu selten ist, reinen tugendhaften Menschen in dieser Kolonie zu begegnen. — Obwohl nun die Schwarzen wenig Neigung zeigen, das Kristenthum an zu nehmen [oder] sich über religiöse Gegenstände den Kopf zu zerbrechen, so sind sie doch häufig recht klug nd verschlagen. Dieß ist besonders der Fall, wenn Kinder sehr jung von den Eltern entfernt werden und nun getrennt einige Erziehung erhalten. Die Stämme der Schwarzen sterben indessen sehr schnell aus und oft sind nur noch 2, 3 von früher 100tn starken Stämmer übrig. Sie thun an einigen Orten der Ansiedlern großen Schaden, indem sie die Kühe mit ihren Speeren erstecken oder Schaafheerden davon treiben ud Schäfer tödten. — Ungefähr 100 Meilen gegen Norden scheiterte vor vielleicht 8 Jahren ein Schiff welches einige Ochsen an Bord hatte. Ein Ochse schwamm an das Ufer nd gedieh herrlich auf der reichen Weide. Die Schwarzen hatten nie ähnliche Thiere gesehen und der gewaltige Ochse erregte ihr Erstaunen. Von fern und nah kamen sie, das unbekannte Ungeheuer in Augenschein zu nehmen und viele Pläne wurden geschmiedtet, es zu tödten. Doch sobald sie näher kamen, rannte der Ochse auf sie zu und verjagte sie. Endlich kamen die Weißen Ansiedler zu jenen Gegenden. Sie kannten das fette Thier sehr wohl und eine Kugel war hinreichend, es zu Boden zu strecken. Es war ausserordentlich fett und ein Herr welcher ein beefstake von dem Fleische kostete, wußte seine Zartheit nicht genug zu rühmen.

Ich weiß nicht, ob ich dir geschreiben, daß ich einst in der

größten Lebensgefahr war, indem ein wilder Ochse mich angriff. Ich was allein, im Walde — nd hatte just Zeit mich hinter einen Baum zu retten. Er griff mich mehrere Male an; ich versetzte ihm einen Schlag mit dem Hammer, dieß schien ihn zu beruhigen und ich rettete mich, indem ich von einem Baume zu andern schlich. — Ich verlor dabei meinen schönen Hammer, der mich durch Frankreich, die Schweiz und Italien begleitet — was indessen natürlich besser war, als das Leben. —
Meine Gesundheit ist so fest nd fester denn jemahls. Indessen sagen mir die Leute, daß ich älter aussehe, als früher — und die weißen Haare werden häufiger. Vielleicht würde ich weniger die Beschwerden der Reise ertragen haben, hatte ich mich nicht stets bestrebt einen reinen ⌜keuschen⌝ Lebenswandel zu führen, und deiner meine gute Mutter, immer würdig zu bleiben. Hattest du den Knaben nd den Jungling lieb, der dir doch mitunter, obgleich gegen seinen Willen, Kummer verursachte, so würdest du den Mann mehr lieben, in welchem sich allmählig die Leidenschaften der Jugend beruhigen und der dennoch die kindliche Herzens frische zu erhalten verursacht hat. — Laß uns hoffen, uns unter dem alten Himmel noch einmal wieder zu sehen. Mag Gott dir bis dahin nd länger Gesundheit und Heiterkeit schenken. Ich möchte so gern von Euch hören — und wahrscheinl. erwärten mich Briefe von Euch in Sydney. Doch das Innere des Landes ist nicht durch regelmäßigen Postenlauf mit Sydney verbunden, un da ich beständig meinen Aufenthalt wechsele, ist es unmöglich Briefe zu empfangen. Doch selbst heut höre ich vielleicht von Euch, da just ein Dampfschiff von Sydney nach Moreton bay gekommen ist, mit welchem ich Briefe von meinem Freunde in Sydney erwarte. Ich habe glücklicherweise einen recht wachern Freund in diesem Lande gefunden, der mich nach allen seinen Kräften unterstützt und der gleichsam Williams Stellvertreter ist, den der liebe Gott mir zu sandte, als William in Europa zurück blieb. Auch von William, dem theuern Freunde un Bruder habe ich lange nicht gehört. [Ich sende ihm

Moreton Bay

indessen diesen Brief um ihn zu Euch zu schicken. —⌐ Nun lebe wohl liebe Mutter. ⌐Grüße mir alle unsere Lieben⌐ von deinem herzlich dich liebenden Sohn[e,]

Ludwig

Moreton Bay, 27th June, 1843

Dearest Mother,
I can now tell you that I have already made long journeys in this colony—I've seen a great deal and I've been through a great deal. I've often been quite alone, on the tops of mountains or in the remoteness of the bush, and, as I lay by my fire at night, wrapped up in my blanket, my thoughts would lead me back to you, and to the rest of the family, so far away over there, at home. But your son is having a wonderful time. Poor though he is, he's doing his best to cope with circumstances and whenever the prospect has been gloomy somebody has offered a helping hand. Think of a young farmer in jacket and trousers, riding a small black horse, with a roll of blankets and a satchel strapped across the saddle, with a heavy hammer hanging from the pommel and a lighter one in his pocket, and you'll have a fair idea of what your good son looks like as he goes riding through the Australian forests. Often my only companion, apart from the horse, is a white pointer. When it's important to keep moving, and the country is not very interesting, I make as much as 5-6 German miles a day. But if the country is interesting I halt, give my horse a spell, and go wandering about, observing and collecting anything worth while. ⌐*The settlers on the whole are very hospitable. Most of them are young, unmarried men, with 500 to 1500 head of horned cattle, and are trying to increase their herds as fast as they can. The stations are about four [German] miles apart, and consist of no more than a few slab huts, but they give adequate protection from wind and weather in such a mild climate as this.—Some of the cattle are fine beasts though, and I never ate better beef or saw fatter animals during my journeys in Europe.—Other parts of the country are covered with flocks of sheep. There are thousands and thousands of sheep on the hills of New England and the plains of the Darling Downs. These districts are particularly well suited for sheep, which can there be run in flocks of 2 to 3 thousand beasts—a great advantage, considering the high wages*

paid to shepherds. These fellows get from 140 to 190 rix dollars a year. Don't forget, though, that they find necessities correspondingly dear. In the interior of the colony meat is the staple diet. On sheep and cattle stations they eat no other meat but beef or mutton, with a kind of unleavened bread called damper. It's made from well kneaded dough baked in the hot ashes. They drink unheard of quantities of tea, and a billy can of tea appears at all meals—breakfast, dinner and supper. It seems to take the place of soup. Although pumpkins do well here and are very floury, they're seldom eaten. Even though people can have just as much meat as they can possibly eat, and they eat nearly 2 lbs each a day (they're allowed 12 lbs a week), they don't, on the other hand, get the vegetables that you all enjoy so much at home. Potatoes and cabbages are luxuries because the heat of Summer so often kills the plants.⌐ I have met several Germans, and am glad to say that Germans are generally liked here because they work hard and are modest in their demands. They quickly manage to become independent. Some of those I met were thought of highly as shepherds. Another was a wine grower. Here at Moreton Bay I'm living with the German Mission. It consists of good, zealous people who have been sorely tried in their most disappointing efforts to convert the blacks. They are all married and ⌐now⌐ number 7 families with 22 well-behaved children. I feel very contented here among them, because their simple goodness is something one encounters far too seldom in this colony.—Although the blacks show little inclination to accept Christianity or to bother about religious matters, many of them are highly intelligent and very astute. This is most evident with children who have been removed from their parents when very young and have then been taught apart [from the blacks]. The little clans, however, are dying out, and in many instances clans that were once 100 strong can now muster only 2 or 3 individuals. In some places they cause the settlers great loss by spearing cattle or by driving sheep away and killing shepherds.—About 8 years ago there was a ship wrecked 100 miles or so farther North, which had some bullocks on board. One of them swam ashore, where he waxed fat on virgin pastures. The blacks, who had never seen so huge a creature before, were wonder struck. They came from near and far to gaze at the unknown monster, and many a plan was forged for his undoing. But whenever they approached him

Moreton Bay

the bullock charged, and scattered them. At length the whites appeared on the scene. They understood the ponderous wonder, and it fell to their first shot. It was extraordinarily fat, and one gentlemen who tried a steak from the carcase was never tired of telling how deliciously tender it was.

I can't remember if I told you that my own life was once in danger when I was attacked by a wild bullock. I was alone in the forest, and I dodged behind a tree just in time to save myself. He went for me several times so I dealt him a blow with my hammer, which seemed to take the fight out of him, and I darted back from tree to tree.—It meant the loss of the hammer, which I had carried through France, Italy and Switzerland —but better that than my life, of course.—

My health is as good as, if not better than ever. Yet people tell me that I'm looking older—and the grey hairs are coming faster. I might not have stood the hardships of travel so well had I not always held to a clean respect for my body and striven to honour the training which I owe to my good old mother. If you were fond of the boy and the youth who sometimes gave you a lot of trouble—though he never really meant to— you'd rejoice even more in the man, who has come slowly through to a calmer state of mind yet wants to retain the glad heart of childhood.— Let us hope that we shall meet again under the same old sky. Until then and thereafter may God grant you health and happiness. How I long for news of you all—and perhaps there's a letter waiting for me in Sydney. Regular postal services, however, do not extend [far] into the interior from Sydney, and, as I am constantly moving on, letters can't be forwarded to me. Still, I might hear from you this very day, as a steamer has just arrived at Moreton Bay from Sydney and I expect letters from my friend by it. I'm fortunate in having found a really staunch friend in this country, who supports me to the limits of his ability. It's as if he were William's deputy, providentially sent out here ahead of me when William decided not to come. And not even from my brotherly William have I had any news for a long time. ⌈I'm sending him this letter, just the same, so that he can forward it to you.—⌉ Well, Mother, good-bye for now, ⌈and give my love to the rest of the family.⌉

Your affectionate son,
Ludwig.

84 [TO MRS MARLOW, Sydney, N.S.W.

German Mission Station, Moreton Bay, 13 July 1843]

My dear friend

I should fear you had forgotten long time ago your fellow passenger, whose weak point is to roam through the wilderness, to pick plants and to knock at the rocks, if I was not sure that Mr Lynd had given you from time to time some account of my strange doings. It was through him, that I heard you had changed your residence to the Surrey hills and that you were living as quietly and as happily as Sydney could permit to one who had enjoyed the comforts and the social life of home.

Whenever I wrote to Mr Lynd I begged to be remembered to you, as I feared you might forget the wanderer through the bush who homeless and solitary has [sic] he is, felt so happy in thinking, you took a motherly interest in his welfare. Mr Lynd gave me also in his letters intelligence of you, but I was often for months and months without any news, as I was far beyond the regular postal communications and as I moved continually from place to place, even not knowing beforehand, what acci‑ dent might change my preconceived plans. I never dreamt to extend my travels over the Hunter and when I was in Sydney and Newcastle even the next mountain range appeared to me as an insurmountable barrier. Alas, it is with my science and with travelling, as it is with the desire of making money. The more you have, the more you wish to have. It is not like the social pleasures of life, which are only pleasant in moderation: as one toils up to a range of blue mountains, others confine again the horizon, one toils on again, to obtain an illimited view—and comes never to an end. You will see how much I have to tell, when I return to Sydney. No soldiers napsac was ever fuller of dry biscuits, as mine is with strange visions of the bush. Far beyond the Liverpool Range I met a young man, who told me he was a particular friend of Ben Marlow; they were cronies together. His name is Murray Prior[a] of a good

Irish family, educated in Brussels and living for the greater part of his life in England. I forget now, wen he saw you—if I am not mistaken in the South of England after your return from the Bermudas, as he told me, that Captain Marlow was staying at the time at those islands.—You know yourself how pleasant it is to talk about absent friends and I did not fail to tell him, how pretty a girl the little Miss Maryanne had become, and how the whole family had changed, as he told me of his friend Ben Marlow and Miss Jane.—I thought very little at the time that you were visited by so heavy calamities and that Capt Marlow and Miss Marlow had to suffer so severely. I heard of it at my arrival in Moreton bay through Mr Lynd and believe me I was deeply afflicted. I wished to write to you; but I was so unsettled and had so frequently to change my quarters, that I could not accomplish my desire before the departure of the steamer and in consequence I had to wait for the next arrival. Mr Lynd wrote me in the letter of yesterday, that Captain Marlow is well again and I saw the signature of his name on the official papers sent to the Barrack Serjeant of Moretonbay, so firm that I think it even an ocular demonstration of the good health of the distant friend. Miss Marlow, I am glad to hear is recovering and I hope and pray, that she may soon be again in the full enjoyment of her health. May I find you and your family as well as I left you, may I find you content and happy. My regards to Capt. Marlow and Miss Marlow to John Kathe Sophy and little Gussy.

      Believe me ever to be Yours truly
        Lud. Leichhardt.

85 HELENUS SCOTT Esq: / Glendon

      Bunya Bunya Mtains the 24th July 1843
My dear Sir

 My wanderings have been extended far beyond my original intentions, as allways some opportunity was offered for visiting new parts of the colony, and I am almost afraid, I shall not be in Glendon in time, to put your wine into new casks, which should be done as soon as the temperature increases. If you however paid attention to the syphons and to the sweetness of the water, I would propose to let the syphon remain, till the blossoming time and the fermentation movement is over and to draw the wine off immediately afterwards, should I not be able to return during the interval. I have seen a great part of the colony and I am now almost at its northern extremity, just ready to start for Wide Bay, where Mr Eels [sic] has been humbugged into a new sheep station. Of all the country from Newcastle to the Condamine, none could be equalled to the Darling downs, wherever they are sufficiently watered. The country between the Coast Range and the sea which forms a belt about 100 miles wide, has the advantage of plenty of water, though the rather dense forest and the abundance of scrubs renders them less fit for sheep farming. There are however very generally 1500 sheep in a flock which one shepherd is well able to attend to. The vegetation is rich, the turf is denser, than in any part of the colony I traversed ⟨you know I did not get to see New England⟩; the grasses are various though the Kangooroograss ⟨which might perhaps be different in species from that which covers the mountains of the Glendon Estate⟩ predominates. Mr Kent[a] told me, that 12 years observations have not shown the slightest decrease in the density of the turf nor in the variety of the grasses. The run of Mr Thomas Bell[b] at Laidleys plains is remarkably rich and favourable: I never saw fatter sheep nor tasted finer mutton; and the cattle which was brought from there to be salted at the salting establishment at Moreton bay gave a pattern of beef. I stayed for some time at Brisbane and roamed through the brushes which cover the banks of that river and of its numerous creeks. Three days ago I set out for the Bunya Bunya district[c] and past a country not occupied by any settler within 50 miles from the settlement, which would give rich pasturage to thousands and thousands of cattle. You did perhaps know, how Mr Th. Bell did obtain the run of Laidleys plains, to which the country between the Pine River and Mr Archers station is by no means inferior.

Moreton Bay

Everyone was afraid of settling there, as it was within the 50 miles. Mr Bell was not so easily frightened, he took the run and will have the benefit of it for at least 10 years to come. It occurred to me, my dear Sir, that you would derive a great advantage from making use of a part of this country, if you had a man, on whose integrity and activity and knowledge you could rely and to whom you could give in charge a certain flock of cattle. Moretonbay is only 40 miles distant; every head of cattle would become at least 100 lb heavier than at Glendon —it would loose nothing by being brought down to the market, either to be sold to the butcher, or to be salted at the salting establishment of Mr Campbell[d] which I visited and which is well conducted. When I arrived at the station of Mr David Archer,[e] a gentleman to whom I was introduced at Moreton bay and with whom I had opportunity of being intimately acquainted, my mind was again occupied with similar considerations and I asked Mr Archer, whether he would not be inclined of taking cattle on thirds on similar conditions as Mr Otley ⟨has your cattle at the Gwydir⟩.[f] As he was willing to do so, I made him write down his terms and I send them to you, that you may think of them.—You have here a rich country, amply supplied with water which *never* fails and with the richest pasturage which perhaps New Holland is able to afford. You have a market 40–50 miles distant, a town, which at present consists of some scattered houses, but which must rapidly increase with such an extensive and almost unequalled backcountry as the Downs to the inland. The Bar, which at present prevents ships from coming up the River, will soon disappear, as the population becomes richer and denser. For there was many a River in the United States of America, which had three times as difficult bars as the Brisbane at their entrance. You have the possibility of salting your beef here, if the butchers would not take it. Every beast killed here will weigh 100–150 lb heavier—and perhaps even more so than at Glendon or at Cassilis. At last you have a consciencious man, who had charge of cattle beforehand, who has much colonial experience,

Leichhardt's Letters

A Scotchman too, actif and mindful. Be not frightened by the accounts given of the savage disposition of the blacks. It is true, they killed several shepherds even within the coast Range, but they never rushed cattle. Mr Mackenzie[g] lost sheep and shepherds, but his cattle was never troubled. I must however acknowledge, that the blacks would be the only drawback against your entering into my proposal. There are however several points, which lessen the danger from the blacks considerably. They have become much more tractable.—Should you think my proposal worth your consideration, I should suggest, that you execute it as speedily as you can, particularly, as some other gentlemen appear to be on the scent. The present time is very favourable for driving cattle. They find plenty of food from Glendon or Cassilis up to the Downs and to this part of the country after the fine rains, with which the colony seems to have been universally favoured; the season is equally favourable, as the temperature is not to high. I should say, send as many cattle as you can and you will derive great benefit from it. Your wild cattle of Glendon estate would soon become tame, as the tameness increases so wonderfully with the increased fatness and under the careful eye of consciencious gentlemen superintendents. As the communication between Glendon and Moretonbay is so long, I should like to lay everything, necessary for you to form a judgement, at once before you.—The terms of Mr Archer are inclosed in the letter. You have to send the cattle here; Mr Archer would meet them at the downs and pilot them down the coast Range to the Bunya district. The road from the Gwydir to the Severn is without difficulty, as there are stockyards at every station; the road from the Severn to the Condamine (about 90 miles) is level but scrubby, without stations or stockyards. The passage down the coast range would be perhaps the most difficult.

Now I conclude my letter with the earnest wish, that you may think myself as mindful of your interest, as it is in the power of a poor mortal to be. If I knew of any danger, having seen the locality and living at present in it, I should certainly

Moreton Bay

mention it. But nature is alltogether favourable—and what is much rarer: man is favourable too.

May I mention to you that I gave the following orders on you during my journey: To Capt. Wilkinson 4 sh. To Murphy (head of Gwydir ⟨near Mr Rusdens stat.⟩)ʰ 2 sh. To Mr Otley 16 sh. (summa 1£ 2 shll.

I join my congratulations to you and to Mrs Scott on her happy confinement and I hope and pray that she herself and her little baby may be in the full enjoyment of good health. Many kind regards to Mrs Scott to Dr and Mrs Glennie to Sarannah and Helenus.—And believe me, my dear Sir, ever to be

Yours truly Ludwig Leichhardt.

[P.S.] Mr Archer refers you for his character and his ability of managing cattle to Mr Francis Walker ⟨of M[aitland]⟩.

He has sheep stations here; but 2 of his brothers are living with him. He has therefore full time to superintend cattle.— On my long journey I would have recommended only two men to you without hesitation, though I saw a good number of them and these are Mr Hentig and Mr Archer.—I do not mention Mr Otley, because you know him even better than I do.

86 [TO LT. ROBERT LYND, Barrack Master, Sydney, N.S.W.]

From Head Station of Mr Eales, Wide Bay River.

7th August, 1843

[My dear friend,]

... ...

In travelling from Archer's and Mackenzie's we had to cross a very high range of sienite, which had been broken through by basaltic rock. This range divides the district of the Brisbane river from the Wide Bay river district.

... ...

These Glasshouses are very remarkable from their abrupt and fantastic shape, and they are visible for a great distance.

... ...

The Bunya-Bunya tree is noble and gigantic, and its umbrella-like head overtowers all the trees of the Brush. It seems to be a brush tree, but covers, probably like Araucaria Cunninghami, more open ranges. … …The Bunya Bunya is by no means a yearly regular crop; it gives rather a feast to the black fellows than food. … …Many tribes come at that time to the district, and fight day after day, while the women collect the cones and seeds of the tree, and prepare dinner. … … … …

87 [ADDRESS NOT PRESERVED.]

Moreton Bay District 27tn Aug. 1843

Meine theuerste Mutter

Immer noch hat dein Wandernder Sohn keine ruhige Stätte gefunden, sondern irrt durch Wald nd Steppe, über Berg nd Thal, schläft nun in seine Wollendecke gehüllt unter freiem Himmel vor einem warmen Feuer, nun in der Hütte des gastfreien Ansiedlers — ja bisweilen unter einer Decke mit dem transportirten Verbrecher, dessen frühere Sünden ich über die gegenwärtige Gastfreundschaft vergesse. Nachdem ich einige Zeit in Moretonbay gelebt und die dortigen Gebüsche durch sucht hatte, welche die Ufer des Flusses und der Bäche bedecken, ritt ich mit zwei Pferden, denn meine Familie hat sich um eine junge Stute vermehrt, zu einer weniger bekannten Gegend, welche gewöhnlich Bunya Bunya gegend genannt wird. Es wächset nämlich hier in den Gebüschen, welche die Rücken der Berge bedecken ein mächtiger Baum, gleich der Tanne in Deutschland, mit großen Tannenäpfeln, zwischen deren Schuppen mehlige süßliche Kerne liegen, die von den Schwarzen sehr geschätzt werden. Dieser Baum giebt nun nach der Aussage der Eingeborenen alle 3 Jahre eine Erndte und zu dieser kommen die Stämme fern nd nah um sich für ungefähr 3 Monate mit den nahrhaften Bunya kernen zu mästen. Hier fechten denn die Männer, während die Weiber ausgehen, die Früchte zu sammeln und mancher Schwarze kehrt nicht lebend zu seiner Heimath zurück. Seine Anver wandten verzehren indessen den gefallenen Bruder, reinigen

Moreton Bay

seine Gebeine Schenkel u Armknocken nd Schädel, welche die Weiber in einem kleinen Netze mit sich tragen. Einige Stämme haben den sonderbaren Glauben, daß die Stärke des Gefallenen auf den übergeht, der ihn verzehrt und ihn doppelt so stark macht. Es herrscht über diesen Genuß des Menschenfleisches nicht der mindeste Zweifel, Männer, die lange Jahre mit den wilden gelebt, so wie die Missionäre, haben es beobachtet. Es ist keine Hoffnung, die gegenwärtige Generation zu Kristen zu machen nd wahrschl. wird diese Generation die letzte sein, da die Schwarzen schnell aussterben, wo sie mit Weißen in Berührung kommen. Diese schwarzen Kinder des Busches sind indessen in vielen Beziehungen recht interessante Geschöpfe. Es fehlt ihnen durchaus nicht an Scharfsinn. Wo die Natur ihnen den geringsten Vortheil bot, haben sie sich seiner bemächtigt nd sind deßhalb in Bezug auf ihren Unterhalt so reich an Entdeckungen, als wir, die wir lernten, aus Waizen n Roggen Brod zu backen, Gemüse zu zubereiten, Fische zu fangen, Wild zu schießen, den Bienen ihren Honig zu rauben und gewiße Pflanzen als Heilmittel an zu wenden.

— Sie sind dabei abergläubig, glauben an Gespenster, haben dunkle Vorstellungen gewißer Gottheiten und einige dieser Vorstellungen sind recht sonderbar. Ja es scheint selbst daß ihre nächtlichen Tänze, während welcher sie sich mit weißem Thone mit Streifen bemahlen, oft zur Verehrung ihrer Gottheiten oder zur Besänftigung ihres Zornes angestellet werden. — Ihre Weiber behandeln sie wie Sclaven un Lastthiere nd die armen Geschöpfe haben für die trägen Männer Wurzeln zu suchen, die Kinder und Netze zu tragen, in welchen sie ihre Habseligkeiten fortschaffen. Jeder Stamm hat einen gewißen Bezirk. In diesem wandern sie beständig herum, um die hinlängliche Nahrung zu finden. Oft ist der ganze Stamm beisammen, oft sind sie zu zwei, 3 nd 4 Paaren zerstreut. Ihre Hütten od. Humpies, wie sie sie hier nennen, machen sie aus Stöcken nd Baumrinde, indem sehr viele Bäume leicht ihre Rinde ab streifen lassen. Am Meere sind die Männer gewöhnlich größer, hier sind sie kleiner, doch untersetzt nd wohlgenährt. Fast jeder Stamm hat seine eigene Sprache, oft haben selbst Familien eine Menge abweichender Worte; doch selbst fremde Stämme verständigen sich leicht. — Sie sind gewöhnlich verrätherisch, nd man muß sich vor ihnen, selbst wo sie sich freundlich bezeugen, in Acht nehmen.[a] In Wide Bay hatten sie just vor meiner Ankunft 5 Schäfer ermordet nd hier auf Hr. Archers Station suchten sie einen Schäfer mit ihren Speere zu durchbohren; doch glücklicher weise war es nur eine oberflächliche Schulter wunde. — Es ist näturlich, daß die Weißen sich zu rächen suchen und daß mancher Schwarze sein Leben verliert. Der Haupt fehler scheint in der unzulänglichen Polizei zu liegen. Es ist sehr schwierig, Euch eine richtige Idee dieses Mißverhältnisses zu geben. Die Regierung scheint mir in so vielen Beziehungen irrig zu handeln, daß es schwer wird zu sehen, wo eigentl. der Hauptirrthum liegt. Größtentheils sind die Weißen, welche auf den verschiedenen Stationen beschäftigt werden, transportirte Verbrecher, fast ohne Ausnahme oder mit wenigen Ausnahmen unverheirathet, ohne die geringsten moralischen Grundsätze nd Gefühle. Diese Männer haben nun häufig Verkehr mit den schwarzen Weibern, welche sich natürlicher Weise leicht zu den Hütten halten, da sie gute reichliche Nahrung finden, während sie im Busche gar oft hungern müssen. Die schwarzen Männer, obwohl durchaus nicht so genau mit ihren Weibern, wenn sie nur Tabak nd zu Essen erhalten, wollen es doch nicht leiden, ihre Weiber ganz zu verlieren nd ⌜so⌝ fangen ⌜sie⌝ an mürrisch zu werden, zu drohen und endlich sich zu rächen. Dieß thun sie indem sie entweder das Rindvieh un die Schaafe tödten, oder selbst die weißen Männer angreifen, wo sie nur ihrer Herr werden können. Ihr mögt fragen, warum die Besitzer der Schaaf n Viehheerden so schlechte un moralische Personen in Dienst nehmen? Die Antwort ist: die freien Einwandere haben eines Theiles nicht Erfahrung genug, um sich nützlich zu machen, andern Theils fehlt ihnen der Muth, entweder Einzeln oder mit Weib un Kind oft Hunderte von Meilen ins Inland zu gehen, um dort Schäfer oder Kuhhirten

Moreton Bay

zu werden. Die transportirten Verbrecher, welche man hier 'old hands' Alte Hände nennt, waren dazu gezwungen, lebten lange Zeit an demselben Orte nd machten so viel Erfahrung. Es ist wunderbar, was ein solcher Mann nicht alles zu thun versteht. Er ist Schäfer, Kuhwächter, Hütten, bauer, Säger, Holz spalter, Tischler, Schmidt; er versteht die Krankheiten der Pferde, der Schaafe, versteht Brod zu backen, zu kochen, zu nähen — kurz alles was zum Leben im Busche gehört. Dabei ist er ausserordentl. ausdauernd, kann lange Märsche unternehmen und Hunderte von Meilen in einem Zuge reiten. — Das Buschleben ist in dieser Beziehung sehr belehrend; der Mann lernt seine eigenen Kräfte kennen nd ich habe sehr viel Erfahrung in dieser Beziehung gewonnen, da ich gegenwärtig ein ziemlich guter Buschmann bin.[b]

Da meine Mittel zu Reisen allmählig schmelzen, werde ich bald ernstlich darauf denken müssen, mein Brod zu machen; denn ich will Australien nicht verlassen, ehe ich es nicht quer durchreist habe. Ich hatte mehrere Vorschläge; doch habe ich mich noch nicht entschlossen. Ich habe fast den ganzen Nördl. Theil der Kolonie gesehen und hoffe nächstes Jahr den südlichen Theil zu besuchen. Von Moreton bay kehre ich nun bald nach New England, zum Hunter nd nach Sydney zurück wo ich in 3 Monaten an gekommen zu sein hoffe; denn mein Reisen geht so langsam vor sich, ich habe so vieles zu sehen nd zu suchen, daß ich gar mühsam vorwärts komme. In Moretonbay sah ich die deutschen Missionäre, von denen einige von Berlin un Pommern kommen. Ich glaubte fast wieder in meiner Heimath zu sein, als ich ihrem sonntäglichen deutschen Gottesdienst beiwohnte. Es sind 7 Familien, zwei Prediger,[c] und die ganze Gemeinde hat jetzt 22 kleine Kinder, die viel, leicht besser erzogen sind, als irgend ein anderes Kind in der Kolonie. Früher waren sie von der Regierung unterstützt, gegenwärtig leben sie von Gartenbau, bis ihr weiteres Schicksal entschieden sein wird. Sie sind in so fern nützlich gewesen, als sie die Schwarzen mit Menschen freundlichkeit nd Milde behandelten nd so den Ansiedlern befreundeten. Doch in

Bezug auf Lehre un Kristenthum konnten sie leider wenig ausrichten. Katholische Missionäre sind hier vor wenigen Monaten angekommen. Ich weiß nicht ob der äusserliche Gottesdienst der Katholiken mehr geeignet sein wird, die Schwarzen an Einem Orte zu fesseln und ihre Aufmerksamkeit von den allein körperlichen Bedürfnissen auf geistige nd göttliche zu richten. Könnte man die Kinder von dem Stamme entfernen und sie in der Ferne regelmäßig erziehn so liesse sich wohl etwas hoffen. Doch dieß wird von der Englischen Regierung als Eingriff in die Rechte brittischer Unterthanen, als welche sie die Wilden betrachtet, angesehen—nd so schwindet denn auch diese Hoffnung eines künftigen Erfolges. —

⌜Ich hoffe, Ihr habt meine Briefe erhalten: ich habe nur einen Einzigen von Euch nd ich bitte Euch durch William an mich zu schreiben.⌝ Lebet Alle wohl nd behalte ⌜Ihr alle, besonders du meine gute Mutter, mich lieb nd⌝ in Andenken.

Dein dich herzlich liebender Sohn
Ludwig ⌜Leichhardt.⌝

Moreton Bay district, 27th Aug., 1843

Dearest Mother,

Far from having found a quiet retreat your restless son is still wandering through the forests and plains, over the mountains and across the valleys, and sleeping sometimes under the open sky, wrapped in his blanket with a warm fire burning near him, sometimes as a guest in a settler's hut—and sometimes even sharing a bunk with a transported convict, whose present kindness makes me forget his criminal past. After spending some time at Moreton Bay and making a close study of the dense vegetation along the river and the creeks there, I rode on, with two horses—for I've added a young mare to my family—to a less well-known part of the country, commonly called the Bunya Bunya district. It's so called after a huge tree that grows in the bush along the ridges of the mountains. It resembles the pines of Germany, and bears big cones, between the scales of which lie sweetish, floury kernels that are highly relished by the blacks. These people say that the trees bear a heavy crop

Moreton Bay

once in 3 years, which draws the clans from near and far. They assemble for about 3 months, to feast on the nourishing bunya kernels. During that time the men settle their differences whilst the women go out to gather the pine cones, and there's many a black who never returns to his own part of the country alive. His kinsmen proceed to devour the fallen brother, but they clean the bones of his trunk, limbs and skull, and the women carry them about in a small net. Some clans have a strange belief that the dead man's strength passes into whoever eats him and makes the latter twice as strong [as before]. Nobody has the slightest doubt that the blacks enjoy eating human flesh. [White] men like the missionaries, who have lived amongst them for years, have been witness to it. There is no hope of converting this generation to Christianity, and this generation will very likely be the last, for the blacks are rapidly dying out where they are in contact with the whites. These black children of the bush are nevertheless highly interesting creatures in many ways. They're not in the least lacking in perception. They have held their own in natural conditions that offered hardly anything in their favour, and in doing so have discovered as many things contributing to their support as did we when we learned how to make bread from wheat and rye, how to cook, how to prepare vegetables, to catch fish and to shoot game, to deprive bees of their honey, and to make use of certain plants as remedies.—At the same time, however, they're superstitious, believing in ghosts and having obscure ideas about certain aspects of divine power, some of which are decidedly odd. It seems, in fact, that even their nightly dances, for which they paint themselves in stripes with pipe clay, are often performed as acts of worship or propitiation of divine power.—They treat their women as slaves and beasts of burden, and the poor creatures have to dig roots for the lazy men, and to carry the children and the nets in which they transport their belongings. Every clan has its own circuit, within which it moves about continually, to find enough food to live on. The whole clan is often united, but they're often scattered in groups of two, 3 or 4 couples. Their huts, or humpies as they call them here, are built of sticks and bark, as the latter can easily be stripped off many of the trees. Most of the men at the seaside are taller than those here, but they're thick-set and sleek enough here. Nearly every clan has its own dialect, and even many families use a number of words peculiar to themselves; but strange clans easily come to understand each other.—As a rule they're treacherous, and you have to be cautious even if they express friendship. Just before I reached Wide Bay they had murdered 5 shepherds there. They tried to spear a shepherd here on Mr Archer's station, but fortunately inflicted no more than a superficial wound on his shoulder. Naturally enough, the whites try to avenge themselves, so that many a black man loses his life. The fault seems to be mainly in the inadequacy of the police force. It's very difficult to give you people over there a proper idea of the disproportion [between police and population]. In my opinion the Government errs in so many of its dealings that you can hardly say with certainty what the fundamental mistake is. The majority of the whites who go to work on the various stations are transported convicts, and almost without exception, or with few exceptions, they're unmarried, without a vestige of moral principle or feeling. There's a pattern of easy relations now, between these men and the black women. The latter easily get permission to go to the huts, because they're given plenty of good food there and they have quite often to go hungry in the bush. Although the black men are not, as a rule, too strict with their wives so long as they return with tobacco and provisions, they won't tolerate being deprived of them; so they turn sullen, begin to threaten, and sooner or later avenge themselves. They do so either by killing the cattle and sheep, or by attacking the whites when they can get the better of them. You may wonder why the holders of sheep and cattle stations employ such wicked and immoral people, but the explanation is this: the free immigrants are, on the one hand, too inexperienced to be of much use; and on the other, not courageous enough either to go alone or with wives and children, perhaps hundreds of miles inland, to become shepherds and stockmen. The transported convicts, whom they call 'old hands' here, were compelled to do so. They've lived where they are for a long time and have gained considerable experience. It's wonderful to see how many things a man like that can turn his hand to. He's a shepherd, stockman, builder, sawyer, timber cutter, carpenter and blacksmith; he can treat sick horses and sheep; he can bake bread, cook and sew—in short, he can do everything that's called for if you live in the bush. Besides all this, he has remarkable endurance, can walk for long distances or ride hundreds of miles in one stage.—Life in the bush is very

Moreton Bay

instructive. A man discovers his own powers, and I've had a great deal of this kind of experience myself, which means that I'm a fairly good bushman[b] by now.

As the means that have enabled me to travel are quietly melting away, I shall soon have to think seriously about earning my daily bread; for I will never leave Australia until I have travelled right across it. I had several proposals, but I've not yet made up my mind. I've seen practically the whole of the northern part of the colony and am hoping to visit the southern part next year. I shall soon be on my way from Moreton Bay, by New England and the Hunter, back to Sydney, where I hope to arrive in 3 month's time; for, having so much to see and to seek, I have to proceed slowly, and I move laboriously forward on my journeys. At Moreton Bay I saw the German missionaries, some of whom are from Berlin and Pomerania. Attending their Sunday services, which they conduct in German, I could almost imagine myself back at home. There are 7 families, with 2 pastors,[c] and the community can now muster 22 small children—and they're about the best-behaved of any children in the colony. The missionaries used to be maintained by the Government, but they're now living by market gardening whilst the question of their future is being settled. Their work was not quite in vain, for they dealt kindly and humanely with the blacks, which made things easier for the settlers. It's a pity, however, that they could accomplish so little in regard to education and religion. Catholic missionaries arrived here a few months ago. I don't know whether the pageantry of the Catholic service would be more effective in keeping the blacks in the locality, and in diverting their attention away from mere bodily needs towards mental and devotional interests. If the children could be taken away from the clans and brought up properly quite apart, there might be hope for them. But the British Government regards this as interference with the rights of British subjects, which it considers the blacks to be—so even this hope for a better future fades away.

⌈I hope that you've been getting my letters. I've had just one solitary letter from the family, so do write to me through William.⌉ Good bye to you all, and keep me ever warmly in mind⌈—especially you, my dear old mother⌉.

Your affectionate son,
Ludwig ⌈Leichhardt.⌉

88 [TO LT. ROBERT LYND, Barrack Master, Sydney, N.S.W.]

Archer's Station, Bunya Bunya, 4th September 1843

[My dear friend,]

............Last Saturday, I returned from a three days' excursion to the Glasshouses.[a] These mountains are very remarkable. Out of low ranges they rise like needles, like castles, like those isolated rocks in the Ocean, to which sailors have given similar names. The highest of them, Biroa, (or Birwah,) is about 1000 feet high, and is composed of a rock entirely different from the surrounding mountains. I have seen similar mountain features in the neighbourhood of Clermont-Ferrand in Auvergne, and geologists have called the rock domite, because of its affecting generally the form of a dome. This Domite belongs to the Trachytic group. It is rather an earthy paste with some scattered crystals of feldspar. The mountain is extremely steep and its sides almost naked rock; but wherever a hollow or a depression has allowed the accumulation of some soil and of moisture, a rich vegetation appears—single [sic] but full high bushes of a broad leaved Boronia, a dendrolobium with red blossoms, a shrub belonging to the Tiliaceae, Zieria, Calytrix, and several old friends. There is no doubt in my mind that the sea heaved once round these mountains: they are surrounded by sandstone ridges of a coarse grain, and the soil is composed of pure sand, slightly mixed with vegetable mould. The grass-tree grows in thousands, casuarina, the apple tree, chorizema, three species of Banksia, whilst only one Banksia is found inland.

............The clouds are gathering again, and it seems as if no change of the moon could pass without some days' rain. During rainy weather it is warm and close, and vegetation advances rapidly; but once a strong dry westerly wind sets in, the nights get cold, the days very hot, the soil drys rapidly and the young plants suffer.

............I cannot omit mentioning, that I found the Moreton Bay passion-flower in full blossom, near a water-hole, in a rather swampy place, with tea trees, (melaleuca) and coarse grasses.
....

89 [TO LT. R. LYND, Barrack Master, Sydney, N.S.W.

Archer's Station, Bunya Bunya,
c. end of September, 1843]

My dear friend,

............The black-fellow, in his natural state, and not yet contaminated or irritated by the white man, is hospitable and not at all devoid of kind feelings. We had a striking instance of the honesty of these men. A native dog which they had tamed, came during our absence and took our meat provisions. When we returned, one of the black-fellows came and brought back a piece of bacon and the cloth in which it was. The ham had been devoured by the dog, but the blacks brought even the bones which still remained. For about three figs of tobacco they provided us two days with oysters and crabs. They are a fine race of men,[a] tall and well made, and their bodies, individually, as well as the groups which they formed, would have delighted the eye of an artist. Is it fancy? but I am far more pleased in seeing the naked body of the black-fellow than that of the white man. It is the white colour, or I do not know what, which is less agreeable to the eye. When I was in Paris, I was often in the public baths in the Seine, and how few well made men did I see! There is little fat in the black-fellow, but his muscles are equally developed and their play appears on every part of the body, particularly on the back, when you are walking behind him and he is carrying something on his head. The Bunya Black, who lives on the food which the brushes yield to him, is shorter but sturdy and thickly set. As much as I was able to observe there is nothing in the nature in which they live which they have not discovered. They make fine baskets of the leaf of Xerotes, and ropes and nets of the bark of Hibiscus: they make vessels of the sheath of the leaf of Seaforthia, or hollow out pieces of wood. They are quite as particular about the material of their wommerangs, their spears, nullah-nullahs, and helimans, as a European artist. They make little canoes of the Stringybark tree, which they call Dibil palam. Some of their discoveries are very singular. They prepare for food, for instance, the tubers and the stem of Calladium, which is so hot that the smallest bit chewed will produce a violent inflammation and swelling. How is it that they were not frightened by the first feeling of pain, but went on experimenting? Some particular circumstances must have assisted them in this discovery. Their resources for obtaining food are extremely various. They seem to have tasted everything, from the highest top of the Bunya tree and the Seaforthia and cabbage palm, to the grub which lies in the rotten tree of the brush, or feeds on the lower stem or root of the Xanthorrhaea. By the bye, I tasted this grub, and it tastes very well, particularly chewing the skin, which contains much fat. It has a very nutty taste, which is impaired, however, by that of the rotten wood upon which the animal lives. They are well aware that this grub changes into a beetle, resembling the cockchafer, and that another transforms into a moth. Particularly agreeable to them is the honey with which the little stingless native bee provides them amply. You have no idea of the number of bees' nests which exist in this country. My black-fellow, who accompanies me at present, finds generally three or four of them daily, and would find many more if I gave him full time to look for them. They do not find these nests as the black-fellows in Liverpool Plains; they do not attach a down to the legs of the little animal; but their sharp eye discovers the little animals flying in and out of the opening—even sixty and more feet high. 'Me millmill bull,' (I see a bees' nest) he exclaims, and, so saying, he puts off his shirt, takes the toma-hawk, and up he goes. If in a branch, he cuts [it] off the tree and enjoys the honey on the ground. Is it in the body of the tree, he taps at first with the tomahawk to know the real position, and then he opens the nest. The honey is sweet, but a little pungent. There is, besides the honey, a kind of dry bee-bread, like gingerbread, which is very nourishing. The part in which the grub lives is very acid. The black-fellow destroys every swarm of which he takes the honey. It is impossible for him to save the young brood.......

90 [TO LT. R. LYND, Barrack Master, Sydney, N.S.W.]

Mr Bigge's Station, Mount Brisbane,
19th October 1843
[My dear friend,]

.... I am now moving again, after having been compelled, by rain and sore backs (of the horses), to stay about six weeks at Mr Archer's station, at the foot of a spur of the Bunya-Bunya range. I was not idle, however; for, as I wrote you in my last letter, I made an excursion to the coast,[a] and had a treat of oysters with my friends, the Nynga-Nynga blacks. You could have imagined to enter a primitive village—their bark huts in a circle round a fire and irregularly scattered over a sandy flat, with a swamp, which provided them with fresh water. As I came to them, the one brought me a handful of oysters, the other some crabs—a species of lupaea, the finest crab which Mr Archer or myself ever tasted. During low water the sea leaves a very extensive flat dry, and this gives as many fine oysters to the black-fellows as they wish to collect.

I went from Mr Archer's to Mr Mackenzie's, who has a sheep and cattle station almost due west from Archer's. Here I visited several brushes and found several brush-trees in blossom. The rock is slate, primitive, and one point trachytic (at least belonging to the group). From Mackenzie's I turned to the south, travelled for some miles over sandstone, and entered, at Mr Bigge's again, primitive and igneous formations. Mr Bigge's Station[b] lies under the southern head of the Brisbane, which is here dioritic—that is, the rock is composed of feldspar and hornblende—now the one, now the other predominant. To the east, a rock like that of Biroa and the Glass-houses is found —crystals of feldspar in a feldspathic rather earthy paste.

On Mount Brisbane I found a little shrub, belonging to the family of the Mallows, in blossom. The flower is tubular and red, about an inch long, and very showy. The same brush grew on the sandstone hills of Wide Bay. Calothamnus, a little tree about twenty feet high, with drooping branches, somewhat like the weeping willow, is in full blossom, and adorns the creeks most elegantly with its numerous cylinders of red blossoms. It is extremly rich. Several species of Clerodendron, smaller or bigger trees, enter into blossom. One of the finest sights I had was that of a Glycine, a climbing shrub, which is now in full blossom. The flower is a pale violet, the inflorescence long grapes, which form the most elegant festoons from tree to tree along some creeks. The blacks call this brush Birri or Birrwi, and eat the fruit on the pod when young. I was so struck with the beauty of the sight that I almost forgot, in gazing, to take specimens. I took these at last; but I am sorry to say that many of the blossoms dropt off, and that they lose considerably in colour. It is in general very difficult to dry brush plants—even simple branches. The brilliant green which they generally possess fades away; the leaves shrink, if not under pressure of a press, and get sickly and ugly.

The settlers have treated me very kindly. Mr Archer, Mr Mackenzie, and Mr Bigge are all well educated men. It is remarkable how many of these settlers have been in Germany. This makes their conversation even in so far agreeable, as it frequently recals [sic] to my mind my country and its customs. Few of them take any interest in my pursuits, but they assist me as much as I require. It is very hot, and yet I make excursions generally, up hill and even up mountains! ....

91 [TO LT. ROBERT LYND, Barrack Master, Sydney, N.S.W.]

Mr Bigge's Station, 8th Nov. 1843

[My dear friend,]

.... I have seen some forty miles more of the District,

Moreton Bay

and the more I see, the more I feel convinced that it is eminently fit for small settlers. .... I went down to Lockyer's Creek, which is surrounded by extensive plains. Out of one of these plains, on the station of Mr Wingate, a little conical hill rises, and excites by its unexpected presence, the curiosity of the passer-by. I ascended it, and found that it is composed of a curious variety of basaltic rock, with the fragments of evidently broken pillars like those of the Giant's Causeway.[a] They are much inclined, and their heads lie towards south by west. This hill is called Tarampa, and its vegetation is quite as curious as its geological composition. I gathered a great number of shrubs in blossom and fruit, one of which struck me equally from the fragrancy and from the colour (white and yellow) of its flowers. The scrubs, between Lockyer's Creek and Limestone, gave me also some good plants, and I observed here the rosewood tree, which is nothing else but an acacia, the flowers capitate and axillary. .... The brushes of the south head of Mount Brisbane or Brisbane Range, are rather poor. The Araucaria Cunninghami (Moreton Bay pine) is frequent, as in all the mountain brushes of the district. .... In travelling from the Brisbane Range downwards to Limestone, you soon leave the primitive rock behind you, and pass over Puddingstone, which seems to be made up of pebbles from the Primitive Ranges towards the north. As soon as you cross the river, and go towards Lockyer's Creek, you are on sandstone, which composes ridges of small elevation, between which extensive plains exist. In some of the creeks which I had to cross, I found pebbles of carbonate of lime. This substance occurs frequently in small concretions of the size of the hazel-nut in a loose black soil, which is equally found on Darling Downs, and on Liverpool Plains. Here very large grass-trees grow on it. Thus the different species of Xanthorrhoeaceae indicate either an extremely poor sandy soil, or very rich black mild soil, containing much carbonate of lime, and being very fertile, provided rain and water is frequent. Besides this description of soil, there are many flats between the Primitive Mountain ranges and ridges, which are very stiff and clayey. A bed of clay lies generally one and a half to three feet, below the surface. This clay is probably the produce of the decomposition of the feldspar contained in the dioritic, sienitic, and granitic rocks, which form the principal mountains intervening between the heads of the Brisbane and the Wide Bay River. ... ... The forest ground resembles at present one uninterrupted oat or rye field in harvest time. Antistheria Australis is almost the only predominant grass. On the black soil, several others are found, but they are rare. The Antistheria grows here about three to four feet high; in lower situations much higher; and as you ride or walk through it, you feel sorry that such a fine grass does not produce a fuller and more nutritive grain. In a few weeks it will be burnt, to have fine fresh shoots for the sheep and cattle. This burning off of the grass, with its seeds, has been blamed as injurious to the density of the turf; but though Count Strzelecki himself partook in this opinion, and made the Home Government send out orders against the burning of the grass, it is altogether erroneous. The soil is only able to nourish a certain number of stems or of tufts of this perennial plant. These tufts increase with the richness of the soil, and decrease and become scattered as the soil gets poorer. The fire does not destroy the roots, and as the ashes form a good manure, it rather contributes to enrich the soil and enlarge the tuft. A rich turf would be formed, if other species of grasses were sown in the interstices of the tufts of Antistheria; for the soil, though incapable of producing more of *one* species of grass, would still yield nutriment enough to other species. This I explained to several of the advocates of the not-burning system, which is extremely foolish and injurious to the live stock. Even the black-fellow seems aware of its usefulness. He burns early in the year, whenever time is favourable, small patches, which afterwards attract the kangaroo by the sweetness of the young grass. ... ...

Moreton Bay

92 [TO LT. R. LYND, Barrack Master, Sydney, N.S.W.]

Archer's Station, 24th Nov. 1843

[My dear friend,]

... ... I went to the Station of Mr Scott, which is on the banks of the river, below the junction of Stanley Creek[a] and Archer's Creek. A high Range rises two miles behind the hut, and the banks of the river rise towards the south-east, into another less elevated Range. The forest is formed of silver-leaved ironbark (Eucalyptus Glaucus?) real Ironbark and Angophora Cardifolia. Singularly enough some few box trees, which, in the interior indicate a good soil, grow here on a poor soil, almost entirely devoid of lower vegetation. Between Conglomerate and Pudding Ranges (which rock seems to fill the whole basin between Mount Brisbane, Mount Esk and D'Aguilar's Range) small flats intervene here and there with a rich black soil, and with concretions of Limestone. These flats have a richer grass, a greater variety of herbs, and are generally embellished by scattered Xanthorrhoaeas eighteen to twenty feet high, and one foot in diameter. It is very remarkable that this species, which I first met on the Darling Downs, grows always on a rich, mild soil, whereas the smaller species, which I think identical with that near Sydney, indicates invariably the poorest sandy soil. The water-holes and the banks of the river are often densely covered with reeds twelve feet high and upwards. Near the river, and particularly where brushes commence, these reeds are accompanied by nettles often as high as themselves, and by different Cyperaceous Plants which form an almost impenetrable jungle. There was a little brush opposite Scott's, with outskirts of that description. After much pains I got into this brush, and here all lower vegetation ceases at once. Every thing strives to get to the light. The climbers ascend to the tops of the trees, and display there their rich foliage and blossoms. The trees themselves rarely form branches under forty to fifty feet high. At that elevation, the crown of one tree touches that of the other—all interwoven with vines, and spread over with sheets of white, purple, or red blossoms. Over all these, the Moreton Bay Pine and the Bunya-Bunya raise their towering heads, both differing in shape, and the former perhaps the more picturesque of the two.

.... After losing a day again in search of my horse, I went to Mount Esk, about two miles from Mr Scott's. The rock is a feldspathic porphyry, resembling a little that of Mount Biroa. In a gulley which comes down from this mountain, I found an entirely new Hibiscus, with leaves of the Platanus, but thickly covered with down. From Mr Scott's, I went to Mr M'Connell's, a gentleman who came out with considerable property, which he laid out in sheep and cattle. He is considered to have the finest *run* in the district. He has built a very comfortable house, which he keeps considerably free from fleas—the curse of the bush—as the people do not take the trouble to air and dust their blankets regularly. He took more than common interest in my pursuits, and I spent a very agreeable time during my stay with him. From his place I visited several points. In his brush I found several shrubs in blossom. On his creek, Castanospermum Australe, the Moreton Bay Chestnut entered into blossom about the 13th November, and a species of Melaleuca, with small linear lanceolate leaves, was covered with a sheet of snowy flowers. Thus the creeks of this interesting district change their dress or their ornaments almost every fortnight. From the 3rd October to the 20th, creek, river, and water-holes were adorned with the scarlet blossoms of Calothamnus, which hangs with drooping branches like the weeping willow over the waters, looking on its own beauty, like Narcissus, and dropping blossom after blossom into the murmuring stream, as man drops his blighted hopes into the stream of life. Melaleuca opens its flowers about the 25th October, and is in its richest colours about the 10th of November. Castanospermum forms its flowers on the leafless part of the branch; the large dark-green pinnate leaves being collected at the end of the branch. The flowers are almost like those of Kennedia Rubicunda, in short grapes. The impression is

Moreton Bay

exceedingly rich. Another myrtle is not far from blossoming; it is called Kulu by the black-fellows; is a considerable tree of dark-green foliage, and is found at almost every creek. I think I have mentioned in a former letter the beauty of the Birwi or Bread-fruit—a climber, with grapes of violet blossoms, which I found at the creek of Nurrum-Nurrum. Nor shall I forget to mention Calladium, which enters at present in blossom. The plant rises on a short cabbage-like stem, and has large rich green leaves, with long fleshy stalks. The flowers are collected in a cylinder—the females below and the male ones above—and this cylinder is surrounded by a large green Spathe. It smells very strong, like violet, or rather like scented soap, for it is almost too strong to be pleasant. At a certain time of the year it forms tubers at its roots, which are eaten by the black-fellows after three days' soaking and pounding. As I am convinced that it is a plant very similar to the Taro (?) of the islands of the Pacific Ocean, I tried whether the poisonous properties could not easily be destroyed by boiling; but though the green leaf had lost considerably of its strength, it still made the tongue raw after half-an-hour's boiling.

.... I ascended a sugar-loaf, east of M'Connell's, which is connected with the Brisbane Range. At the foot conglo- merate; at the top a dioritic rock (feldspar and hornblende); in a saddle a beautiful augitic porphyry, as I remember having seen two fine vases made of in the Vatican in Rome. About the 11th November, I continued my journey to Mr Balfour's, whose station is in the highest on Stanley Creek.[a] There is no brush nearer than twelve miles, and consequently, after having examined the geological composition of the neighbouring mountains,[b] I bent my way to the east, crossed the Brisbane Range, and returned to Mr Mackenzie's.

.... It is now fifteen months since I left Sydney, and last October I finished my 30th year. The rapidity of time reminds me of the shortness of my life, and it is not so much for me to think of dying, as to think of dying without leaving something behind me that may speak for me when my ashes are driving in the wind. Even in assisting people with my medical knowledge, I feel a great satisfaction, for I am an immediately useful member of society. ... ...

93 SIR THOMAS MITCHELL / Sydney

Mr Archers Station Bunya Bunya the 11th Decbr. 43

My dear Sir

I received your letter of the 23th Octbr last night and I hasten to thank you for your kind remembrance. I feel the deepest interest in the success of the proposed expedition and I should be happy, could I contribute to make it useful to science. I shall hasten down to Sydney as soon as I hear that the expedition is decided. A week will be sufficient to make the necessary preparations for myself—I should not like to leave this part of the country before I am certain of the expedition, as I should have to return in case of its not taking place which would be an expence too heavy for my limited means. May I therefor charge once more your kindness, may I ask you the favour of writing a few lines, whenever the question is settled? I receive my letters immediately after the arrival of the steamer and your letter did not find me because Dr Simpson[a] thought I had left Mr Archers Station.

I remain my dear Sir
Your most obedient humble servant.
L. Leichhardt

94 [TO GAETANDO DURANDO, Paris.]

Bunya Bunya district Archers Station le 6 Janv. 1844

Mon cher ami

⌈J'ai reçu votre lettre du 25 Mars 1843 avec les petites lettres de Mons. Rouault de Mons. Witherill et avec les notes de Mons Brogniart et de Mons. Cordier. Pourquoi vous dire que je les avalai avec la plus grande avidité. Elles me trouvèrent

Moreton Bay

dans les montagnes de la Bunya Bunya, 60 miles anglaises de Moreton Bay. La distance de lieu est si grande et des accidents se peuvent faire si facilement pendant un voyage de 3-4 mois, que je commençai à craindre que ma lettre ne fût pas perdue et que toute communication entre nous ne fût pas détruite. Grace à Dieu, je me suis trompé et nous causerons encore bien de fois ensemble. Il faut que je vous donne une histoire abrégée de mes voyages dans ce pays-ci[1]. Je quittai Sydney après avoir étudié la botanique de Botany baie par 6 mois, j'avais Rob. Brown et les 7 volumes de Decandolle. Il y avait plusieurs familles de plantes, que je ne puis étudier, parce que les livres nécessaires me manquaient par exemple les Euphorbiacées. Dans l'étude des autres j'avancai rapidement et je me trouvai bientôt en état de faire quelques excursions publiques, les premières faites dans la colonie et de donner un cours sur la botanique, dans lequel je tachai d'expliquer l'organisation de ces familles, qui excitent particulièrement l'attention du peuple sur leurs promenades, p. example des Myrtacées, des Rutacées, des Proteacées, des Epacridées, des Cycadées. — C'était à la fin d'Auguste 1842 que je quittai la capitale de New South Wales et que je me portai à l'Hunter, pour étudier la géologie et particulièrement la position de la Houille. L'embouchure de l'Hunter n'est pas du tout si riche en plantes, que les environs de Sydney; mais il y a quelques unes d'une beauté remarquable aux montagnes 10 miles de Newcastle et dans les marais tout près de la ville. Doryanthes excelsa, qui atteint quelques fois la hauteur de 12-16-18 pieds se trouve près du lac Macquarry et sur la montagne nommée le Sugarloaf entre Newcastle et Maitland, où elle vit ensemble avec Xanthorrhœa hastilis entre des blocs immenses de Poudingue. Blandfordia à grandes fleurs d'écarlat habite les marais, qui sont pourtant généralement au sec. Plusieurs Melaleucas (the teatree) et Calothamnus avec ses belles fleurs rouges, plusieurs éspèces de Leptospermum, Fabricia, Baekia se trouvent ici en grand abondance. Après deux mois de séjour je quittai Newcastle et j'ascendai la Rivière. Après avoir visité plusieurs localités intéressantes pour la géologie, par example Harpers hill, où se trouvent beaucoup de coquilles fossiles, et Blackcreek où j'examinai l'étendue et la position d'un calcaire ⌜probablement lacustre,⌝ je m'établi pour quelque temps à Glendon, une ferme très grand appartenant aux MM. R. et H Scott, qui tachaient de rendre mes recherches aussi faciles que possible. La flore change ici et même à Harpers Hill; les plantes de la zone maritime nous quittent et celles de l'interieure se montrent. Un grand nombre de petites legumineuses herbacées, la petite Mimosa (M. terminalis) végétent sur un sol généralement argileux et riche après chaque pluie, qui est pourtant si rare, que les propriétaires se trouvent forcés d'abandonner l'agriculture. La roche prédominante est encore le poudingue et un grès d'un grain plus ou moins gros, qui change souvent en poudingue. La houille se trouve dans beaucoup de localités de Newcastle jusqu'à la Liverpool Range et elle se fait voir généralement dans les bancs de la rivière ou dans des ruisseaux, qui s'y embouchent. Ainsi on la trouve dans le Glendonbrook, Westbrook, Fallbrook, Foybrook et à Bengalla et à la montagne brûlante Mont Wingen[a] elle même n'est autre chose qu'une masse de Houille brûlante au dessous du grès. La houille est remplie d'empreints de fougères. C'est particulièrement celle à fronde lanciolée, d'une pouce jusqu'à 2 pieds de long–de $3'''$ [?"] jusque à $2''$–$2\frac{1}{4}''$ de largeur; mais il y a plusieurs autres qui se rapportent plus ou moins aux genres de fougères fossiles déjà connus. Le grès renferme aussi des fougères; des equisetacées, des calamites, des troncs d'arbres, qui résistent mieux à l'atmosphère, que celle[s?] du schiste argilleux, qui tombe facilement en poussière. — Au NEst de Glendon il y existe un système de collines et de montagnes d'une origine different. C'est un porphyre feldspa‑ thique et pyrossenique dont je ne me souvient [sic] pas avoir vu des échantillons dans le Musée. ⌜Les collines formées par cette roche ont généralement une direction de NOuest à Sudest ou de NEst à Sud ouest.⌝ Il paraît, qu'elles se sont élévées par le poudingue, par le grès et par un conglomerate, que la roche igneuse a frequemment rendu extrêmement dur. Vers le Nord,

Moreton Bay

36 miles anglaises de Glendon on trouve encore le poudingue ⌜et le conglomerate, ci⌝ et ⌜là interrompus par du porphyre et alors le grès qui forme le talon⌝ des montagnes ⌜qui separent la Nouvelle Angleterre du bassin de l'Hunter.⌝ Ces montagnes se composent de Basalt, qui renferme frequemment ⌜du peridot et⌝ des Zéolithes, dont je n'ai pas pourtant vu des cristaux parfaits. Je visitai Mount Royal un épéron (spur) de ces montagnes, qui s'élève jusqu'à 3000 pieds et qui est une des plus hautes montagnes de cette partie de la colonie. Mt Royal descend de la Nouvelle Angleterre du Nord au Sud en ligne droite. Son centre et sa masse prominente est basaltique, pendant que ses flancs sont formés du grès. Sa flanc orientale est couverte d'une forêt très épaisse d'une nature toute particulière, que l'on appelle Mountain brush dans la colonie. Il me paraît que ce mt brush, dont je fais à présent un étude particulier, n'est autre chose que la forêt vierge de l'Amérique Méridionale. La belle description que Monsieur Waterton en donne dans ses Wanderings in South America correspond mot par mot au Mt brush de Mount Royal et plus encore au Mt brushes de la Bunya Bunya. Une grande variété d'arbres qui s'élancent à une hauteur considerable avant de former des branches se trouvent sur un espace limité, liés par des vignes egalement variées, qui ascendent à la couronne des arbres pour y étaler leur foliage et leurs fleurs. Les plantes herbacées n'existent qu'à l'exterieur du brush ou dans les espaces libres, où l'air et la lumière peuvent entrer. Les fougères sont particulièrement riches et variées et c'est dans les petites ravines sous la montagne, qui se sont remplies d'une terre végétale mélangée avec le basalte decomposée, que l'Alsophila croit de 15′ de hauteur et de 10″ d'épaisseur. J'ai fait une assez bonne récolte dans ces brushes; mais je n'ai pas encore déterminé mes plantes, parce que je n'avais pas de livres. — De Glendon je continuai mon voyage à Liverpool plains 150 miles anglaises plus à l'interieur. Me voila arrivé sur la cime de la Liverpool Range, le bassin de l'Hunter et du Goulbourn à mes pieds vers l'Orient, des immenses plaines vers l'Ouest. La Liverpool Range forme un

immense arc de Basalt autour le bassin de l'Hunter, qui est rempli de grès, de poudingue et de conglomerate, au dessous des quels s'étend une immense couche de Houille. ⟨A New׳ castle se voient quatre couches de ce minéral separées de schiste et de psammite et de poudingue! La 3me et 4me sont exploi׳ tées.⟩ On voit plusieurs dykes basaltiques sur la côte; ils ont une direction du quartier Est–Sud au quartier Ouest–Nord; il n'est pas difficile de les réunir avec l'arc de la Liverpool Range, et de chercher un centre des actions igneuses anti׳ diluviennes dans la mere pacifique. — Les porphyres feld׳ spathiques et pyrosséniques forment un système subordonné, plus rapproché de la mer. Il y a des montagnes contenant du calcaire au Paterson, et on dit qu'il y a même des Trilobites, que je n'ai pas encore vus, bien que j'ai trouvé beaucoup d'empreints de coquilles et des Orthoceratites dans un grès ou psammite calcaire au pied de Mt Royal. ⌜J'ai mentionné le calcaire lacustre de Black Creek, qui se trouve encore sur la ferme de Glendon; un calcaire concretionné, renfermant des Limnaeus et des Cyclas et des Cerastinum se trouve dans les lits des ruisseaux de Cassilis, qui viennent de la Liverpool Range.⌝— Je n'ai pas visité le bassin du Goulburn, qui est limité vers l'Ouest par des montagnes de granit: j'éspère de les visiter à mon retour de Moretonbay.

Liverpool plains sont encore une localité pas du tout épuisé pour le botaniste. Quand je passai ces plaines singulières, un grand nombre de composées etaient en fleur. Je fis une petite collection, parce que je n'était pas pourvu d'autre papier, que de journaux que les Messieurs, que je visitai, voulurent bien me donner. ⌜Voila votre ami voyageur demandant quelques journaux pour faire une collection de plantes.⌝— Liverpool plains ont été probablement le lit d'une chaine de lacs du sein desquels beaucoup d'îles s'élévaient, généralement formées de grès et couvertes d'une forêt ouverte composée de differentes espèces d'Eucalyptus. The Cypress pine (Calitris) croît ici frequemment et on l'employe pour la construction des bush׳ huts. Je passai le Mokkai, le Namoi le Peele la Manilla pour m'établir pour quelque temps à la source du Gwydir.[b] Toutes ces rivières n'ont que des étangs dans leur lits et vous pouvez les passer sans mouiller vos pieds. Les lits sont pourtant très larges et remplis de cailloux roulés. Ils montrent en effet deux lits: l'un dans lequel se trouve généralement un peu d'eau et qui est bordé par une haie épaisse de Casuarina (Swamp oak) l'autre rempli de sable et de cailloux roulés ⟨très large⟩, ça et la avec un grand gumtree, qui a su résister à la violence des torrens après un orage or après une pluie continuée par quelques semaines. Entre le Peel et le Namoy la forêt se change et au lieu de voyager au dessous des spotted gums, de Box, d'Iron⟋bark à feuille étroite (narrowleaved ironbark) on entre dans une forêt d'Ironbark à feuilles d'argent (silverleaved Ironbark), dont les feuilles sont d'un gris verdâtre. Entre le Namoy et le Gwydir, aux sources de Rocky Creek, d'un des ruisseaux qui s'em⟋bouchent dans le Gwydir, il y a un système de montagne[s] trachytiques,[c] que j'ai visité et examiné tant que les moyens limités et les sauvages m'ont permis. Vers l'Ouest on voit de ces montagnes les plaines sans limites par lesquelles le Big River va joindre le Namoy qui s'appelle plus bas le Bavan et encore plus vers son embouchure le Darling. ⌜Après le Gwydir on passe le Big River (Bandarra dans la Nouv. Angleterre) the Mac⟋Intire (Macdonald) la Severne, le MacIntire brook et Canal creek.⌝ Toutes ces rivières viennent du plateau de la Nov. Angleterre, qui est situé vers l'Est, et vont joindre le Bavan. Toutes se cheminent d'abord par des montagnes de grès, de basalt, de granite, de quarzite et contiennent assez d'eau dans leur course superieure; mais quand elles quittent les montagnes pour les pleines occidentales leurs eaux disparraissent et leur lit vide ne montre que le grand chemin des eaux pendant les longues pluies. Il est extrêmement interessant de voir que les pluies qui tombent sur le plateau de la Nouvelle Angleterre pas 90 miles anglaises de la mer orientale, se portent plus de 1000 miles pour se joindre aux vagues de la mer méridionale. Entre la Severne et le Condamine le pays est plain et couvert d'une forêt quelques fois très èpaisse, que les colonistes ont appellé

Bricklow scrub.[d] Le Bricklow est une espèce d'Acacia à phyllode long raide grisâtre; cette acacia est accompagnée de Casuarinas (forest oak) et de beaucoup de brousailles, d'Iron‑bark ⟨à feuilles étroites⟩ et d'une éspèce de Banksia, la seule, qui quitte la zone maritime. Le sol est très sablonneux avec l'exception de petites plaines près des ruisseaux, où le sol contient plus d'argile et de terre végétale. Sur ces plaines Angophora lanceolata (the Appletree des colonistes) croit bien. Comme le Bricklow caractérise cette partie du pays, le Myall, une autre espèce d'Acacia caracterise les plaines de Liverpool, du Gwydir, du Big River. Le Myall (Acacia pendula) a des phyllodes et des branches souples, pendantes comme Salix Babylonica. Son bois est très dur et d'un parfume exquis de violette. Les noirs en font leurs bommerangs, qui paraissent voyager de main en main sur tout le continent. — Le Con‑damine est la première rivière qui n'appartient pas au système du Bavan (Darling) mais qui paraît se tourner vers le Nord et dont les eaux vont peut être au Golf de Carpentaria[e] dans une curve semblable à celle du Bavan. — Du Condamine les Darling downs commencent.[f] C'est un pays onduleux, sans forêts, d'un sol noir, riche, basaltique, couvert de differentes graminées, dont une éspèce d'Antistheria (the oaten grass) forme la nourriture principale de nombreux troupeaux de brebis, qui se multiplient rapidement dans un pays si favour‑able. Un nouveau gumtree (the Moretonbay ash) est frequent sur les collines. la partie inferieure du tronc de cet arbre est couverte d'une écorce à écailles grosses noires, pendant que la partie superieure est blanche grisâtre et lisse. Dans les plaines on voit ça et la quelques Xanthorrhœas, d'un caractère bien different de X. hastilis. Cette Xanth. croit sur un sol riche, pendant que X. hastilis ne se trouve que sur un sol pauvre sablonneux; elle est 10–15 pieds de hauteur et presque un pied en épaisseur. ⟨Son comma (flowerstalk) est plus court que celui de l'autre éspèce, duquel les noirs font leurs lances.⟩ Dans un des ruisseaux (Hudsons Creek) on voit un lit de Houille, dont la position est bien remarkable dans un terrain basaltique.

Moreton Bay

Dans l'alluvium des vallées et dans les bancs des rivières, particulièrement du Condamine on trouve des ossemens fossiles. Je tacherai d'en obtenir quelques unes. Il n'est pas de doute, que les ossemens fossiles d'éléphans ont été trouvé; mais je voudrais bien, qu'on ne se hâte pas trop tôt de faire la conclusion que cet animal a vécu dans la Nouvelle Hollande. Il me paraît, que l'élévation du système basaltique ⟨la formation de Liverpl. Rang et de sa continuation dans la Nouv. Angleterre et dans la Coast Range de Moreton Bay⟩ a contribué considérablement pour élever le continent hors des eaux et il est très probable, que les ossemens ont été déposées, quand le continent etait encore sous les eaux — ainsi, que les ossemens pourraient avoir été apportés des Indes ou des grandes îles entre les Indes et la Nouv. Hollande. Les Darling Downs sont 1450′ au dessus de la mer; les nuits étant très froides, même pendant Septembre et Octobre. — La descente vers Moreton bay est très rapide, comme celle de la Nouvelle Angleterre vers Pt Macquarry.[g] Aussitôt que l'on entre le bassin du Brisbane, la végétation devient plus vigoureuse, les arbres plus hauts et plus epais. Les flancs des montagnes, les bancs des ruisseaux et de la rivière sont couverts d'un Brush presque impénétrable ⟨Mountain and River *brush*⟩. ⌜Au pied orientale de la Coast Range, qui est basaltique, on trouve le grès, la poudingue et le conglomerate; 20 miles de la mer à Limestone on trouve la houille au dessous du grès et quelques collines sont composées de calcaire probablement déposé ⟨des eaux⟩ de sources, qui s'élévaient par un terrain basaltique. A Moreton bay la roche est un Talkschiste pénétré de Quarz, qui remplit toutes les fissures de la roche. —⌝ Je me portai de Moreton bay 60 miles vers le Nord au Bunya Bunya district, ainsi nommé de l'Araucaria Bunya Bunya, qui croit ici dans les Mt. brushes. Me voici arrivé sur la scène de mon activité présent. Dans un quart d'heure je puis me porter à la forêt vierge, à un des Creek brushes, qui viennent des Bunya Bunya Mountains. Ces montagnes séparent le bassin du Brisbane de celui de Wide Bay River.[h] Leur direction générale est de l'Ouest à l'Est; elles se joignent à la Coast Range; elles envoient beaucoup de branches vers le Sud, entre lesquelles les differens ruisseaux descendent au Brisbane. La carte de Dixon[i] est tout⸝a⸝fait erronée dans la partie Septentrionale de Moreton bay. ⌈Je vous donne une representation grossière pour aider votre imagination. [*Small diagram*, $1\frac{1}{4}''$ by $\frac{3}{4}''$.] ⌉ La roche diffère beaucoup. C'est bientôt un Sienite ⌈à élémens également dévéloppé⌉ d'une couleur blanchâtre; bientôt mica entre dans le mélange et forme un granite ⌈de la même couleur; bientôt les élémens sont très petite et la roche devient obscure, un peu bleuâtre. Très souvent, particulièrement⌉ vers l'est de la chaine et à la Mt Brisbane Range le quarz paraît manquer entièrement ⌈et alors c'est un Diorite (Hornblende et feldspar).⌉ Souvent c'est un porphyre de Hornblende, quelques fois une Hornblende massive. Il y a encore une roche feldspathique qui se trouve souvent dans les ruisseaux de ces montagnes et qui appartient probablement à l'époque des Glasshouses. Ces montagnes sont des pics isolés, fantastiques au Sud de la Bunya Range, 15–20 miles anglaises de la mer. Elles sont composées d'une roche friable à pâte feldspathique avec des cristaux de feldspath plus ou moins nombreux. ⌈La montagne principale s'appelle Birroa (Birwa). Elle est peutêtre 12–1500' de hauteur. Une aiguille vers l'Orient de Birroa s'appelle Gnarra nurrui (Kumaunum [?]); une autre plus au Sud Dunbabola (Paidarr bolé) (avec deux pics $\frac{dunba}{paidarr}$ montagne $\frac{bola}{boledo}$ two.) Une quatriéme Birrwaman, une 5ième Waiamurrum pp.⌉ La roche du Puy de Dome, celle du Sarconi et d'une troisième montagne dont je ne me rapelle pas le nom ressemblent beau⸝ coup aux variétés des roches des Glasshouses; et il est singulier, que leur physiognomie générale est presque la même. ⌈⟨Les deux roches au Mt Dore (la Tuillière?) rappellent encore la forme singulière du Gnara nurui, bien que la roche en diffère plus.⟩ Ces montagnes sont entourées d'un grès à grain gros contenant beaucoup de limonite.⌉ N'est il pas [étrange?] que je n'ai pas trouvé la moindre indication de metaux, ⌈avec l'exception d'un

Moreton Bay

morceau de Massicot sur le ter[rain] basaltique de Cassilis, ou de pierres precieuses? On m'a montré souvent du Mica jaune d'or comme de la poussière d'or mais jamais d'autre métal, comme si la science ne veut pas permettre, que je serve à autre maître qu'à elle seule.

Mais que vous dire du Bunya Bunya brush? Que vous dire de la Bunya (*Bodné* des noirs), de cet arbre majestueux [dont] le tronc parait former un pilier de la voûte du ciel, sur les fruits duquel les noirs viennent 100 miles et plus pour fêter pendant 2-3 mois (Janvier, fevrier, mars). Que vous dire de ce nombre de plantes, d'arbrisseaux et d'arbres rares qui croissent dans cette brush, qui couvre les montagnes pour 50 miles anglaises de longueur et 10 de largeur[? Le] Castanospermum australe ⟨duquel vous me parlez⟩ ne se trouve pas seulement dans les brushes de la rivière et des ruisseaux, mais il est en[core] très frequent dans le brush des Bunya Mts. C'est souvent un arbre de 80' de hauteur. Ses fleurs rouges et jaunes poussent du bois même en petites grappes, pendant que ses feuilles composées d'un vert foncé se rangent vers la pointe des branches. J'ai trouvé un autre arbre de la même famille hors du brush sur la montagne, [dont] le bois est très epongeux, les noirs en font leurs Helimans. L'écorce est couvert de tubercules de liège. Les fleurs sont rouges, grandes; les feuilles sont ternées; les feuillets petiolés triangulairs, les angles arrondis. — Je c[rois que] c'est une Erythrina. Il y a deux autres arbres legumineux dans le brush, dont l'un est orné de riches racèmes de fleurs jaunes; l'autre me paraît appartenir aux Mimosées. Ses feuilles sont bipinnées, les feuillets éllipticolanceolés [ces] près de la pointe plus grandes, que ces près de la base; ses legumes tournés en tire-bouchon. 4 arbres appartiennent aux Proteacées, ⟨Les noirs distinguent les arbres suivants du brush appartenans aux Proteacées par des noms particuliers:⟩ Wairum à feuille raide très longue pinnatolobé; the Silver oak [Gre]villea robusta; Dulabi à feuille lanceolée ⌜2-3″ long 3/4″ large⌝; Dinangerumbin à feuille de Wairum, mais la surface inférieur d'une belle couleur d'argent. Les médullaires sont très visibles d[ans] le bois de ces 4 éspèces. Il y a plusieurs arbres singuliers appartenants aux Malvacées et aux Sterculiacées. L'un ⌜est appellé Kayar ou Kēr⌝ (Bottle tree des colonistes, parce qu'il a la forme d'une bouteille se gonflant 3–4 pieds au dessus du sol.) L'écorce est très dure, mais l'interieur est mou et epongeux et rempli de suc. Je n'ai pas encore vu la fleur, mais le fruit est une capsule très semblable à celle de la Sterculia. Les noirs mangent les semences. Une autre Sterculiacée s'appelle Banni Banni; c'est aussi un arbre très considerable à bois épongieux et à écorce épaisse. Les feuilles sont $\frac{5}{4}'$ de longueur $1'$ de largeur oblong⌜, accuminées, un peu cordées à long peti[oles]⌝. L'écorce contient un suc gélatineux transparent, s'attachant aux doigts. Un petit arbre ou arbrisseau a des fleurs tubu[laires] d'écarlate; il croit sur les montagnes entre les roches. Une autre Malvacée se trouve au bord de la mer. C'est encore un petit arbre à feuilles de figuier, comme le précédent, à fleur large d'Althaea et d'Hibiscus. Le bois [est] dur et le cœur du bois est d'une belle couleur de velour jaune foncé. J'ai trouvé deux autres éspèces d'Hibiscus, Hibiscus heterophyllus (le Courad⟋ jong) se trouve par toute la colonie et son écorce tenan forme de belles cordes nat[urelles]. Cette éspèce à des fleurs blanches rougeâtres avec la partie inferieure des petales et les étamines d'un pourpre foncé. L'autre éspèce a des fleurs jaunes et la troisième a les feuilles de figuier avec des fleurs rosâtres. Araucaria Cunninghami (Gunam: Warrall) the Moretonbay pine se trouve dans tous les brush de la rivière et des ruisseaux. C'est un arbre d'une grande hauteur et sa couronne se voit bien au dessus de tous les autres. Une troisième éspèce croit dans les brush de Wide Bay River et les noirs l'appellent Dandajōm. J'ai entendu qu'il y a encore une 4th vers les côtes de la mer. Le Cypress pine (Calitris) se trouve frequemment sur un sol sablonneux au voisinage de la mer. ⌜Ma lettre devient trop longue; je vous écrirai bientôt une autre.⌝ J'enverrai 2 collec⟋ tions de plantes de la brush ⌜a Dr Nicholson, qui vit à present à Newcastle upon Tyne en Angleterre. Dr Nicholson en avertira le Musée.⌝ J'enverrai aussi une collection des bois du

Moreton Bay

brush et de la forêt de Moreton bay. ⌜Le prix sera un peu plus haut pour couvrir les dépenses de voyage; mais je pense qu'il n'y existe pas une collection semblable en Europe.⌝ J'ai mis les fruits de la saison dans de l'eau salée et je les enverrai avec la collection des bûches. J'ai fait une collection de roches, ⌜mais les échantillons sont ⟨plus petits que Mons Cordier les veut avoir. Ils sont 2″ de largeur 3″ de longueur. En effet les dimensions des grandes échantillons de Mons Cordier demandent tant de temps qu'il m'était impossible de commencer une telle collection. J'ai essayé pourtant, mais après avoir taillé bien long temps pour faire un échantillon je me suis désespéré de réussir. La collection de petits échantillons sera pourtant très utile et ses numeros seront en rapport avec la description geologique que je vais faire du departement de Morteon bay, du Système trachytique du Gwydir et des porphyres feldspathiques de Glendon.⟩ 3″ de long et 2″ de large; ce sont des échantillons des roches de Moreton bay, du Système trachytique du Gwydir et du système de porphyres feldspathiques de Glendon.⌝ Il me paraît que les geologues ne pensent pas qu'il y existe une telle variété de roches dans la Nouvelle Hollande. ⟨Je ne sais comment faire avec les empreints de fougères de Newcastle; la roche se decompose très facilement et je crains, que les échantillons n'arriveront pas en bon ordre en Europe. Je prendrai pourtant toute precaution. Il n'y a pas de coton, mais il y a beaucoup de laine et je les emballerai en laine aussi bien que possible.⟩ [*Last page of holograph lost. From incomplete copy in Durando's handwriting:* ... ... on a proposé de faire une expedition de Sidney à Port Essington aux côtes septentrionales de la Nouvelle Hollande, mais le gouvernement est trop pauvre; j'espère qu'une belle expedition sera faite un jour soit⸝il par le gouvernement, soit⸝il par les habitans de la colonie. ... ... Nous avons vu la Comète[f] dont vous parlez du 3 mars jusqu'au 1 [*for* 11] avril 1843. Tout le pays par le quel j'ai voyagé, avec quelques exceptions, est occupé par des proprietaires de brebis, et de bestiame: les stations sont 20–30 miles anglaises l'une de l'autre. Souvent j'ai voyagé seul sur ma bonne jument de

Valparaiso accompagné d'une chienne (pointer)…… Souvent j'ai campé seul dans la forêt, ou sur la montagne en même temps cuisinier, groom, washerwoman, naturaliste.…… Les hommes sont pourtant bien hospitaliers, et moi j'etais bien heureux de faire la connaissance de quelques personnes très respectées dans la colonie. Souvent j'ai cru que je serais forcé de laisser mes études et de chercher mon pain. Votre lettre m'a donné beaucoup d'espoir et de plaisir.…… Il est bien singulier qu'il y a ici une petite plante très semblable à celle que vous m'avez envoyé des marais de la Toscana (Hypericum quinque‑ nervium Walt. Sarothra Blentinensis Sav.). Je mets dans la lettre la fleur de l'arbre que j'ai mentioné, à feuille terné, à bois très mou et épongeux dont les noirs font des vaisseaux et des Helimans (leur shield).…… Ces noirs même sont des êtres très interessants sur les quels j'ai fait beaucoup d'observations, vivant presque toujours entre eux. Ce sont à present des tribus puis‑ santes, qui seront bientôt anéanties par la civilisation. La philanthropie s'en fâche, mais on voit que ces pauvres noirs ne peuvent être sauvés de la destruction par les differents moyens qu'on a employés. On ne peut accuser ni les Europeens, ni le gouvernement, ni les noirs. Il paraît être destiné que ces races disparaissent devant la race Caucasienne bien que les mêmes passions, et le germe des mêmes vertus etaient données à toutes.…… d'après tout ce que je vois je ne puis croire que le Noir de cette partie de la colonie soit un être stupide incapable d'une éducation. Cette éducation ne se peut faire en 2–3–10– 20 ans. C'est éducation par des siécles que cette race demande, mais hélas! 10 ans la reduiront à peu de chose par les moyens des maladies veneriennes, qui se montrent déjà ici, et par les liqueurs spiritueuses.

Mr Leichhardt donne ensuite quelques coupes géologiques, et l'analyse du fruit du Bunya Bunya tree (Bödne of the black‑ fellows.]

Moreton Bay

Bunya Bunya district Archers Station 6th Jan., 1844

My dear friend,

⌜I've received your letter of the 25th of March 1843, together with the short letters from Mons. Rouault and Mr Witherill and the notes from Mons. Brogniart [sic] and Mons. Cordier. I need hardly say that I simply devoured them. They found me in the Bunya Bunya mountains, 60 English miles from Moreton Bay. We're so far apart, and things can so easily miscarry during a passage of 3 or 4 months, that I was beginning to think that my letter had gone astray and that we had lost all touch with each other. Thank Heaven I was wrong, and we shall be able to exchange news and views for a long time to come. I must give you a short account of my proceedings in this country.⌝ After studying the botany of Botany Bay for 6 months I left Sydney. I had Robt. Brown and de Candolle's 7 volumes, but there were several families of plants, such as the Euphorbeaceae, which I was unable to study through lack of works of reference. I made rapid strides in studying the others, and was soon able to conduct public excursions—the first to take place in the colony—and to give a course [of lectures] on botany, in which I tried to explain the organisation of the families [of plants] that arouse most interest when people go out walking, e.g., the Myrtaceae, Rutaceae, Proteaceae, Epacridae and the Cycads. It was at the end of August 1842 that I left the capital of New South Wales and transferred myself to the Hunter, to study the geology, in particular the position of the coal. The mouth of the Hunter is not nearly so rich in plants as the country around Sydney, but there are some of remarkable beauty in the mountains 10 miles from Newcastle and in the swamps quite close to the town. Doryanthes excelsa, *which sometimes attains a height of 12 to 16 or 18 feet, is found near Lake Macquarie and on the mountain called the Sugarloaf between Newcastle and Maitland, where it grows in association with* Xanthorrhoea hastilis *between immense blocks of pudding stone.* Blandfordia, *with big, scarlet flowers, grows in the swamps, most of which happen to be dry. Several* Melaleucas (*the tea tree*), Calothamnus *with its fine red flowers, several species of* Leptospermum, Fabricia *and* Baekia, *are also very common here. I stayed at Newcastle for two months, and then proceeded up the valley. I visited several places of geological interest, such as Harper's Hill,* where fossil shells are plentiful; and Black Creek, where I studied the extent and position of ⌜what is very likely a lacustrine⌝ limestone. I stayed for some time at Glendon, a very large farm belonging to Messrs R. and H. Scott, who did all they could to make things easy for me. There is a change in the flora here which was noticeable even at Harper's Hill. We part company with the plants of the maritime zone and begin to meet those of the interior. Numerous little herbaceous Leguminosae and the little Mimosa (M. terminalis) come up after rain; but it rains so little that the land-holders have had to give up agriculture. The prevailing rocks are still pudding stone and a more or less coarse-grained sandstone which passes in many places into pudding stone. Coal is found at many places from Newcastle right to the Liverpool Range. It crops out mainly in the banks of the river and of the tributary streams. It appears on Glendon Brook, Westbrook, Fallbrook and Foybrook, at Bengalla and at the burning mountain, Mount Wingen,[a] where no more than a coal seam under the sandstone is on fire. The coal is full of impressions of ferns, in particular of one with lanceolate fronds from one inch to 2 feet long and from 3" to 2 or $2\frac{1}{2}$" broad; but there are several others which correspond more or less to genera already known. There are ferns in the sandstone also, besides Equisetaceae, Calamites, and the trunks of trees; and they all survive weathering better than the fossils in the clayey shale, which easily crumbles to dust. To the North-east of Glendon there's a complex of hills and mountains of a different nature. They're formed of a feldspathic and pyroxenic porphyry of which I don't recall having seen specimens in the Museum. ⌜The hills formed of this rock run in general from North-west to South-east or from North-east to South-west.⌝ They appear to represent sills that have risen between the pudding stone and the sandstone and a conglomerate that has been indurated in many places by the igneous contact. Towards the North, 36 English miles from Glendon, the pudding stone ⌜and the conglomerate⌝ appear ⌜again intruded here and there by the porphyry; and farther on is the sandstone at the base of the mountains which separate New England from the Hunter River valley. These⌝ mountains ⌜are⌝ composed of basalt rich in ⌜olivine and⌝ zeolites, though I never found any of them well crystallised. I visited Mount Royal, one of the spurs of these mountains. It rises to about

3000 feet and is one of the highest mountains in that part of the colony. Mount Royal extends due South from New England on a straight line. Its heart and the conspicuous mass of it are basaltic but its flanks are formed of sandstone. Its eastern slope is covered with dense vegetation of quite an unusual character, called 'mountain brush' in the colony. To my mind, this brush, of which I happen to be making a special study just now, is the same thing as the virgin forest of South America. Waterton's fine description of it in his Wanderings in South America applies quite aptly to the Mt. Royal brush, and even more so to the Bunya Bunya mountain brushes. Within a small area you find a great variety of trees, thrusting up to a considerable height before branching, and bound together by just as great a variety of vines that have run right up to spread their leaves and flowers over the crowns of the trees. Herbaceous plants are found only on the margins of the brush, and in spaces open to the light and air. Ferns grow with particular luxuriance and in great variety; and in the little ravines at the base of the mountain, where a mixture of vegetable mould and decomposed basalt has accumulated, Alsophila grows to 15' in height and 10" in thickness. I've a satisfactory collection from these brushes but have not yet determined my plants as I have no books with me.—From Glendon I went on towards the Liverpool plains, 150 English miles farther inland. Just imagine me on the crest of the Liverpool Range, with the whole basin of the Hunter and the Goulburn at my feet as I gaze eastwards, and the immense plains behind me to the West! The Liverpool Range forms a sweeping arc of basalt around the basin of the Hunter. The latter is filled by sandstone, pudding stone and conglomerate, which overlie an immense deposit of coal. ⟨At Newcastle you can trace four seams of this substance, separated by shale, sandstone and pudding stone. The 3rd and 4th seams are being worked.⟩ There are a few basaltic dykes to be seen along the coast, striking from South-east to North-west, and it's not hard to relate them to the arc of the Liverpool Range, or to connect them with an antediluvian centre of volcanic activity out in the Pacific Ocean. The feldspathic and pyroxenic porphyries constitute a subordinate series closer to the sea. There is limestone in the mountains of the Paterson, which is said to contain trilobites, but I have not seen any, though I've found plenty of impressions of shells and of Orthoceratites at the foot of Mt. Royal. ⌈I mentioned the lacustrine limestone of Black Creek, which appears again at the farm at Glendon; and there's a concretionary limestone containing remains of Limnaeus, Cyclas and Cerastium, which occurs in the beds of the streams that descend from the Liverpool Range at Cassilis.—⌉I have not visited the basin of the Goulburn which is bounded on the West by mountains of granite, but I hope to do so on my return from Moreton Bay.

The Liverpool plains are one of those areas that still hold out the hope of something new to the botanist. When I crossed this strange region there were numerous Compositae in flower. I made only a small collection of them because the only paper I had was the newspapers that my several hosts had been kind enough to give me. ⌈Just imagine your peregrine friend begging for newspapers so that he might collect plants!⌉ —The Liverpool plains have probably been the bed of a chain of lakes that contained numerous islands formed of sandstone and covered with forests composed of many different species of Eucalyptus. The Cypress pine (Calitris) grows here in many places, and the bush huts are built of its timber. I followed the Mokki, the Namoi, the Peel and the Manilla, to spend some time at the source of the Gwydir.[b] These rivers were no more than chains of ponds and you could cross them without wetting your feet. Their beds, however, are very wide and are full of water-worn stones. In fact, they expose two beds: the one, which usually retains a little water and is lined by a dense hedge of Casuarina (swamp oak); and the other which is full of sand and ⟨very big⟩ water-worn stones, with here and there a sturdy gum tree that has been able to withstand the violence of the torrents that follow after storms or after weeks of rain. Between the Peel and the Namoi the [character of the] forest changes, and instead of riding between spotted gums, box and narrow-leaved ironbark you enter a forest of silver-leaved ironbark, whose leaves are greenish grey. Between the Namoi and the Gwydir, on the headwaters of Rocky Creek (one of the streams that join the Gwydir) there is a system of trachytic mountains[c] which I visited and studied as closely as my means [of investigation] and the wild blacks would allow. From these mountains the view extends across the boundless plains through which the Big River flows to join the Namoi. The latter is called the Barwon farther downstream, and the Darling still farther down towards its mouth. ⌜After the Gwydir you pass the Big River (Bandara in New England), the Macintyre (Macdonald), the Severn, Macintyre Brook and Canal Creek.⌝ All these rivers flow from the plateau of New England, which lies to the East; and they flow into the Barwon. They all rise in mountains of sandstone, basalt, granite and quartzite, and they contain enough water in their upper courses; but when they come out of the mountains on to the western plains they lose their water, and their empty beds merely mark the highways of torrents that flow after rain sets in. It is most interesting to observe how rain water which falls on the plateau of New England not 90 miles from the East coast, runs for more than 1000 miles to lose itself in the waters of the Southern Ocean. Between the Severn and the Condamine the country is flat, covered with forest which is very dense in some places, known to the colonists as 'brigalow scrub'.[d] The brigalow is a species of Acacia with long, stiff, greyish phyllodes. It is associated with Casuarinas (forest oak), and a good deal of brushwood, ⟨narrow-leaved⟩ ironbark and a species of Banksia—the only one found far from the coast. The soil is very sandy except on small flats near the streams, where it contains more clay and vegetable mould. The 'apple tree' of the colonists, Angophora lanceolata, grows well on these flats. Just as the brigalow gives character to this part of the country, another species of Acacia, the myall (Acacia pendula) characterises the Liverpool plains and the plains of the Gwydir and the Big River. The myall has phyllodes and supple, drooping branches like Salix Babylonica. Its wood is very hard and has a delightful smell of violets. The blacks use it for making boomerangs which, it seems, are passed on from hand to hand throughout the continent.—The Condamine is the first river which does not belong to the system of the Darling (Barwon), but seems to turn towards the North; and its course may curve like that of the Barwon, but to lead its waters into the Gulf of Carpentaria.[e] From the Condamine you are on the Darling Downs,[f] undulating, open country, with rich, black, basaltic soil covered with numerous grasses of which one species of Antistheria, the oaten grass, is the staple food of numerous flocks of sheep. The flocks are rapidly becoming more numerous in country that suits the sheep so well. A new gum tree, the Moreton Bay ash, is common on the hills. The bark of the lower part of its trunk *forms big black scales, but the upper bark is greyish white and smooth. Here and there on the flats you see a few* Xanthorrhoeas *of a kind quite different from* X. hastilis. *This* Xanth. *grows on rich soil, whereas* X. hastilis *is found only on poor sandy soils. It grows to be 10–15 feet high and nearly a foot thick.* ⟨*Its comma* (*flower stalk*) *is shorter than that of the other species, from which the blacks make their spears.*⟩ *In one of the streams* (Hudson's Creek) *there's a bed of coal. Its occurrence in what is basaltic country is quite extraordinary. Fossil bones are found in the alluvium of the vallies and in the banks of the river, particularly in those of the Condamine. I shall try to get some of them myself. There's no doubt that the fossilised bones of elephants have been found, but I hope that people will not be too ready to assume that these animals once existed in New Holland. It seems to me that the eruption of the basaltic system, which forms the Liverpool Range,* ⟨*and the extension of the activity to New England and the Coast Range of Moreton Bay,*⟩ *had a lot to do with the emergence of the continent from under the sea; and it's quite likely that the bones were deposited before the land was uplifted—for they could have been transported from the Indies, or from the greater islands between these and New Holland. The Darling Downs are 1450' above sea-level and the nights are very cold, even in September and October.—The descent to Moreton Bay is abrupt, like that from New England to Port Macquarie.*[g] *When you enter the basin of the Brisbane you are at once aware of the greater vigour in plant-growth. Trees are taller and they grow closer together. The flanks of the mountains and the banks of streams and of the river are overgrown by almost impenetrable brush* ⟨⟨*mountain brush and river brush*)⟩. ⌜*At the eastern foot of the Coast Range, which is basaltic, sandstone, puddingstone and conglomerate appear. 20 miles from the sea, at Limestone, coal is found under the sandstone; and there are several hills formed of a limestone, which was probably deposited by spring waters that rose through basaltic formations. The rock at Moreton Bay is a talc schist invaded by quartz which fills all the fissures in the rock.*—⌝*From Moreton Bay I went on to the Bunya Bunya district 60 miles farther North. It takes its name from the* Araucaria '*Bunya Bunya*', *which grows here in the mt. brushes. So here I am at last, at the scene of my present activities. In a quarter of an hour I*

Moreton Bay

can get to the primeval forest, one of the creek brushes that extend downwards from the Bunya Bunya Mountains. These mountains part the waters of the Brisbane from those of the Wide Bay River.[h] Their general direction is from East to West, where they join the Coast Range; and they send a number of spurs to the South, between which run a number of streams that flow down to the Brisbane. Dixon's map[i] is quite wrong for the northern part of Moreton Bay ⌜, so here is a crude sketch that will help you to form a better idea of it: [small diagram $1\frac{1}{4}''$ by $\frac{3}{4}''$]⌝. The rock formation is far from uniform. Here, it may be a whitish syenite, ⌜its constituent minerals equal in amounts; there, it contains mica and becomes⌝ a granite ⌜of the same light colour; elsewhere, it may be very fine-grained, bluish in colour, and its classification uncertain. In many places, particularly⌝ towards the East [end] of the chain and in the Mt. Brisbane Range, quartz seems to be quite lacking ⌜and it becomes a diorite (hornblende and feldspar)⌝. In many other places it forms a hornblende porphyry, and in still others it passes into massive hornblende. There's another feldspathic rock to be found in many of the streams in these mountains, which probably belongs to the epoch of the Glasshouses. These mountains form fantastic, isolated peaks to the South of the Bunya Bunya Range and 15 to 20 English miles from the sea. They are composed of a friable rock of feldspathic groundmass and more or less numerous phenocrysts, also of feldspar. ⌜The most important of them is called Birroa (Birwa) and is perhaps 1200 or 1500′ in height. A spire to the East of Birroa is called Gnarra nurrui (Kumaunum [?]), another, more to the South, is Dunbabola (Paidarr boledó)—with two summits: $\begin{Bmatrix} \text{dumba} \\ \text{paidarr} \end{Bmatrix}$ mountain $\begin{Bmatrix} \text{bola} \\ \text{boledo} \end{Bmatrix}$ two). A 4th is Birrawaman and a 5th Waiamurrum.⌝ The rock of the Puy de Dôme, that of Sarconi, and of a third mountain (I can't recall its name) are very like the kinds of rock that form the Glasshouses. It's remarkable that these mountains all look alike in a general way. ⌜⟨The two peaks at Mont Dore (la Tuillière?), in fact, remind me of the odd appearance of Gnara nurui, although [the former are] composed of a rock rather different from the other rocks of [the series?].⟩ These mountains are surrounded by a coarse-grained sandstone that contains a lot of limonite.⌝ Isn't it strange that I've seen not the least sign of metals or precious stones, ⌜except for a lump of massicot in the basaltic soil at Cassilis⌐? I've often been shown yellow mica that looked like gold dust, but never any other metal. It's as if Science had ordained that I shall serve no other gods but her alone.

But what am I to say about the Bunya Bunya brush? Or about the bunya (the bodné of the blacks)? About this majestic tree whose trunk looks like a pillar supporting the vault of Heaven? About its fruit, which brings the blacks for a hundred miles and more to feastings that last for 2 or 3 months (January, February and March)? What am I to say about this multitude of plants and shrubs and rare trees that grow in the brush that covers the mountains over an area 50 English miles long and 10 miles wide? The Castanospermum australe ⟨that you mention⟩ is found not only in the river and creek brushes, being even commoner in that of the Bunya Mts. It's a tree that often reaches a height of 80'. Its red and yellow blossoms grow straight out of the wood in little clusters, whilst its dark green, compound leaves are confined to the ends of the branches. I've found another tree of the same family up on the mountains outside the brush. Its wood is very spongy and is used by the blacks for their helimans. The bark is covered with corky tubercles; the blossoms are large and red; the foliage is ternate, the leaflets being petiolate and triangular with rounded angles.—I think it's an Erythrina. There are two other leguminous trees in the brush, one of which bears rich racemes of yellow blossoms. The other looks to me like a Mimosa. Its leaves are bipinnate, the leaflets elliptico-lanceolate, those at the end larger than those at the base, and the pods are twisted like corkscrews. ⟨The blacks distinguish the following Proteaceous trees by name:⟩ wairum, which has stiff, very long pinnato-lobate leaves; the silver oak, Grevillea robusta; dulabi, which has lanceolate leaves ⌐2 to 3″ long and $\frac{3}{4}$″ broad⌐; and dinangerumbin, which has leaves like wairum but with beautifully silvery under sides. The medullary rays show very clearly in the wood of these 4 species. There are several curious trees belonging to the Malvaceae and the Sterculiaceae. One of them is ⌐called kayar or kēr⌐ (the bottle tree of the colonists, because it is shaped like a bottle, swelling out 3 or 4 feet above the ground). The bark is very hard, but the inside is soft and spongy and full of sap. I've not yet seen the blossom but the fruit is a capsule like that of Sterculia.

Moreton Bay

The blacks eat the seeds. Another of the Sterculiaceae is called banni banni. It too is a tree of some size, with spongy wood and thick bark. Its leaves are oblong, $\frac{5}{4}'$ long and $1'$ broad, slightly chordate, ⌜acuminate, and with long petioles⌝. The bark contains a transparent, gelatinous juice that sticks to the fingers. There's a small tree or shrub with tubular, scarlet blossoms, that grows between the rocks on the mountains. There's another malvacaous plant to be found by the sea side. Its leaves, like those of the former, resemble fig leaves, and it has large, broad flowers like Althaea and Hibiscus. The wood is hard and of a lovely deep yellow, velvety colour at the heart. I've found two other species of Hibiscus. Hibiscus heterophyllus (the Couradjong) is to be found all over the colony. Its strong bark makes excellent natural rope. This species has pinkish white flowers, the base of the petals and the stamens being deep purple. The other species has yellow flowers, and there's a third with reddish flowers and leaves like fig leaves. The Moreton Bay pine, Araucaria Cunninghami (gunam: warrall) is found in all the river and creek brushes. It's a tree so tall that it towers above all the rest. There's a third species growing in the brush at Wide Bay, which the blacks call dandajōm. I've heard tell that there's even a 4th out towards the coast. The Cypress pine (Callitris) is common on sandy soil close to the sea. ⌜But my letter is getting too long so I'll write to you again soon.⌝ I shall be sending 2 collections of brush plants ⌜to Dr Nicholson, who is now living at Newcastle-upon-Tyne in England. He will let the Museum know about them.⌝ I shall also send a collection of timbers from the brush and from the forests at Moreton Bay. ⌜I shall have to put a good price on it to cover the cost of shipping it, but I don't think that there's a similar collection in Europe.⌝ I've preserved such fruit as was in season in brine, and shall send it with the collection of woods. I've also made a rock collection⌜, but my specimens are ⟨smaller than Mons. Cordier would like to have them. They are 2" broad and 3" long. The truth is that large specimens of the size Mons. Cordier requests take such a time to prepare that I could not even begin to collect them. I tried, but the trimming of a specimen took so long that I gave up in despair. The collection of small samples will nevertheless be very useful, and they are numbered to correspond to the account of the geology of the Moreton Bay district, and of the trachytic formations of the Gwydir and the feldspathic porphyries of Glendon, which I intend to write.⟩ 3" long and 2" broad. They represent the rocks of Moreton Bay, the trachytes of the Gwydir and the feldspathic porphyries of Glendon.⁷ It seems to me that the geologists have no idea that so many kinds of rock are to be found in New Holland. ⟨I don't know what to do with the impressions of ferns from Newcastle. The rock crumbles so easily that I fear that my specimens may not reach Europe in good condition. However, I shall take all precautions. There's no cotton but there's plenty of wool, so I'll pack them in wool as well as I can.⟩ [Last page of holograph missing. From incomplete copy in Durando's handwriting: ... ... There's been a proposal to send an expedition from Sydney to Port Essington on the northern coast of New Holland, but the government can't afford it. I hope that a proper expedition really will be sent out some day, either by the government or by the people of the colony. ... We saw the comet[j] you mentioned, from the 3rd of March until the 1st [for 11th] of April 1843. All the country through which I passed has been taken up, except for a few places, by owners of sheep and cattle, and their stations are 20 to 30 English miles apart. I often travelled alone, on my good Valparaiso mare with a bitch (pointer) for company. ... ... I've often camped alone, in the forests or on the mountains, and have been cook, groom, washerwoman and naturalist all together. But the people are very hospitable, and I am pleased to have made the acquaintance of several men of high standing in the colony. I've often thought that I would have to give up my studies and turn to earning my living. I enjoyed your letter, which has raised my hopes considerably. ... It's odd that there's a little plant here very like the one you sent me from the marshes of Tuscany (Hypericum quinquenervium Walt. Sarothra Blentinensis Sav.). I am enclosing herein the flower of the tree I mentioned, that has ternate leaves and soft, spongy wood that the blacks use for their utensils and helimans (shields). ... ... The blacks themselves are very interesting beings about whom I've been recording a lot of information, as I've been living amongst them nearly all the time. There are some strong tribes in existence, soon to disappear before civilisation. Philanthropists deplore the state of affairs, but one can see that these poor blacks can't be saved from destruction by any means of salvation that have so far been applied.

Moreton Bay

Neither the government nor the Europeans nor the blacks themselves can be blamed. It seems to have been ordained that these races are to disappear before the Caucasians, although the human passions and the possibilities of human virtue are common to all men. ... ... As I see him, I can't believe that the black man of this part of the colony is a stupid creature incapable of education. But he can't be educated in 2, 3 or 10 or even 20 years. To educate his race would take centuries; but 10 years, alas, will work its degeneration, through venereal disease, which has already taken a hold here—and strong drink.

Mr Leichhardt then gives several geological sections and the analysis of the fruit of the bunya bunya tree (Bödne of the black-fellows).]

95 [TO LT. ROBERT LYND, Barrack Master, Sydney, N.S.W.]

Durrundur, Archer's Station, 9th Jan. 1844

[My dear friend,]

... ... The 31st December I returned from my expedition to the Bunya feast of the black-fellows, and well have I feasted myself. Mr John Archer and Mr Waterstone[a] accompanied me, with three black-fellows, who carried our provisions, as the dense brushes did not permit to take horses with us.

... ... I have travelled again in those remarkable mountain brushes, out of which the Bunya-Bunyas lift their majestic heads, like pillars of the blue vault of heaven. I measured several, and their circumference six feet high was 17 to 20 feet. The black-fellows go up to the top of these giants of vegetation with a simple brush-vine and which they put round the tree and which they push higher with every step they take upwards. They break the cones, almost a foot long and $\frac{3}{4}$ in diameter, and throw them down. They whistle through the air and their fall sounds far through the silence of the brush. Those trees the fruit of which we gathered had 4, 5, or 6 cones. Every cone contained, perhaps, 40 to 50 fertile scales; many, and particularly those at the top of the cone, are not fertile. The black-fellows eat an immense quantity; and, indeed, it is difficult to cease, if one has once commenced to eat them. If you find a favourable tree, and if the circumstances are favourable too— for instance, if the day is cool, in the morning and evening— the kernel of the Bunya fruit has a very fine aroma, and it is certainly a delicious eating; but during a very hot day, or from an unfavourable tree, the fruit is by no means so tasteful as I hoped to find them generally. The black-fellows roast them, and we tried even to boil them; the fruit lost, however, in flavour in both cases. Besides, it did not agree with my stomach. The black-fellows thrive well on them, but Mr Archer told me that the young people return generally with boils all over the body, and I witnessed myself some cases.

There is a little valley, an open plain, in the midst of these brushes which cover, perhaps, an extent of 50 miles long and 10 miles broad. This plain they call Booroon,^b and it seems the rendezvous for fights between the hostile tribes who come from near and far to enjoy the harvest of the Bunya. I went to Booroon. A creek, fed by the shady gulleys of the brushes, passes it; low ranges, all covered with thick brush, all of igneous formation, surround it: many a Bunya-tree looks down on the capers of the sable children of the forest and brush, of plain and mountain, of the sea-coast and of the country inland. They intended to have a fight, but their antagonists did not appear. I saw some black beauties—young unmarried women, about 14 to 17 years old, or perhaps still younger. They were very regularly formed; their movements free and graceful; they were full of mirth and joke, and examined us with all the *naïveté* of youth. Their faces could not be beautiful, for their flat broad noses will never permit it; but the proportions of the body in woman and man are as perfect as those of the Caucasian race, and the artist would find an inexhaustible source of observation and study amongst these black tribes. The weather was very unfavourable; we had during the whole journey showers of rain, which made my collecting plants very difficult and precarious. The Bunya was not yet entirely ripe, and, according to the statements of the black-fellows, it was no rich harvest. It would have been in vain to collect cones to send them to Sydney or to Europe; they would have been rotten in less than a week. I took several kernels with the woody case: I took every precaution and dried them in the sun and in the shade; but alas! they are already gone! It seems, indeed, that these trees bear very good fruit only every three years or in such period, though there are trees in bearing every year.

......I have been all the time collecting specimens of wood, and become familiar with the various trees of the bush. I have at present about 120 different trees (varieties of timber). ......

96 [TO LT. R. LYND.]

[Durundur, Archer's Station,] 19th January 1844.

[My dear friend,]

......The rainy season has commenced—powers of rain have poured down; the rivers and creeks were filled to the highest brim, and the adjacent flats and hollows were extensively inundated. The waters, falling on the steep slopes of the Bunya Range and of its spurs, collected quickly into the gulleys and creeks, and ran off as quickly as they came. The wind blew during the rains from easterly quarters (east and south-east). Last Thursday it changed to the west and fair weather set in again; but even now thunder storms are generally gathering in the afternoon and loose clouds send down occasional showers, particularly towards evening and during the night. The wind during the rains was very slight, and in the morning there was generally a perfect calm. The heat during the sunny intervals is very oppressive, and I think it approaches very much to the description of the moist heat of the East Indies. ......

97 HERREN MALER SCHMALFUSS / Cottbus / Berlin

Archers Station dn 2tn Februar 1844

Mein theuerster Schwager

Du wirst aus meinem Brief, den ich dir mit Missions briefen nach Deutschland sandte, ersehen, daß ich mir über die lang verzögerte Antwort meiner Briefe allerlei Gedanken machte. Wer konnte mir bürgen, daß nicht das Baltische Meer über die Sand ebenen Pommerns nd der Marken hinein gebrochen und alle die Meinigen mit sich fortgeführt, um sie vielleicht auf den Küsten Schwedens ab zu setzen. Hat doch die Prophezeiung daß London in März od. Mai 1840 verschlungen werden sollte,[a] gar manches Herz zittern gemacht — wie sollen die schwachen Köpfe der Antipoden die wachsenden Befürcht⸗ ungen bemeistern. Und wie kann ein Antipode alle die wunderbaren Neuig Keiten auffassen, die du ihm in deinem lieben langersehnten Briefe vom 29tn April 1843 mittheilst. Ihr Alle gesund! Adolph verheirathet! Eisenbahn in Fürsten⸗ walde! Daguerrotype in Kottbus! Mir ist fast, wie jenem 200 Jährigen Schläfer, der beim Erwachen auf seinem Felde einen Dampfpflug heranbrausen sieht, auf welchem der Knecht ein Cigarr raucht und die Zeitungen liest. Obwohl ich dem Grundsatze 'besser hab ich als hätt' ich' hold bin, so laß ich mir doch den Aufschub in der Sendung eines Daguerro⸗ typirten Familien bildes bis zu meiner Zurückkunft nach Sydney gefallen; ich freue mich auf dasselbe, wie auf einen Augenblick des Wiedersehens nd ich hoffe nur, daß Euere Gesichter mit meiner Rechnung übereinstimmen — daß Sor⸗ gen nicht tiefere Furchen gegeben haben, als der Natürliche Lauf der Jahre mit sich bringen. Jenen Cometen, welchen Ihr in Deutschland beobachtet, haben wir auch hier gesehen; es war eine der herrlichsten Himmelserscheinungen, die mir in meinem Leben zu Theil geworden; und ich konnte nicht satt werden, den wunderbaren Fremdling, den schnellen Wanderer, an zu staunen, der auf seinen weiten Bahn durch den un

Moreton Bay

ermeßlichen Raum zwischen den glänzenden Sternbildern dieses reichen dunkelblauen Himmels hindurchglitt nd hinter den Sternen Orions sich in die blaue Nacht verlor. Er wurde zuerst am 3tn u 4tn März gesehen und den letzten leisen Lichtschimmer bemerkte meine angestrengten Augen ungefähr am 11th April. Ausserdem Hauptschweif (a) des Cometen erschien ein Leichter Streif (b), welcher im spitzen Winkel von ihm nach unten nd Westen zu abging; ungefähr so: [*small diagram*]. Er war viele Male länger, als der Schweif; doch bemerkte ich ihn nur den 2tn Abend. Ich bin neugierig zu hören, ob man diesen Streifen auch in Deutschland beobachtete nd wie ihn die Astronomen erklären.[b] Ich bin immer noch in dem Bezirke von Moretonbay, obwohl gegen meinen Willen, indem der beständige Regen, und die angeschwollenen Bäche und Wasser mich von meiner Heimkehr zurückhalten. Du wirst aus meinen frühern Briefen ersehen haben, daß dieser Bezirk sich wesentlich von der Uebrigen Colonie unterscheidet. Die Leute haben hier fast zu viel Regen, viel zu viel für Schaafzucht — während die andern Theile der Kolonie an Dürre leiden. Du mußt indessen wohl den Strich Landes unterscheiden, welches sich in eine Breite von 6 Meilen am Meeresufer hinzieht und den regelmäßigen Meereswinden ausgesetzt ist, die gewöhnlich von 10–11 Uhr bis 3–4 Uhr die mit Wasserdünsten geschwängerte Luft des Meeres, über die benachbarte Landschaft westlich führen. In diesem Küstenstriche sind die Erndten sicherer, obwohl auch sie in sehr trockenen Jahren leiden. Die Gärten geben gewöhnlich einen reichen Ertrag von Pfyrsichen, Feigen, Granatäpfeln, Apricosen, Trauben. — Die Rebe gedeiht ausserordentlich, obwohl die Qualität des gewonnen [*sic*] Weines sich noch nicht als sehr gut bewährt hat. Es ist eine Freude, von dem Monotonen graugrün der Waldung das Auge auf dem frischen Grün der Reben u Pfyrsichbäume aus zu ruhen. Der Hunter scheint besonders für Weinpflanzungen geeignet und ich machte voriges Jahr an fast 2000 Quart Wein un mehr — welche nach dem Urtheil der Sachverständigen zu den besten Weinen gehören, welche in der Kolonie gemacht sind. Die Kartoffel gedeiht nicht so gut; die besten werden von Van Diemens^c land eingeführt; doch habe ich im Moreton bay district schöne Europaeische Kartoffeln gegessen. Während der Regenzeit gedeihen alle Gemüse ganz vortrefflich. Sie wuchern vielleicht zu sehr. Melonen von verschiedener Güte, Wassermelonen, Kürbisse, Gurken wachsen überall in Fülle — ja der Kürbiß ist so zu sagen die Kartoffel der Kolonie und man sagt ein Scherz: man könne ein eingeborenes Mädchen (Currency) von einem Englischen eingewanderten Mädchen (Sterling) unter‑ scheiden, wenn man ihr Kartoffeln un Kürbiß vorsetze. Das Hiergeborene Mädchen wird zuerst vom Kürbiß essen, während die Engländerin die Kartoffel vorzieht. Die Kürbisse sind ausgezeichnet schön, groß nd mehlig und halten sich länger, und wachsen weit im Innern um die Hütte des Hirten nd Schäfers, wenn alle andern Gemüse unter der Dürre erliegen. Man baut in Moretonbay 'die Süsse Kartoffel' (Convolvulus Batatas — eine Art Winde mit Kartoffeln od. Knollen. Diese Pflanze giebt sehr reiche Erndten, schmeckt wie ein etwas wässrige süßliche Kartoffel nd wird oft 17 Pf schwer. Die Banane nd das Zuckerrohr gedeihen hier gleich‑ falls; doch letzteres erfordert zu viel Aufmerksamkeit un Arbeit, um fürs Erste für die Kolonie nützlich werden zu können, da Arbeit hier so ausserordentlich theuer ist. — Ich erhielt vor einiger Zeit vom französischen Museum eine Aufforderung eine Sammlung der verschiedenen Holzarten der Kolonie nach Paris zu senden. Ich habe versucht dieser Aufforderung nach zu kommen nd habe ungefähr 130 Holzstücke 1′ lang ud 1–3″ dick in diesem Bezirke gesammelt. Wenn du überlegst, von wie wenigen Baumarten unsere deutschen nd eingebornen Wälder gebildet sind, so wirst du dich nicht wenig wundern, wenn ich dir sage, daß ungefähr 120 dieser Bäumen fast in dem Durchschnitt einer halben Meile zu finden sind. 100 von diesen gehören den dichten fruchtbaren Berg nd Fluß gebüschen an, während 20–25 den offenen Wald bilden. Dieser Wald läßt sich nicht, wie bei uns

Moreton Bay

nach einem vorherrschenden Baume Kiefer wald od. Tannen wald od. Eich. u Buchenwald nennen, sondern jene 25 verschiedenen Waldbäume sind gleichmäßig gemischt — indessen mit Vorwalten einiger Arten nach der verschiedenen Beschaffenheit des Bodens; der Boden unter den Bäumen, welcher in unsern Eichwaldungen häufig mit Preißelbeer un Heidelbeergesträuch bedeckt ist, nährt hier besonders eine Art Graß, welche man Kangooroo graß genannt hat, sei es, daß es die vorzüglichste Nahrung der Kangooroos bildet, oder daß es dem Korn nd Hafer gleich zu eine Höhe aufschießt, über welcher die weidenden Kangooroos wenn sie sich auf Schwanz un Hinterfüßen aufrichten, just mit ihren Köpfen hinübersehen. — Im October u Novembr. wird dieß Gras reif un der Waldgrund erscheint dann wie ein weites un unter brochenes Haferfeld. In Novembr. u Decembr. wird es trocken und Buschfeuer finden statt,[d] welche oft Meilen weit durch den Wald sich hinziehn und den Boden von Gras un trockenem Holze reinigen. Die zurück bleibende Asche bildet den Dünger unter welchem beim ersten Regenschauer das junge süße Gras hervorschießt, welches von Kangooroos, von Schaaf un Rinderheerden mit großer Begierde beweidet wird. Hunderte von Meilen stehen während der heißen Jahreszeit in Feuer; es beginnt, wo der Schwarze sein Nachtlager aufgeschlagen, indem er beim Weitergehen just den glühendsten Brand herausnimmt und ihn gegen ein Stück Rinde anglimmen läßt, um sogleich Feuer zu haben, wenn er sich wieder niederläßt, oder der reisende Weisse setzt das benachbarte trockene Gras in Brand; oder die Ansiedler brennen das alte Gras systematisch, um junge Weide für ihre Schaaf u Rinderheerden zu haben. Man hat mir gesagt, daß Feuer entstehe bisweilen durch das Reiben zweier Baumstämme od Zweige gegeneinander, wie sie vom Winde bewegt werden. Ich habe es nie gesehen nd glaube, daß es nur sehr selten statt findet, wenn es wirklich der Fall ist. Die Schwarzen wissen Feuer durch Reiben od. vielmehr schnelles Drehen eines Stockes in der Höhle eines andern hervor zu bringen; doch macht es ihnen zu viel Mühe nd sie ziehn es vor, stets Feuerbrände mit sich zu führen. Die Buschfeuer sind oft un besonders während der Nacht sehr mahlerisch. Eine lange wellige Feuerlinie, hinter welcher der dichte Rauch empor wirbelt, bewegt sich lang⸚ samer od. schneller in ihrer ganzen Ausdehnung gegen den Wind, welcher ihr den nothwendigen Sauerstoff zuführt; wie sie einem Strauche begegnet rennt sie knisternd an ihm hinauf: sie frißt in die alten Bäume hinein, welche gewöhnlich im Innern hohl sind, nd wie sie (vielleicht im Laufe mehrere Jahre) endlich zur Höhle eindringt, treibt sie, wie in einem Ofen, durch die ganze Höhe des Baumes nd der Rauch wirbelt aus den Enden der abgebrochenen Äste hervor, bis das Feuer selbst zu diesen hinansteigt und nun wie aus einem Hohofen aus ihnen hervorbläßt. Zweige fallen brennend nied[er], benachbarte Bäume werden von der Flamme ergriffen, welche schnell über das Geblätter hinlauft und in eine Feuer masse auflodert, endlich verliert der Baum das Gleichgewicht und stürzt prasselnd nieder, die hohlen Zweige brechen nd zersplittern, und wie die Aussere Luft freiern Zutritt zur verkohlten innern Masse gewinnt; wird jeder abgebrochene Zweig der Mittelpunkt eines unabhängigen lebhaften Feuers. Ich habe mich in der Nacht in der Nähe nd selbst in der Mitte dieser Feuer befunden. Bäume stürzten nach allen Seiten, von allen Seiten kam die Flamme heran gekrochen. Wenn das Gras nicht zu hoch ist, kann man über sie hinweg⸚ springen; bei hohem Grase, wenn die Flamme hoch auflodert wird es bedenklich. Die Gefahr bedroht besonders von den fallenden Bäumen, und in der Nacht während des Schlafes. — Während von Ende Novembr. bis zu Aufang Januar das Feuer im Walde regiert, haben während Ende Januar nd Februar die Fluthen die Herrschaft. Schon vor Waihnachten treten häufige schwere Gewitterstürme ein, welche meistens mit heftigen Regenschauern begleitet sind. Diese Gewitter kommen von west. Quartieren (besonders zwischen S West–N West). In Januar folgen den Gewittern oft für 3 Tage allgemeine leichte Landregen von schweren Schauern .trbrochen. Der

Moreton Bay

Wind weht entweder leise von Osten oder es ist vollkommen Lufstille. Die Schauer werden häufiger, heftiger mit dem Karakter tropischer Regengüße und dauern 5-6 Tage un Wochenlang. Schnell laufen die Wasser von den Bergabhängen in die Schluchten, diese schwellen nd führen es zu den bis dahin fast wasser losen Bachbetten, die von hohen Ufern umgeben sind. Das Wasser steigt bis zu den höchsten Rändern ⌐nd darüber⌐ hinweg nd breitet sich über die benachbarten Ebenen un Vertiefungen. Der kaum rinnende Bach ist jetzt ein staatlicher tobender Fluß, welcher [jetzt] um dicken Stämme der im Flußbette wachsenden Bäume lärmt, sie untergräbt und oft umstürzt. Doch da der Fall des Wassers groß ist, läuft es schnell ab, wie es kam nd in 2-3 Tagen kehrt es zu einem mittlern Stand zurück. Das Reisen ist nun sehr beschwerlich. Der Boden ist aufgeweicht; die Pferde sinken tief ein, die Lastwagen wühlen bis zu den Achseln in Koth nd heftige Windstöße entwurzeln leichter hohe Waldbäume, deren Wurzeln gewöhnlich flach in einem seichten Boden liegen, der von Regen aufgeweicht dem Baume nicht mehr hinreichenden Halt bietet. — Gewöhnlich macht diese[m] Regen ein plötzlich von Westen her wehender starker Wind ein Ende. Wenn in Europa der Wind von Westen nach Norden Osten u Süden wechselt, so folgt er in Neu holland der entgegen gesetzten Richtung von Westen nach Süden, Osten, Norden. — Dieß ist gewöhnlich der Fall und weicht er ab, so springt er gewöhnlich wieder zurück. In Moretonbay u Wide Bay habe ich indessen bemerkt, daß östliche Winde plötzlich mit westlichen wechseln, als wenn 2 Luftströme über einander hin flössen, von denen sich der *Obere* westliche .tr gewissen Umständen senkt.

Man hatte die Absicht eine Expedition von Sydney nach dem Nordwestl. Theile Neuhollands (nach Pt Essington) zu senden, um das Innere des Landes zu .trsuchen; Ich hätte diese Expedition natürlich begleitet, doch die Regierung, oder vielmehr der gegenwärtiger Gouverneur hat sich wegen fehlende Mittel dagegen opponirt. Man schreibt diese Opposition indessen persönlichen Motiven zu, indem Sir Thomas Mitchell, welcher die Expedition leiten sollte, kein Freund des Gouverneurs Sir George Gibbs ist. Ich weiß gegenwärtig nichts nähreres, da ich von Sydney entfernt bin. Wahrschl. wird eine solche Expedition früher od. später stattfinden. Man hat mich überall .ßerordentlich gastfreundschaftl. behandelt nd mein gegenwärtiger freundschaftlicher Wirth ist Hr. David Archer, welcher hier Schaafstationen hat. Er ist von Norwegen, wo sein Vater lebt, der indessen Schotte ist. Ich habe viele junge Männer gefunden, welche deutsch sprechen nd sie sind fast ohne Ausnahme von guter Familie nd wohlgebildet, obwohl das einsame Leben im Busche manche von ihnen feinen Sitten fremd macht. Die Häuser sind gewöhnlich Bretterwohnungen mit Baumrinde bedeckt. Einige Waldbäume spalten sich sehr leicht nd ihre Rinde schält sich in großen Platten (6' lang 6' breit) ab, so daß die 2 nothwendigsten Artikel zur Errichtung der Wohnungen überall od. fast überall zur Hand sind. — Ich kehre im 8 Tagen nach Sydney zurück nd. befinde mich gegenwärtig in bester Gesundheit. ⌜Eine meiner Stuten hat gefohlt; meine Pointerhundin hat 2 Jungen.

Grüße mir Mütterchen, Adolphen und meinen neuen Schwägerin, Barths nd die Freunde die sich meiner in Kottbus erinnern. Grüße Fettchen u deine Kinder⌝ von deinem herzlich dich liebenden Schwager

Ludwig Leichhardt.

— *Schmalfuss, Esq., Artist / Cottbus / Berlin*

Archer's Station, 2nd February, 1844

My dear brother-in-law,

You'll see from the letter I posted to Germany with the mail from the Mission, that this long, long waiting for answers has been causing me to imagine all sorts of things. For all I knew, the Baltic Sea might have come surging across the sand plains of Pomerania and the Marches and swept you all away, to deposit you perhaps on the coasts of Sweden. If the prophecy that London was to be engulfed in March or May, 1840,[a] *was enough to make many a strong heart beat faster, how could weak*

Moreton Bay

mortals in the Antipodes overcome their growing apprehension? And how is a poor Antipodean to form any idea of the extraordinary innovations you reported in your most welcome but long awaited letter of the 29th of April 1843? So everybody's well! Adolph married! The railway as far as Fürstenwalde, and Daguerrotype has reached Kottbus! It almost makes me feel like the man who fell asleep in his own field and woke up after 200 years to see a steam plough come roaring towards him with the ploughman puffing a cigar and reading the newspaper. Although I believe in the principle that a bird in hand's worth two in the bush, I shan't mind if you postpone sending me a Daguerrotype family group until I get back to Sydney. I'm looking forward to it as if I were going to have a real peep at you all, and I only hope that you'll look like I think you do—that the wrinkles and furrows are no deeper than the simple passage of time must have made them. The comet that you saw in Germany was visible to us here too. It was one of the most magnificent of Heavenly spectacles that it has ever been my lot to behold; and my eyes were never tired of gazing nor my mind of wondering at this astounding stranger in swift passage as it sped silently on its huge orbit out between the brilliant constellations of this deep, crowded, azure sky to be lost to sight in the darkness beyond the stars in Orion. It was first seen on the 3rd and 4th of March, and, by straining my eyes, I could just make out the last faint glimmer of it about the 11th of April. As well as the usual tail (a), this comet had a thin secondary tail (b), leaving the former from the under side at an acute angle and extending westward rather like this [small diagram]: It was many times longer than the main tail, though I noticed it only on the second night. I would like to know if this secondary tail was seen in Germany too, and how the astronomers explain it.[b] I'm still in the Moreton Bay district, where I'd prefer not to be, but incessant rain and the swollen creeks and rivers have been stopping me from setting out for home. You will have inferred from my previous letters that this district differs in important respects from the rest of the colony. They get almost too much rain here—far too much for sheep rearing, whilst other parts of the colony are suffering from drought. You must of course except the belt of country about six [German] miles wide that extends along the coast, and is exposed to regular sea breezes. The latter, which blow as a rule from 10–11 o'clock until 3–4 o'clock, carry sea air impregnated with water-vapour westward across the neighbouring country. In this coastal belt harvests are better assured, except in very dry years. The orchards and vineyards generally bear heavy crops of peaches, figs, pomegranates, apricots and grapes. Vines do extraordinarily well, though the quality of the wine produced has so far not been very good. After the monotonous grey-green of the bush the lively green of the vines and peach trees is a great relief to the eye. The Hunter seems to be particularly suitable for viticulture, and last year I made as much as 2000 quarts or more of wine which experts consider to be equal to the best the colony can produce. Potatoes don't do quite so well. The best are those imported from Van Diemen's Land,[c] but I've eaten very good European potatoes in the Moreton Bay District. All vegetables grow vigorously in the rainy season—in fact, they even make too much growth. Melons of a range of quality, watermelons, cucumbers and pumpkins grow everywhere in profusion—the pumpkin, in fact, is the potato of the colony, and they've a joke about it: 'You can tell a currency (colonial-born) from a sterling (English-born) lass if you serve her with pumpkin and potato. The colonial girl will eat her pumpkin first, but the English girl will prefer the potato.' The pumpkins are superb, big and floury, and they keep longer [than ours]. They grow far inland, around the stockmen's and shepherds' huts, when all other vegetables succumb to the drought. In Moreton Bay they grow the sweet potato (*Convolvulus batatas*)—a kind of bindweed with potatoes or tubers. This plant bears prolifically, and tastes like a sweetish, rather watery potato, and may weigh up to 17 lbs. Bananas and sugar cane do just as well here, but the latter calls for too much attention and labour to be of much advantage to the colony at present, labour being so uncommonly dear here.—Some time ago I had a request from the French museum to send a collection of the various kinds of colonial timber to Paris. I've been trying to do what they want, and I've collected about 130 specimens of wood from this district, each 1' long and 1–3" thick. When you consider how few different kinds of trees go to make up our German woods and indigenous forests, you'll no doubt be astonished when I say that about 120 of these trees are to be found within a radius of a quarter of a mile. 100 of these belong to the dense, rich mountain and river brushes, whilst 20–25 form open forest. This forest, unlike ours, can't

Moreton Bay

be named after any predominating tree, the way we speak of pine or fir forest, oak or beech forest, for the 25 kinds of trees are evenly mixed, except where the soil favours some rather than others. The ground under the trees, which would be covered with blueberry and whortleberry bushes in our oak forests, is mostly covered with kangaroo grass here, so called either because it's the favourite food of kangaroos or because it grows like wheat and oats, and gets just high enough for them to peep over when they sit up on their hind legs and tails whilst they're browsing. This grass ripens in October and November, when the ground under the trees looks like an even, sweeping field of oats. In November and December the weather gets dry and bushfires break out.[d] They're often a mile wide, and they clear the ground of grass and dead wood as they sweep through the bush. The ashes left behind are like manure to the sweet, tender grass that shoots up as soon as it rains and is eagerly consumed by kangaroos, flocks of sheep and herds of cattle. During the hot months fire can rage over hundreds of [square] miles. It starts [either] where the blacks have been camping for the night, as all they do when moving on is to pull a brightly burning stick out [of the fire] and keep it smouldering against a piece of bark, so that they can light a fire at the next camp; or, [where] white men have set the dry grass on fire [when] passing on; or [where] settlers have been systematically burning off the old grass to obtain young pasture for their sheep and cattle. I've been told that fire is sometimes caused by the friction between trees and boughs rubbing together in the wind, but I've never seen it and it must happen rarely if at all. The blacks know how to produce fire by friction, particularly by rapidly rotating one stick in a hole in another; but it takes too much trouble, so they prefer always to carry fire-sticks with them. Bushfires are often very picturesque, especially at night. A long, wavy line of fire, behind which the smoke swirls upward, advances, here quickly, there slowly along its whole length into the wind, which brings it the necessary oxygen. If it encounters a bush, it runs crackling up through it. The flames eat into the old trees, most of which are hollow; and when, perhaps after several years, they do at last find their way into the cavity, the whole tree draws like an oven and smoke comes pouring out at the ends of broken branches until the fire itself gets that far, when they blaze like blast furnaces. Burning branches come crashing down, trees near by catch fire, the flames reach their foliage and envelop the whole tree, which finally loses its balance and comes toppling down ablaze; hollow branches break off and burst, and, as the open air is now in freer contact with the mass of inner embers, every bough that has snapped off now becomes the centre of a strong, separate fire. I've been close to—in fact, in the midst of—fires like this at night, with trees falling on all sides and the flames creeping in from every direction. If the grass is not too tall you can jump over it out of the way, but in tall grass with the flames leaping high it's risky. The greatest dangers are from falling trees and from being overtaken at night during sleep.—Though bushfires are the danger from November until the beginning of January, floods [are likely to] prevail late in January and in February. Even before Christmas heavy thunderstorms begin to break, accompanied as a rule by heavy showers of rain. They come from westerly quarters (particularly S West–N West). In January these storms are often followed by 3 days of light, general, continuous rain punctuated by heavy showers. Either the wind blows gently from the East or there's a complete calm. The showers then come oftener, they get heavier, like tropical downpours, lasting for 5–6 days or even a week. Down comes the water from the mountain slopes and runs into the gorges, which choke, and lead it farther down into what have been almost dry watercourses between high banks. The water rises to the very tops of the banks, ⌜overflows them⌝ and spreads out over the neighbouring flats and lowlands. What was a mere trickle of water has become a noble, turbulent river, swishing past the stout trunks of the trees growing in its course, undermining them and often hurling them down. Nevertheless, the gradient of the stream being steep, the water runs away as quickly as it came, and in 2 or 3 days it returns to its usual level. Travelling is now very fatiguing. The ground is sodden, the horses get bogged, the drays sink down to the axles in the mud, and heavy gusts of wind easily uproot tall trees in the bush. The trees as a rule are flatly rooted in light soil, which affords insufficient hold when softened by the rain. A strong westerly wind usually brings these rains to a sudden end. If in Europe the wind changes from West to North, East and South, the corresponding change in New Holland is in the opposite direction, from West to South, East and North.—This is the rule, to which there is quick return if it be broken. In both [the]

Moreton Bay

Moreton Bay and Wide Bay districts, however, I have observed that easterly winds suddenly change to westerly, as if there were two streams of air flowing in, one above the other, the upper of which, the westerly, descends under certain conditions.

They were intending to despatch an expedition from Sydney to the north-westerly part of New Holland (to Port Essington) to examine the interior of the country, and I would naturally have gone with it, but the government, or rather, the present Governor, has opposed it on account of the expense. People are saying that the reasons for his opposition are personal, because Sir Thomas Mitchell, who was to lead the expedition, is no friend of the Governor, Sir George Gipps. That's all I know about it just now, being so far from Sydney. But an expedition will most likely be sent out in that direction sooner or later. I've been treated with extraordinary kindness everywhere, and my friendly host at the moment is Mr David Archer, who has sheep stations here. He comes from Norway, where his father lives although he's a Scotsman. I've met a number of young men who speak German, and practically all of them are well educated and come from good families, though their lonely existence in the bush deprives them of the refinements of life. Most of the houses are slab huts roofed with bark. Some of the bush timbers are very easily split, and bark can be peeled off the trees in big sheets (6' long 6' wide), so that the 2 essential materials for building dwellings are right at hand everywhere—or nearly everywhere.—I'm returning to Sydney in a week's time and am feeling very well just now. ⌈One of my mares has foaled and my pointer bitch has 2 pups. Love to mother and Adolph and my new sister-in-law. Regards to the Barths and to any friends in Cottbus who remember me, to Fettchen and your children, from⌉

Your affectionate brother-in-law,
Ludwig Leichhardt.

98 [DR WILLIAM NICHOLSON

Archer's station,] den 6tn Februar[, 1844]

Mein theuerster William

Der Regen dauert fort nd hält mich nicht nur in dem Bezirke von Moreton bay fest welchen ich nun gern verlassen möchte, sondern [hält] mich auch in das Haus, wo ich mit Lesen und mit Grübelei die trägen Stunden verkürze. Ich habe eine Sammlung der hiesigen Holz arten gemacht, die ich indessen nicht für vollständig halte, da ich in den Fluß un Meeres gebüschen mehrere Bäume fand, von denen ich noch nicht kein Specimen besitze. Das Sammeln dieser Specimen ist mit große Mühe verbunden als du dir vielleicht vorstellst. Der Karakter der Rinde ist häufig so unbestimmt, besonders in verwandten Baumarten, daß es nöthig ist den Baum zu fällen um sich von der Verschiedenheit zu überzeugen. Ist ein Schwarzer zur Hand so steigt er hinauf un wirft Zweige nieder; doch die Schwarzen haben wenig Neigung, mich oft auf meine ermüdende Excursionen zu begleiten und es hält .srordentlich schwer sie zum häufigen Besteigen der Bäume zu bewegen. Sie sind in dieser Beziehung wunderliche Geschöpfe. Werden sie ein Opossum oder Honig auf der höchsten Spitze eines 180′ hohen Baumes bemerken, sie würden auf der Stelle sie zu erreichen streben; doch da sie nicht verstehen weßhalb ich Zweige un Früchte verlange, steifen sie sich [solange] wie möglich, ehe sie sich zum Ersteigen eines Baumes entschliessen. — Du hast mich geboten dir Sceletons zu senden. Auch diese sind nicht so leicht erhalten wie du glauben möchtest. Ich selbst bin kein Jäger, die Thiere zu schießen nd hänge deßhalb vom guten Willen meiner Wirthe un der Schwarzen ab. Meine jetzigen Wirthe sind gl.flls keine passionirten Jäger und die Schwarzen haben auch zu guten Appetit, um was sie erlegt, auf zu geben; und sie essen alles, was sie nicht ißt. — Sie erlauben dir indessen die Haut eines Thieres nd begnügen sich mit dem zurückgelassenen Fleische. Koche ich das Fleisch die Knochen leicht ab zu lösen, so verlangen sie nichts desto weniger das Fleisch, welches sie von den Knochen selbst abessen; da sie indessen den Werth der kleinen knochen nicht verstehen, gehen häufig mehrere davon verloren — Hierzu kommt, daß alles mir auf den Händen liegt. Ich hatte meine Pflanzen zu trocknen, mein Papier aus zu spreiten nd zus. zu bringen welches für 3–400 Pflanzen keine Kleinigkeit ist. Ich hatte in den Busch zu gehen nd Bäume zu fällen, Specimen zu sägen und nun, wenn die Schwarzen zufällig ein Thier brachten, dieses zu häuten, zu kochen nd das Scleton zu machen. Diese letztere litten fast immer nd ich schäme mich fast dir ausgestopfte Thiere nd die Knochen zu senden, wenn ich dir nicht zeigen wollte, wie gern ich deinem Wunsche nach kommen möchte. Sollte die Zukunft mir Mittel in die Hände geben einen Mann als Beistand nd Diener zu miethen, so sollst du den Unterschied bald wahrnehmen. — Das beständige Regenwetter in einer Gegend, welche zum Theil mit üppige Vegetation bedeckt ist konnte für die Gesundheit der Bewohner nicht ohne Nachtheil sein. So hatten wir mehrer Fälle von 3 tägigen Fieber und der Knabe, an welchem ich die Hydrocoele operirte[a] hatte eine Krankheit von ganz eigenthüml. Natur. Es ist nun 2 Monaten daß die Operation gemacht wurde nd während dieser Zeit war das Kind vollkommen gesund. Vor 14 Tagen hatte es Aphthen, welche es sehr quälten — vor 8 Tagen verschwanden diese; doch eine wasserhaltige Schwel' lung zeigte sich wiederum im Vorlaufe des Saamenstranges. An der Orte wo die Milz legt, auf der linken Seite .tr den Falschen Rippen zeigte sich [nämlich] eine harte Schwellung, der Puls war sehr schnell nd hart nd das Kind lag besinnungs' los mit blauen Lippen, offenem Mund im Schooße der Mutter. Man fürchtete jeden Augenblick seinen Tod; man rief mich, ich setzte seine Füsse in heißes Wasser, legte ein Senfpflaster auf die Gegend der Schwellung nd ließ das Kind zur Ader. Es erholte sich ein wenig; doch dauerte die Krankheit fort. Inner' lich wurden Castor öl gegeben. Anfälle derselben Art nur schwächer fanden jeden Tag statt nd vor 3 Tagen stellte sich Kälte vor den eigentl. Anfall dar. Dieß ließ mich die Krank' heit als tägliches Fieber .f fassen nd ich fing an das Kind .t Sulph. of Quinine[b] zu behandeln, welches schnell einen großen Wechsel der Erscheinungen hervorbrachte. Das Kind wird lebhaft, gewinnt Appetit, den ich indessen nur mit Reis un Reiswasser befriedige und die Anfälle werden schwächer. Die Schwellung selbst scheint abzu nehmen, während das

Wasser an die Hoden nicht zunimmt. Im allgemeinen ist dieses Land für Kinder sehr günstig, indem sein mildes Clima und die Trockenheit der Luft sie vor Erkältungen schützt. Besonders sind Viehstationen, wo die Kinder so viel Milch haben, wie sie nur trinken können wahre Treibhäuser; Schaafstationen sind un günstiger, und ich sehe hier, wie beschränkt man ist im Falle daß Kinder krank werden. Obwohl ich die Anlage von Gärten nd die Cultur der Gemüse auf den Stationen der Squatters, als Zeichen ihrer Industrie .t Vergnügen bewerte, so bin ich doch geneigt zu glauben, daß Stationen mit Fülle von Vegetabilien sich eines geringen Gesundheits zstds. erfreuen, als solche, auf. lchen die Bewohner fst .s schließlich auf animalische Nahrung nd auf Brod od Damper und Thee angewachsen sind. Wahrscheinl. liegt die Ursache nicht in der Gegenwart der Vegetabilien, sondern in dem Uebermaße ihres Gebrauches. — Man hat vielfach den uebermäßigen Genuß des Fleisches in dieser Colonie getadelt. Das wöchentl. Quantum für jeden Mann ist von 10–14 Pfd. (also fast 2 Pfd. per diem). Dieß ist .serordentlich viel. Doch diese Leute halten gewöhnl. mehrere Hunde, welche an den Rationen bedeutend Theil nehmen; nd meine Erfahrung zeigt, daß die Leute sich mit dieser Nahrung einer guten Gesundheit erfreuen, während ich .f Stationen .t viel Vegetabilien, wo indessen die Fleischrationen fast ebenso groß waren, häufiger Klagen über Unwohlsein gefunden habe. Ein auffallender Beispiel giebt die Lutherische Mission in der Nähe von Brisbane; die Leute essen vorherrschend vegetabilishe Nahrung und sie sind fast alle bloß siech und ungesund. Im Innere dagegen, auf den Downs pp. haben sie keine Gärten oder nur Kürbisse nd die Leute sind sonnig nd frisch. — Die Consumption des Thees in dser. Col. übertrifft die in England nd in den vereinigten Staaten bedeutend. Gewöhnlich trinkt jeder Mann im Innern 1 Quart Thee am Morgen 1 Quart zu Mittag, 1 Quart am Abend. — Die Ration für jeden Mann wöchentl. ist 1/4 Pfd. un brauner Zucher 2 Pfd. Auch gegen dieß Getränk hat man gepredigt un geschrieben. Meine persönl. Erfahrung hat mir gezeigt, daß Thee das gesundeste un erfrischendste Getränk ist, welches man in der Kolonie hat und ich würde jeden Mann, der sich viel bewegt un bedeutend .s dunstet rathen, sich an die 3 Quarts zu halten. Je wärmer es ist, je weniger verlangt man Zucker un ich habe oft, wenn der Schweiß aus allen Poren drang, Thee ohne Zucker dem Thee mit Zucker vorgezogen. — Je wärmer der Thee ist, je mehr man während des Trinkens schwitzt, je mehr kühlt un erfrischt er nachher. Milch stark mit Wasser verdünnt ist gleichfalls ein gutes Getränk, welches nur indessen man auf Viehstationen haben kann. Es ist für die Gesundheit der Leute im Innern von größere Wichtigkeit daß sie keine Spirituösen Getränke haben. Doch wenn sie zu den Städten un zu den öffentl. Häusern kommen geben sie sich dem Trunke gewöhnl. so lange hin, daß sie krank od. erschöpft zur Station zurückkehren, auf welcher vollige Enthaltsamkeit ihre Gesundheit wieder herstellt. In dieser Beziehung sind Herr un Knecht oft völlig gleich — und beide gleichen dem Matrosen, welcher nach einer langen See[fahrt] ans Land kommt und sich nun den wildesten Bachanalien fast besinnungslos hingiebt. Die gentlemen Squatters, welche man in Brisbane Jackooroos (wie die wilden Schwarzen) nennt begeben Streiche, wie sie nur immer die Tollheit deutscher Studenten oder auch wohl Englischer Oxfordmen erdachte. — Es versteht sich, daß diese nur die Ausnahme sind nd daß der größter Theil der Squatter im Becken des Brisbane wohlgebildete enthaltsame Männer sind, die jeder Gesellschaft Ehre machen würden. — Brisbane town, welches während der Wollschur der Sammelplatz der Squatter ist, indem sie hierher ihre Wolle bringen, um sie ein zu schiffen ist der [anerkte] Schauplatz nächtlicher Trunkenheit Hurerei nd Tollheit. — Auf ihren Stationen sind indessen diese wilden Jünglinge, in deren Adern sich ein elektrisches Blut .f zu häufen scheint, welches sich bei Berührung mit der Gesellschaft entladet, zuvorkommend, gastlich nd gesetzt. Da die Stationen oft 20 nd mehrere Meilen von einander entfernt sind, so findet selten nur selten Berührung zwischen ihnen statt. Diese ist durchaus nicht so freundschaftl. wie man in solche Zurückgezogenheit erwarten sollte. Die Grenzen ihrer Runs und go betweens sind oft die Ursache von Streitigkeiten, in welchen der einsamlebende Mensch weniger willig nach zu geben scheint, als der gesellschaftliche, dessen Aufmerksamkeit sich leichter von selbstsüchtigen Interessen ablenken laßt. Wie überall, so sind die Leute auch hier geneigt über ihre Nachbarn Beurtgn. zu machen und gewöhnlich die schwachen Seiten am schnellsten zu entdecken; sie geben sich Spott namen, welche oft lächerlich genug sind. So wurde einer meiner Wirthe, ein Superintendent, der 100£ per annum erhielt: cheap and nasty genannt. Die Wohnung eines andern heiß[t] Castle of Misery pp. — Mit unter besuchen sie sich um eine Partei Whist zu spielen, besonders wenn ein Fäßchen Rum od Wein angekommen ist, welches indessen die Knechte während der langen Reise gewöhnlich anstechen nd verdünnen als wohl ohne Umstände .s trinken. — Bei solchen Gelegenheiten sind Schaafe un Rinde nd Pferde die ausschließlichen Gegenstde. der Unterhaltung, nd als man während des niedrigen Preißes der Wolle ein Vorschlag brachte die Schaafe zu schlachten nd das Fett aus zu kochen hattest du von den äussersten Nordl. Station in Wide bay zu Marks Station in Pt Fairy (einen Weg von 1200 Meilen) reiten können, ohne ein anderes Wort als the boiling in^c of sheep zu hören. Dieß ist ganz natürlich, da ja Schaaf un Rinder, besonders die erstene die fast einzige Quelle unabhängiger Reichthümer der Colonie sind. A half penny im Preise eines Pfd. Wolle ist kein Kleinigkeit nd the boiling schien mir wahres Rettungs werk, als jeder sich zischen mußte daß der gegenwärtige Preiß der Wolle nicht hinreiche die Schaafstationen mit den nothwendigen Schäfern zu erhalten. Ob unser Mark zu den Jackooroos gehört? Ich glaube nicht, denn eines Theils ist er verständig nd dann schreibt er mir, daß er weibliche Gesellschaft (eine Anverwandte mit 2 Töchtern[)] bei sich zum Besuche erwartet. Weibliche Gesellschaft ist überall die Retterin guter Sitten! Ich hoffe nun bald zu hören, daß du dich verheirathet hast; denn wie kann ein junger Arzt

Moreton Bay

Zutritt zu Familen erhalten ohne verheirathet zu sein? Du mußt entweder heirathen oder zu mir kommen, in welchem Falle du gleichfalls ein Weiblein mit bringen möchtest. — Ich selbst fühlte gar stark, was die einfaltigen Bibelheiligen Sündhaftigkt. des Fleisches nennen. Mein Moralisches Gefühl, beständiges Verliebt sein und Mangel an Versuchung hat den maiden man in mir erhalten; doch wie ich beständig einen Gegenstand hatte, an welchen sich meine Neigungen reichten, hatte ich auch beständig ein drängendes Verlangen, diesen Gegenstand zu gewinnen. Meine Wissenschaft war indessen ein wenig stärker als meiner fleischlichen Treibe nd Ehrgeiz half vielleicht meiner Wissenschaft; oft schwebte der Weg .f beiden Seiten .f nd nieder nd ich war einen ['']Faulen Streiche[r]' (wie mein Vater zu sagen pflegte) sehr nahe. Dieses Schwanken der Wege habe ich vielleicht niemals so lebhaft gefühlt als in dieser Kolonie, wo mehrere Male die Aussicht auf gesell- schaftl. Unabhängigkt. nd die Möglichkt. ein Weib zu ernähren, recht lebhaft vor mir trat. — Ich fühlte, was ich in meinem Junggesellen stande verliere, es scheint mir, daß ein unverheiratheter Mann kaum dchaus innerlich tugendhaft sein kann, indem sein Auge beständig seine Begierde nährt. Doch soll ich mich durch ein Weib an eine Stelle binden nd die Ausdehnung meiner Studien beschränken? [']L'homme n'est pas digne de la gloire, qui ne peut fouler les plaisirs sous ses pieds' Und so versucht der Ehrgeiz mich zu beruhigen, wenn meine Träume mir das Glück ein Weib zu besitzen vormalen: The treasures of the deep are not so precious, as are the con- cealed comforts of a man locked up in woman's love. I scent the air of blessings, when I come but near the house. What a delicious breath marriage sends forth—the violet bed's not sweeter— Sage mir William, bdauerst du wirklich, daß du dich nicht ausschließlich deiner Wissenschaft zu wendetest nd daß unser gemeinschaftliches Studium uns weit über den Schulkreis der Medecin hinausgeführt? Bedauerst du, daß deiner Geist, wie er frisch nd jung nach allseitiger Bereicherung strebte, mit Wissenschaften nd Künsten schmücktest, anstatt ihn dem einseitigen Nadelkopf malher [sic] gleich vielleicht für die Ausarbeitung einer brillianten medicinischen Dissertation zu suchen? Solltest du jetzt wirklich glauben, daß du deine Jugendzeit nicht wohl angewendet, oder wohl zum Theil nutzlos vergeudet so verspreche ich dir, daß nicht 10 Jahre ins Land gehen sollen, ehe du diesen deinen Glauben vollkommen änderst. Beginne deine Praktis und du wirst bald mit Vergnügen bemerken, daß deine Erfahrung schnell unter dem Einflusse früher gewonnener allgemeinen Ansichten groß wachßt, daß jedes Factum sich schnell nd oft befriedigend zur allgemeine Ideen erweitert und daß die Hülfsquellen in dir selbst für deine Wissenschaft größer waren, als du selbst glaubtest. Beruhigt trittest du, in den Kreis der Gesellschaft, welcher sich täglich erweitert, wie du mit ihm bekannter wirst und nun gewinnen deine accomplishments freier Spiel, welche für den Arzt oft mehr thun, als die nakte Wissenschaft! Nun hast du nicht umsonst der Musik einige Aufmerksamkt. geschenkt, nicht umsonst dich mit Malerei beschäftigt; deine Reisen geben dir Stoffe zu angenehmer Unterhaltung und selbst auf die kalte Schnee fahrt über am Fuße des St Gotthard reagirt Herz nd Geist mit behaglicher Wärme. Wie sich .tr dem schweren Tänzer die Diele so beugt sich vor dem gewichtigen Geiste die Gesellschaft, laß ihn auch immer so anspruchslos in sie eintreten. Ist nur dem klugen Manne Geduld genug gegeben, die ersten schrägen nd dohnenshägenden Blicke ruhig zu ertragen, beginnt er nur mit leisem Tone — bald wird er ihn zu jede Stärke schwellen können und selbst seine Donnerstimme wird dann nicht beleidigen sondern erschrecken. Doch willst du dich auf einmal geltend machen, ohne daß selbst wenige dich kennen, so wirst du die Eitelkeit der Leute beleidigen und sie werden dich vermeiden, selbst wenn sie deine Ueberlegenheit fürchten. Sollte ich ein Symbol wählen, so würde es 'the wedge' sein. Es setzt sich zuerst in einem kleinen Raum ein, den es vorwärts dringend ohne Verletzung weitet. Du tadelst mein unbedingtes Reise leben. Lieber Junge, die einzige Sorge, die auch dabei plagt, ist, daß ich nicht Geld erwerbe, meine

Moreton Bay

große Schuld zu bezahlen! Könnte ich in diesem trivialen Leben etwas besseres thun, als mich dem Studium der Natur nd der Menschen hin zu geben? Ich kann die Armen nicht.tr stützen denn ich selbst bin arm; ich mache mich.t meinen medicinischen Kenntnissen im Busche nützlicher als in der Stadt;[d] denn dort fehlt es an Arzten, während es hier zu viele giebt. Ich suche zu belehren wo ich willige Schüler finde nd belehre auch selbst, wo ich mit klugen Männern zus. treffe. Zugestdn. indessen daß ich an endliche Fixing un Unabhängigkt. denken muß! Geduld — ein Mann mit ['] sound colonial experience' als ich jetzt besitze, [']will never go to the dogs'. Behalte mich lieb nd ich fühle mich.tr der [Sternessitzung?] daß du es thust glücklich. Du wirst wahrscheinl. fr die nächsten 3 Monaten kn. Brf. erhalten. — Denn aber hoffe ich wieder in Sydney zu sein.

Vale vive memento esto mei
L. L.

[*Archer's station,*] *6th February*[, *1844*]

My dear William,

It's raining steadily, so I'm confined not merely to the Moreton Bay district, which I'd now gladly leave, but the very house itself, where I've been whiling the dull hours away in reading and meditation. I've made a collection of the various timbers to be found here, though I don't regard it as complete, for I found several trees in the scrubs by the river and near the sea of which I've as yet no specimens. You can perhaps form some idea of what hard work it is to get such specimens. The character of the bark is often so indeterminate, particularly amongst closely allied species, that you have often to fell a tree to make sure that it differs from another. If you've a black with you he can climb up and throw branches down [to you]; but the blacks don't often want to go out with me on my strenuous excursions, and it's very hard to get them to do much climbing. Yet they've wonderful ability in climbing. If they see an opossum or a beehive right at the top of a tree 180' high, they will at once try to get it; but, because they don't understand why I want branches and fruit, they hold back obstinately as long as they can before they'll consent to climb *a tree [for me].—*You asked me to send you skeletons, but it's not so easy to get them as you might think. I'm a poor shot myself, so I have to depend on the good will of my hosts and the blacks. My present hosts, like me, care very little for shooting; the blacks have too good an appetite to [be willing to] part with anything they kill; and besides, they eat everything that doesn't eat them. They don't mind letting you have the skin of an animal if you let them have the carcase. If I happen to be boiling flesh so that I can separate the bones easily, they still want me to let them pick the bones themselves; and, as they don't understand the importance of the smaller bones, a lot of them get lost.—This means that I have to do everything myself. I've had to dry my plants, to spread my paper out and then to assemble it, which is no small matter with 3 or 4 hundred plants. I've had to go into the bush to fell trees and saw specimens [of timber], and then, if the blacks happened to bring me an animal, I've had to skin it, boil it, and clean the skeleton. The latter nearly always comes to grief and I'm almost ashamed of sending you stuffed animals and skeletons, were I not anxious to show you how much I want to do what you wish. If I can ever afford to employ an assistant you'll soon notice the difference.—Continuous wet weather in country partly covered with luxuriant vegetation cannot but have a bad effect on the health of the inhabitants. In fact, we've had several cases of tertian fever; and the boy on whom I operated for hydrocoele[a] has had an illness of quite a peculiar kind. It's now two months since the operation, and during that time the child has been in the best of health. A fortnight ago it had milk thrush, which caused it great distress, but that subsided 8 days ago. A watery swelling then appeared again at the origin of the spermatic cord. A hard swelling appeared on the left side, under the floating ribs in the position of the spleen; the pulse was hammering quickly, and the child lay unconscious in its mother's lap with blue lips and its mouth wide open. They were afraid that it might die at any moment, so they summoned me. I put its feet into hot water, applied a mustard plaster to the area of the swelling, and bled the child. It recovered slightly, but the disorder persisted. Castor oil was given internally. Attacks like this, though slighter, occurred every day, and 3 days ago the child became cold before each attack. I was thus able to recognise the complaint as quotidian fever, and began to treat the child with sulphate of quinine,[b]

Moreton Bay

which quickly had a beneficial effect. He's now becoming lively and is regaining his appetite, though I allow him nothing but rice and rice water, and the attacks are getting milder. The swelling itself seems to be going, and the water on the testes is not increasing. In general, this country is a good one for children, as its temperate climate and dry atmosphere protects them from catching cold. Cattle stations in particular, where they can drink all the milk they want, are just like forcing houses. Sheep stations are not so favourable, and I have seen here just how little people can do when children get sick. Although I admire the industry of squatters who lay out gardens and grow vegetables on their stations, I'm inclined to think that the people are not so healthy on stations where there are plenty of vegetables as on those where they live exclusively on a meat diet with bread or damper and tea. The difference, I should say, is due not just to eating vegetables but to eating them to excess. You often hear adverse comment on the colonial appetite for meat. Everybody gets from 10 to 14 lbs a week (which is nearly 2 lbs a day). This is an extraordinary amount; but most people keep several dogs, which account for a considerable part of the ration. From what I have seen I can say that the people on this diet are enjoying good health. I've heard more complaints about bad health on stations where there are plenty of vegetables but nearly as much meat is eaten as elsewhere. The Lutheran Mission station near Brisbane provides a striking illustration. The people there live mainly on vegetables and you've only to look at them to see that most of them are neither strong nor healthy. In the interior, on the other hand, as on the Downs etc., where they've no gardens or nothing but pumpkins, they're lively and radiant.—More tea is consumed in the colony than in England or the United States. In the interior everyone drinks a quart of tea in the morning, another at midday and another in the evening. The weekly ration is $\frac{1}{4}$ lb with 2 lbs of brown sugar for each person. The Scribes and the Pharisees have inveighed even against this beverage. From my personal experience I hold that tea is the healthiest and most refreshing drink we have in the colony, and I would advise anybody who is very active and who sweats freely to keep on with his 3 quarts. The warmer the weather the less one wants sugar, and often, when I've been sweating from every pore, I've preferred tea without sugar to tea with it.—The hotter the tea, and the more you sweat while drinking it, the cooler and fresher you feel afterwards. Thin milk and water is also a good drink, but you can get it only on cattle stations. Their lack of spirituous liquors is of greater importance for the health of people in the interior. Yet when they go to town and get to the public houses they give way to such long bouts of drinking that they return to their stations either ill, or spent with dissipation, until total abstinence restores them to health. In this respect there's often no difference between master and man—both act like sailors just back from a long voyage, and indulge in the wildest orgies, hardly knowing what they are doing. Gentleman squatters or jackeroos as they call them in Brisbane (like the wild blacks) play such pranks as it takes the madness of German students or English Oxfordmen to devise.—These of course are the exception, most of the squatters in the basin of the Brisbane being sober, well educated men who would be a credit to society anywhere.—Brisbane town, where the squatters meet at shearing time because they bring their wool there for shipment, is the recognised place for drinking, whoring and folly by night. These wild young men, in whose veins the blood seems to accumulate an electric charge that sends the sparks flying on contact with society, are courteous, hospitable and sober on their stations. As many of the stations are 20 miles and more away from others, it's seldom, very seldom, that they're in touch with each other. Nor is communication between them so friendly on the whole as might be expected from their seclusion. The boundaries of their 'runs' and 'go betweens' are often subject to dispute; and men who live alone seem to be more unyielding in matters of self-interest than those who live in community with their fellows. People here are just as inclined to judge their neighbours as they are anywhere, and very quickly discover their weaknesses. They give each other nicknames, some of which are quite amusing. One of my kind hosts, a superintendent who gets £100 a year, they dubbed 'cheap and nasty'. The dwelling of another they call Castle of Misery, etc.—They'll visit each other now and then for a game of whist, particularly if a keg of rum or a cask of wine has arrived. The liquor has usually been tapped and diluted by the men on its long journey, if they haven't drunk the lot without ceremony.—On such occasions sheep and cattle and horses are the only subjects of conversation; and, as a proposal for slaughtering the sheep and boiling them down

Moreton Bay

for their fat was made when the price of wool fell, you could have ridden from the northernmost station on Wide Bay right to Mark's station at Port Fairy (a distance of 1200 miles) without hearing of anything else but the boiling down of sheep.[c] *This is only natural, because sheep and cattle, particularly the former, are indeed almost the only sources of independent means in the colony. A halfpenny a pound on the price of wool is no trifle; and it seemed to me that the boiling down really saved the situation at a time when everybody was muttering to himself that the prevailing price was too low even to keep the stations with their indispensable shepherds going. Do I think Mark's one of the jackeroos? No. For one thing, he's sensible; and for another, he's written to me saying that he's expecting feminine company (a visit from a kinswoman with 2 daughters). Feminine society is the salvation of good manners wherever you are. I hope that it will not be long before I hear that you've got married: for how can a young doctor [expect to] be admitted into people's homes unless he's married? You must either marry or come out here to me, and if you come you might just as well bring a young wife with you.—As for myself, I used to be keenly aware of what simple-minded advocates of the Bible call 'the iniquity of the flesh'. Through my sense of what is moral, through being continually in love and but little exposed to temptation, I have remained in a state of 'masculine maidenhood', but, since my affections have constantly had an object of attachment, I have just as constantly felt the yearning to win her. My interest in science, however, has been just strong enough to prevail over the promptings of the flesh, and ambition may have given some help to science. My path has wavered from side to side and up and down, and I've come close to being 'a lazy rover' (as my father used to say). Nowhere, perhaps, have I felt so consciously uncertain of my direction as in this colony, where the prospect of social independence, and of the possibility of maintaining a wife, has been vividly before me several times.—I sensed what I was losing by remaining a bachelor; and it seems to me that an unmarried man can hardly be virtuous through and through, since his eyes are constantly adding fuel to his desires. Am I to tie myself to one place though, by taking a wife? Am I thereby to restrict the range of my studies? L'homme n'est pas digne de la gloire, qui ne peut fouler les plaisirs sous ses pieds. Thus does ambition seek to* console me, when my dreams bring intimation of the joy of possessing a wife. 'The treasures of the deep are not so precious, as are the concealed comforts of a man locked up in woman's love. I scent the air of blessings when I come but near the house. What a delicious breath marriage sends forth—the violet bed's not sweeter'—Tell me, William, are you really sorry that you've not turned exclusively to your scientific interests; or that our work together took us far beyond what we could learn in the medical schools? Do you regret that you strove to extend your knowledge in all directions when your mind was young and active, instead of trying perhaps, like a limited painter of miniatures, to elaborate a brilliant medical thesis? If you've really come to think that you did not profit by those years; or worse, that you wasted your time, I'm sure that before 10 years have gone by you'll have completely changed your mind about it. Take up your practice, and you'll soon find with pleasure that your judgment, thanks to general conclusions to which you came long ago, will be maturing quickly; that every fact will rapidly assume its full importance, often in accord with your general ideas; and that your professional resources are greater than you thought they were. You'll take your place easily in a social circle that will expand every day, as you get better known in it; and your 'accomplishments', often more valuable to a physician than his bare qualifications, will enjoy freer play. You'll find now, that the time you spent on music and on painting has not gone for nothing. Your travels have given you something interesting to talk about, and you might even like telling people of that freezing tramp through the snow up to the foot of the St Gotthard. Just as a dance floor yields under a heavy dancer society yields to the man of moment, no matter how unassumingly he make his entry. Only a wise man has the patience to take the first wry, covertly defensive glances calmly. He speaks quietly at first, but is soon able to raise his voice, until, even if he shout, people may be startled but they'll not be offended. You've only to assert yourself once, before at least a few people know you, and their vanity is affronted. They will avoid you, even if they stand in awe of your superiority. If I wanted a symbol it would be the wedge, which widens the narrow space into which it is driven without doing damage. You take me to task for doing nothing but travel. My dear fellow, the only thing I'm worried about is that I'm not earning the money that would [enable me to] pay

Moreton Bay

my heavy debt [*to you*]. *Could I be doing anything better in this trivial existence that to devote myself to the study of Nature and of mankind? I can do nothing to help the poor, being poor myself. My medical knowledge makes me of more use in the bush, where there are no doctors, than in the towns,*[d] *where there are too many. I try to teach when I find willing pupils, and I take instruction myself when I happen to come across able men. Granted that I have to think about eventual 'fixing' and independence. Be patient—a man with 'sound colonial experience' such as I now have. 'will never go to the dogs'. Keep my memory green; and I feel, trusting my lucky star, that fortunately you do. You'll probably get no letters for the next 3 months. By that time I hope to be back in Sydney.*

Vive, vale, memento esto mei.

L. L.

99 [TO A DIGNITARY OF THE LUTHERAN CHURCH, possibly in Germany.

Archer's station, about 2nd March 1844]

Mein verehrter Herr

Da ich in der lutherischen Mission in Moretonbay für einige Zeit lebte nd ihr sechs Monate hindurch nahe blieb, wird es für Sie vielleicht von Wichtigkeit sein, das Zeugniß eines Augenzeugen in die Entscheidung [einer Frage] zu hören, in welche der Karakter nd Ruf eines redlichen Mannes so tief betheiligt ist. Ehe ich nach Moretonbay kam nd die lutherische Mission selbst sah glaubte ich, daß Hr. Schmidt nd Hr. Eipper die geistliche Vorstände der Mission waren nd daß sie die innere nd .ssre Thätigkeit der Anstalt regelten nd leisteten, während die ihnen verbundenen Handwerker als dienende Brüder die körperl. Arbtn. übernähmen um den beiden Geistlichen volle Zeit zur Ausführung ihres Bekehrungswerk zu lassen. Ich glaubte daß sie nur denn an der Belehrung der Schwarzen Theil nahmen, daß sie sie .t ihren respektiven Hdwerken bekt machten und daß sie ihnen in ihren friedlichen religiösen Lebens wandel ein nachahmungs würdiges Beispiel

vorhielten. Als ich zur Mission kam, fand ich die wechsel,
seitigen Verhältnisse ganz anders als ich erwartet hatte. Die
beiden Geistlichen, anstatt sich ungehindert ihren Missions
geschäften hin zu geben hatten vom frühen Morgen bis zum
Abend mit harter Körperarbeit um ihren Lebens .trhalt zu
ringen und sich aller Beschwerden eines gewöhnlichen Tage,
löhners zu .tr ziehen. Ueberdieß .tr richteten sie die Kinder, was
ihnen abwechselnd täglich 4 Stunden Zeit nahm. Wäre dieß
nur für die ersten 6 Monate nach der Ankunft der Fall gewesen,
so würde dieß nicht auffallen, da die erste Ansiedlung gewöhn,
lich mit großen Bedrängnissen verbunden ist; doch es war fast
6 Jahre nach der Ansiedlung. Da ich im Allgemeinen nd
besonders als Preussen an dem Gedeihen meiner Landsleute
lebhaften Antheil nahm, war die Mission sehr häufig Ge,
genstd. meiner Unterhaltung mit Englischen Ansiedlern, die
dieselbe länger als ich in der Nähe beobachtet hatten. Alle, die
der Mission wohl wollten bedauerten mit mir, daß die Geist,
lichen, von deren höherer Erziehung natürlicher Weise auch
höheres erwartet wurde, drch harte Körperarbeit in ihrem
Wirken so .ssr ordentlich beschränkt wurden. Sie vermißten
jenes systematische Zusammenwirken, der verschiedenen
Glieder der Mission, welches in einer Stellung, wie die Ihrige,
so nothwendig ist. Ich würde nach Allem was ich sah,
gemeint haben, daß Hr. Schmidt nd Hr. Eipper keine Gewalt,
ja keinen wirklichen Einfluß auf die übrigen Missions brüder
.s übten. Ich war deßhalb erstaunt zu hören, daß sich die
Brüder Ihnen über Zwang od Unter drückung beklagt hatten.
Hr. Schmidt zeigte mir die Regeln, welche die Committee in
Vorschlag gebracht hatte. Sie schienen mir so vernünftig, so
nothwendig sie waren so übereinstimmend mit den Grund,
sätzen der Englischen Gesellschaft, welche doch so eifersüchtig
über Gewalthaber wachst, daß mir Klagen über diese Regeln
nur auf Mißverstand zu [su]chen schienen. Sie wissen, daß
jede Versammlung von Engländern, welche den Zweck hat
gewisse Punkte zu breitschlagen, sogennante Chairmen od
Praesidenten erwählt, um Ordnung zu erhalten. Meine

Moreton Bay

Preussischen Landsleute waren so eif ersüchtig auf Gleichheit,[a] daß sie selbst Bedenken [trugen] die beiden Geistlichen zu Chairmen zu wählen. Und warum wurde es wahrhaft frommen Männer[n] so schwer, die Vortheile einer höhern Erziehung willig an zu erkennen? Warum scheuten sie sich, die beiden Geistlichen selbst leichte freiwillige Dienste zu thun, aus Furcht, diese Dienste möchten als gzwungene Arbeit gedeutet werden? Eine solche Furcht deutet in jede Gemeinschaft besonders aber unter Missions brüdern auf tiefe radicale Fehler, die ich nicht kenne, auf die ich aber nach aussern Erscheinungen schließen muß. — Würde ich Hr. Schmidt loben, sage ich Ihnen, daß er meinen Idealen eines vollkommen Kristen näher kommt als vielleicht irgend ein anderer Mann, den ich auf meiner Lebensbahn .k lernte, so möchten Sie es wenn nicht für... dichte Schmeichelei, doch für die natürlich Täuschung... halten in einem Freunde das bste zu dnkn. Die Brüder erscheinen in ihrem Verkehr .t den umgebenden Ansiedlern aufrichtig biedere gottes fürchtige Männer nd sie werden überall geschätzt; das Einzige was ich zu tadeln finde ist, daß sie .tr sich selbst nicht so biederliche aufopfernde Gesinnungen hegen, wie man von solchen Männern in einer solchen Stellung, mit einem solchen Berufe erwarten möchte. — Die Mission ist gegenwärtig sich selbst überlassen, jeder schwimmt für sich selbst und sucht sich zu retten.... gewinnt der Egoismus der Menschen natur leichter die Oberhand über den stillen festen Glauben in die göttliche Vorsorge, welcher uns lehrt unserer Brüder auch in eigener Gefahr nicht zu vergessen. Vergeben Sie mein langes Schreiben. Ich hielt es für Pflicht das Meinige zu thun den Guten Ruf eines Freundes vor Ihnen und vor unsern gemeinschaftl. Vaterlde. zu vertheidigen und ich hoffe, daß Sie mein Bestreben Vorurtheils frei zu sprechen anerkennen werden.

Ich verbleibe in tiefster Hochachtung
Ihr gehorsamster Diener
L. L.

[*Archer's station, c. 2nd March, 1844*]

My dear Sir,

Considering that I spent some time at the Lutheran Mission station at Moreton Bay and remained in its neighbourhood for six months, you may find the testimony of an eye-witness of importance in settling a question that so deeply concerns the character and reputation of an honourable man. Before I came to Moreton Bay and actually saw the Lutheran Mission I thought that Messrs Schmidt and Eipper were its spiritual directors, and that they organised and carried out the internal and external functions of the institution, whilst the associated artisans were ministering brethren who took over the manual work, so as to leave the two clergymen free to devote themselves to the task of converting [the blacks]. Only after that [i.e. the manual work] was done, I thought, did they [the artisans] take part in the instruction of the blacks, [and I thought that they did so] by teaching the latter their respective handicrafts and by setting them an example of devout and peaceful conduct. When I arrived at the Mission I found the reciprocal relations to be quite different from what I expected. Instead of being entirely free to devote themselves to their missionary work, the two clergymen had to wrestle with hard bodily labour from early morning until evening to support themselves, and were submitting to all the hardships of common day labourers. Besides, this, they had to teach the children, which they did for 4 hours a day, taking it in turns, day about. Had this been only for the 6 months after their arrival it would not have been amiss, because the foundation [of such an enterprise] is seldom achieved without great difficulty; but it was nearly 6 years afterwards. Because in a general way, and particularly as a Prussian, I was very interested in the welfare of my countrymen, I used often to discuss the mission with English settlers who had been able to observe it closely for longer than I had. Everybody who regarded it with approval shared my regret that the clergymen, from whom something better was naturally expected on account of their better education, were so extraordinarily hampered in their work by [the] hard labour [they had to undertake]. They [the settlers] deplored the lack of that systematic coöperation between different members of the mission which is so necessary in a position like yours. I would have supposed, from all I saw, that

Moreton Bay

Messrs Schmidt and Eipper exerted no authority, and indeed had no real influence, over the other brethren of the mission. I was therefore astonished to hear that the brethren had complained to you about constraint and oppression. Hr. Schmidt showed me the rules which the Committee had proposed; and they seemed to me to be so reasonable, so necessary and in such accord with the principles of the English society (though the latter is growing so jealous of persons in authority), that I think we have to seek the reason for complaint against them in misapprehension. You are aware that whenever Englishmen meet formally to thrash out particular questions, they elect a so-called Chairman or President, to maintain order. My Prussian countrymen, however, were so concerned about equality[a] that they felt compunction even about electing [either of] the clergymen as chairman. And why was it so hard for truly devout men freely to acknowledge the advantages of better education? And why did the two clergymen themselves refrain from giving light, voluntary service? Was it for fear that it might be construed as obligatory service? In any community, but particularly amongst mission brethren, such a fear is a sign of profound, fundamental error, of what kind I do not know but outward appearances constrain me to conclude that it exists. If I may speak in Hr. Schmidt's favour, let me say that he comes nearer to my ideal of the thorough Christian than perhaps anybody with whom I have been better acquainted in my life, though you may regard this, if not as . . .flattery, then as the natural [self-]deception of a friend who wishes to think the best. In their dealings with surrounding settlers the brethren appear to be open, honourable, god-fearing men and everybody thinks highly of them. The only fault I can find is that, between themselves, they do not foster [quite] such generous and devoted convictions as one would expect of men of their calling in such a situation.—The mission has now been left to fend for itself, so that each one of them has had to strike out on his own and try to make sure of a living. . .the selfishness of human nature easily gets the upper hand over that calm, firm belief in Divine Providence which exhorts us to be mindful of our brethren even in our own danger. Please forgive me for writing at such length; but I felt obliged to do what I could to speak in defence of the good name of a friend, before you, and before that Fatherland which is yours and mine. I hope, Sir, that you will accept my assurance that I have tried to speak without prejudice.

I remain, with profound respect,
Your most obedient servant,
L. L.

100 [TO DR CHARLES NICHOLSON, Sydney, N.S.W.

Rosenthal, 26th March, 1844]

My dear Sir,

I received your kind letter in one of the most beautiful and interesting mountain landscapes, I ever saw on my travels in this colony. It was at the foot of Mt Mitchell on the Gap Road to the Darling Downs. I was on my way to Maitland, but the misfortune of losing my pack horse kept me for a week at the eastern side of the coast Range. Though I shall probably be before this letter in Sydney, I shall chance at least an opportunity to answer your letter and to thank you for your ⌜kind⌝ remembrance. This is certainly an interesting district and I should have made a rich harvest but for the incessant rains, for frequent accidents with my horses and for my limited means, which compelled me to be a kind of factotum not only in my scientific pursuits, but even in the necessities of life. After having made a collection of woods in the Bunya Bunya brushes and in the forests near Archers Station, I went down to Brisbane, visited the coal pits near Redbank and the limestone hills near Limestone, I camped at Flinders peak, visited the Mtains near Camerons Station, Mt Edwards, Mt Gravel, Mt Fraser, ascended the coast Range past the Gap, examined the brush which covers the western side of this remarkable locality and went on to the Darling downs, those rich plains bordered by basaltic ranges and covered with the most various herbs peculiar to the rich dark soil intermixed with concretions of carbonate of lime, which accompanies invariably the different varieties of the basaltic rock of this country. The geological relations of this part of the Moretonbay district are simple, but very interesting: they allow perhaps to form surer guesses on the

Moreton Bay

age of the sandstone and the coal of this continent, than the fossils, which we are ⌜not⌝ allowed to range according to the ages of the fossils found in the Northern Hemisphere. We must keep to the principles of the science formed in Europe but we must have a geology and palaeontology of our own. I am inclined to believe, that the igneous rocks correspond much more all over the globe according to their composition, than the sedimentary strata—and as the sandstone and the coal which is covered by it comes in contact with the oldest igneous formations, with those of the old tertiary and younger tertiary series—I have been carried to the conclusion, that those sedimentary formations correspond to the middle tertiary strata—to the miocene and older pliocene period of Lyell. I told you in my last letter that the Glasshouses are composed of a rock almost identick with that of the Puy de Dome and of the Sarconi near Clermont in Auvergne. The same rock forms the interesting Range of Flinders peak in equally bold and fantastic shapes, it is found in Camerons Range, in Mt Edwards in Mt Gravel ⌜in Mt Fraser⌝ and if I am not mistaken in Mt Mitchell and Cordeaux, for it appears on the gap at the foot of these mountains, though the lower ranges are composed of a different rock. Well: round many of these mountains the sandstone lies in horizontal layers, whereas the basaltic rock has broken through it, frequently lifted up the coal and spread even over the sandstone. The sandstone is consequently younger than the rock of the Glasshouses (Domite belonging to the Trachytic rocks) and older than the Basalt, which is considered as belonging to the latest tertiary period. I acknowledge that well founded objections exist even against such an conclusion; but it is certainly not so absurd as to say the deposits which might inclose at the present day Trigonias living near the coast of New Holland are of the same age as the strata of the Northern Hemisphere in which Trigonias have been found—or Trilobites found in New Holland belong to the same geological period to which those of Europe belong. Why—I should not be surprised if our fishermen should bring to light some Trilobites from one of the rocky nooks even in Pt Jackson—not to speak of the unexplored and unknown reefs of Torres Straights.—Was it not in the Indian seas and in the Southern Hemisphere that animals nearly allied to that group of fossil crustaceae have been discovered?

When I came to the Downs I had to change my opinion on the uninterrupted continuation of the basaltic rock from Liverpool plains to New England and from New England to Darling downs.—Connections might probably be found—but there exist large tracts of sandstone country, as in Liverpool plains and my wonder ceased of finding coal on the Downs.—This mineral seems to fill the whole basin of the Brisbane, for it appears here and there from the mouth of the river to the ⌈even⌉ foot of the coast Range, the creeks of which abound with coal pebbles and with those of the blue clayy rock which accompanies the coal.

The limestone Rocks near Limestone owe their origin to springs containing carbonate of lime and silica. Springs of this kind must have existed very generally and they exist even now. The black soil produced by the decomposition of basaltic rock is full of them and they are generally very pure. At the Gwydir and Namoy similar concretions are to be observed, but there they contain much clay and little carbonate of lime; the rock on which they are found there is a clayy rock which breaks into irregular pieces.—I am now looking out for the fossils of the Downs! Alas! I fear I shall little succeed. Alone, with a mare lost, my raiments almost all torn to pieces or worn off and scantily covered with a red shirt—I am now often even ashamed to meet the bushman; but I shall try at least and examine more minutely some of the localities where these interesting fossils have been found.—It is a consolation in all my troubles, that I am schooling myself for the expedition to Pt Essington. The colony and science has you to thank for it, if it takes place and they must aknowledge your good intentions even should it not take place. I hope to see you strifing one day for a little more liberal endowment of a national museum ⌈, though I should not wonder if you thought that the Young Pt Philippian Society was healthier and more mindful of the advantages of such institutions, than the Sydney Society seems to be[1].

You have given me the permission of writing and you see that I get talkative. You are so kind as to offer me your house in case of my returning to Sydney and I should certainly make use of your kindness, if Mr Lynd had not made his house my home. If I can trust his words, and I know him too well to do otherwise, my stay with him cheers rather his solitude and varies a little the dullness of his official occupations—And he is such an excellent man so full of experience and wisdom, that I do not know myself wether his disinterested friendship or the excellence of his caracter or the comminication of his knowledge attracts me more. Should you see Sir Thomas would you do me the favour of presenting my respects to him and to thank him for his second letter[a] in my name.—And

Believe me ever to be my dear Sir
Yours faithfully
Ludwig Leichhardt.

101 [TO LT. R. LYND, Barrack Master, Sydney, N.S.W.]

Rosenthal, 27th March, 1844

[My dear friend,]

...... I visited an extremely interesting mountainous country and beheld some of the finest scenery which this colony possesses. Flinders' Peak Range and several other large mountains, either isolated and forming bold masses of naked rock or in high ranges covered with brush, with some immense precipices, are composed of the rock of the Glasshouses (Domite or earthy Trachyte). Sandstone extends in horizontal layers round their foot; coal outcrops here and there from the mouth of the River to the very foot of the Coast Range. Fossil wood is found in the sandstone in very considerable masses. The brush bears a different character from that of Archer's. It is not so dense, does not contain so many vines, and its principal constituent is the Rosewood Acacia, the wood of which has a very agreeable violet scent like the Myal Acacia (a. pendula) in Liverpool Plains. The ground is dry, but several Malvaceous half-shrubby plants which grow to four or five feet high, make a passage through the brush very difficult. The most interesting tree of this Rosewood brush is the true Bottle Tree—a strange looking unseemly tree, which swells slightly four to five high, and then tapers rapidly into a small diameter; the foliage is thin, the crown scanty and irregular, the leaves lanceolate, of a greyish green; the height of the whole tree is about forty five feet. Several vines of the Asclepiadaceae are in blossom; and some trees which I did not meet before, paid my fatigue.

The finest mountain country I have seen in this colony is the eastern side of the Gap,[a] through which the road passes from Brisbane to the southern part of the Downs. This Gap intervenes between the high mountains—Mount Mitchell and Mount Cordeaux. Sunny ranges, covered with fine grass and open forest, ascend pretty rapidly to the Pass. The coast range forms an amphitheatre of dark steep mountains; a waterfall rushes over a precipice 300 feet high into a rocky valley, which one might take for the crater of an extinct volcano, if the surrounding rocks warranted such a supposition. Bold isolated mountains appear in the distance, in their various tints of blue, and during sunset dimming through a purple mist. Both sides of the mountains have some brushes, particularly the western side, in which many of the trees of the Bunya brushes reappeared. This is the most western point in which the Araucaria Cunninghami has been found. The Seaforthia palm is frequent and high. Both trees are remarkable for the latitude in their conditions of life, as they do not only grow in the lower mountain brushes, and in those which accompany rivers and creeks, but grow equally in the brushes along the sea side. It is, however, observed by carpenters, and men who work the wood, that the mountain pine is by far preferable to the river pine, the grain being much closer.

Moreton Bay

How the eye is pleased at entering again into the open plains of the Downs! Nothing is so agreeable as to see one's way clearly before him. Ranges of middling height now a chain of cones, now flat-topped mountains, covered with brush, now long-backed hills sharply cut at their ends—accompany on each side the plains, two to three miles broad, and many many miles long. The soil is black and yet mild, with many white concretions of carbonate of lime; the vegetation is quite different from that of the forest-ground of the other side of the coast range; the grasses are more various, but they do not cover almost exclusively the ground. They grow more sociably in small communities together, separated by succulent herbs, particularly compositae. The creeks are deeply cut, with steep banks covered with reeds.......

102 [TO LT R. LYND, Barrack Master, Sydney, N.S.W.

Rosenthal?] 13th April[, 1844.]
[My dear friend,]
...... I have returned from my round, and I have been tolerably successful. Isaacs has some beautiful specimens of the lower jaw bone of a gigantic kangaroo, the size of a bullock. Three smaller species must have lived at the same time. The locality is very interesting. I think that the aspect of the country has little changed since these giants disappeared, as the fresh water shells, which live at present, are imbedded with those bones in great numbers. I think it not at all improbable to find the animal still living farther inland, and more into the tropics. An animal of such size, and herbivorous, required much water; the change of climate, which made former lakes dry up, must have destroyed the conditions of life, and the animal either died or retired to more favourable localities. Large plains extend along the River Condamine, and I crossed one of them twenty-five miles broad, and fifty miles long, a true savannah, in the centre of which I saw the sharp line of the horizon, as if it had been on the ocean. Misfortune again! I lost my faithful pointer-

bitch, my cheerful companion, on this plain—knocked up by want of water!

It is probable, my dear friend, that I shall not stay long in Sydney when I come down. I have found young men willing and able to undergo the fatigues of a private expedition,[a] and if I can muster sufficient resources to pay the expenses of provisions for six men, I shall immediately set out for Port Essington. ... ... I know that if I start with these men, whom I know to be excellent bushmen, excellent shots, and without fear, I am sure to succeed. Every one of us has the necessary horses, and all that is required besides, would be six mules with harness for carriage of flour—100 pounds per head—tea and sugar and ammunition. Every one of us has lived weeks and weeks together in the bush, frequently surrounded by hostile blacks, whose character we know, and intercourse with whom we shall always try to avoid. Believe me that one experienced and courageous bushman is worth more than the eight soldiers Sir Thomas intends to take with him. They will be an immense burthen, and of no use.

I start next Monday from the Downs, and I shall proceed as quickly as I can. In three weeks I shall have the pleasure of seeing you again, and I hope in good health. I am afraid my body is so accustomed to the moving life, that close studies will not agree with it. If my friend[b] send me the mountain barometer, I shall be a made man. ... ...

103 [TO WALKER SCOTT, ESQ., Cumberland Place, Sydney, N.S.W.]

Newcastle the 10th Mai 1844

My dear Sir

Here I am again under your hospitable roof after a roaming life of 18 months. I have seen the Moreton Bay district and Wide bay, I made a round over Darling Downs ⟨where I met with some beautifully preserved fossil bones⟩, and returned through New England ⟨and descended the Gloucester Road to Pt Stephens. And so far from being tired of the changing

Moreton Bay

life of a roving bushman, I am full of projects to start again.⟩ The impressions which those different parts of the colony have made are decidedly favourable and I feel sure, that the Moreton bay district will very soon eclipse the Hunters River, if the proprietors at its banks turn not their thoughts seriously to the establishment of vineyards. For such a purpose Moretonbay would be to wet, as heavy rains set generally in when the grape enters into maturity. The district is also not adapted for sheep farming, though there are some excellent runs at the higher and more mountainous parts of it, but cattle will thrive and fatten not only in Moretonbay but also at Wide Bay, which Mr Eales was going to leave with his sheep after having made a sad experience of a year. The Darling downs with their extensive plains, their low ridges, their open forests, their rich variety of grasses and herbs in an elevation of perhaps 1800' above the level of the sea seem formed by nature expressedly for the purpose of nourishing large numerous flocks of sheep.—The people of the Downs commence to make themselves comfortable, to many and to enjoy themselves. They want sadly young ladies who will brave the solitude of the bush: but I hear that several very respectable immigrations are expected, for instance the Mis Gores are expected by some joyful bachelors.

At New England I visited the falls of the Apsley, which must be a magnificent sight when the river is full of water: the chasm is about 300 feet high. Fine scenery is very rare in this colony: the only one, which I met with of the first order is at the eastern side of Cunninghams Gap, through which the road passes from Moretonbay to Darling downs.—The mountain brushes of the Bunya Bunya district are very extensive; they cover perhaps an area of 500 square miles: the Bunya pine is one of their finest ornaments and I never beheld a more majestic tree.

⟨Some young men have joined to make an expedition to the Gulf of Carpentaria, provided they find support enough to pay comparatively trifling expenses. I do not yet know, whether I shall be able to collect some money but I have still some.⟩[a] It is probable that I shall start soon on a private expedition to the Gulf of Carpentaria; this is however to be decided when I come to Sydney, which will be in about a weeks time. You see my dear Sir that it would never have been safe to have anything to do with me in point of business. It is with me as it is with a money making man, the more he obtains the more exorbitant his desire becomes. I cannot however deny that the present comforts after the late fatigues of travelling are very seducing and I am sure that the time will come, when I shall gladly sit down, not to start again, but to ruminate and to digest that which I have seen and learnt.

I have to beg your pardon, that I left my things so long time here in your house. I never thought, that my journey to Glendon and to Bengalla would extend to a line of short of 2500 miles.

When I came to Hexham pond I gave an order on you of 18 pence to the man, as I had no money to pay. I shall pay this to Mr Wallace and I mention it only, in case this order should come into your hand.

May I ask the favour of mentioning my name to Dr and Mrs Mitchell who have had plenty of time to forget it and to Mr Robert Scott and to make my compliments to Mr Kirchner and his family.

I hope to meet you in good health in Sydney; Monday or Tuesday I am going up to Glendon to put my things into boxes that they may be sent down.

G[ood by]e my dear Sir and believe me ever to be

Yours faithfully
Ludwig Leichhardt

104 [TO CARL SCHMALFUSS, SR.]

Newcastle am Huntersflusse den 14 Mai 1844

Mein theuerster Schwager

Ueber 600 Englische Meilen hat mich meine getreue Stute wieder zurück getragen und ich habe auf diesem Wege einige der merkwürdigsten Theile der Kolonie besucht. Von Brisbane

Moreton Bay

town in Moreton Bay besuchte ich einige interessante Punkte, in welchen Kohle zu Tage steht. Sie wird von Sandstein bedeckt und es scheint, daß das ganze Gebiet sich über Kohlen lager ausbreitet. Diese Kohle, nd der Sandstein werden indessen von einigen recht merkwürdigen Bergen nd Bergzügen durchbrochen, welche ich gleichfalls besuchte und welche dasselbe Gestein zeigten, welches ich schon in den Glasshouses fand und welches deutsche Geologen Domite nannten, da es gewöhnlich Berge in Domform bildet. Dichte Gebüsche bedecken den Fuß einiger dieser Ketten. Diese Gebüsche sind indessen von jenen Berggebüschen verschieden, welche ich im Bunya Bunya distrikte untersuchte. Man nennt sie Rosewood scrubs, da vorzüglich eine Acacia in ihnen wächst, deren Holz einen Rosen od. Veilchen geruch hat. Es giebt in den Ebenen gegen Westen eine andere Acacia, welche die Ansiedler Myall nennen (A. Pendula) mit hängenden Zweigen, wie die der Trauerweide; das Holz derselben hat auch den lieblichsten Veilchengeruch, welcher sich oft schon bemerkbar macht, wenn man sich den Myal gebüschen nähert. Ich fand indessen einen andern höchst merkwürdigen Baum sonderling hier. Er wächst ungefähr 45' hoch, ist am Grunde 5/4' dick, schwillt 3' Fuß höher zu 6/4-7/4' an und schießt dann in eine lange Spitze aus. Sein Laub ist sehr dünn, seine Krone unansehnlich. Wie man mit dem Beile in ihn einhaut, ist der Stamm so weich nd schwammig, daß das Beil leicht nd tief eindringt. Das junge Holz ist saftig und wie das Innere eines Kohlstrunks. Man kann es essen nd es ist für den *hungrigen* Magen eine vollkommene Speise. Nach der Beschreibung siehst du, daß die Gestalt des Baumes einer französischen Flasche etwas ähnelt, obwohl der Flaschenhals .sserordentlich in die Höhe gezogen ist. Die Ansiedler haben diesen Baum deßhalb *bottle tree* oder Flaschenbaum genannt. Die Blätter sind bleichgrün, langlanzettlich: ich habe weder Blumen noch Früchte finden können, obwohl ich eigenhändig einen Baum 5' umfang fällte. Aus dem Gebiete des Brisbane stieg ich durch einen hohen Berg pass zu den Darling downs, einem Hochlande 1600′ über dem Meere, auf. Dieser Pass wird 'Cunninghams gap' genannt, da der Botaniker Cunningham der Erste war, der ihn auffand. — Die Aussicht von diesem Passe gegen Osten war .sserordentlich reich nd malerisch. Ich habe in Europa nichts ähnliches gesehen, da die höchst merkwürdigen isolirten Berg gestalten, wie die Sonne hinter der Hauptkette unterging von dem lieblichsten Purpurnebel umflossen wurden, während sich die fernern Bergzüge in schwächern nd schwächern Blauen verloren und die steilen Abfälle der Hauptkette im dunkelsten Schatten wie ein riesenhaftes Amphitheater vor mir lagten. Eine Kaskade stürzte sich 300′ tief in eine Bergschlucht. —

Während der Nacht hatte ich das Unglück meine Stute mit dem Fullen zu verlieren.[a] Sie diente mir als Lastpferd und trug mein Papier und die Felsstücke welche ich sammelte. Acht Tage lang suchte ich nach ihr; doch vergeblich. Ich hatte deßhalb alle meine Sachen auf meine andere Stute zu packen ud zu Fuße weiter zu wandern. — Ich besuchte einen Landsmann Hrn. Friedrich Bracker aus Mecklenburg, welcher die Aufsicht über die Schaafheerden eines Hrn. ⌜Bolton⌝ hat nd machte seine Wohnung zum Mittelpunkt meiner Wanderungen. Die Darling downs sind weite mit Graß bedeckte baumlose Ebenen, von niedrigen Bergzügen begleitet, welche von einem sehr offenen Walde bedeckt werden. Diese Züge un die Ebenen nd das Klima sind für die Schaafzucht, wie auch für Viehzucht ausserordentlich günstig nd Brackers Schaafheerden sind wahre Muster. Seine Schaafe sind Sächsische Merinos, während die Meisten Engländer diese mit einer Englischen Schaafart, dem Leicester schaafe kreutzen. Er behauptet daß diese Kreutzung die Wolle verderbe, obwohl der Körper der Schaafe viel größer nd schwerer wird, da das Leicesterschaaf eine sehr große Art ist. Um dir eine Idee von der Ausdehnung, nd der ausserordentlich schnellen Ausdehnung der Schaafzucht über die Downs un Moretonbay zu geben, führe ich nach der Aussage glaubwürdiger Leute nur an, daß die Ausfuhr in diesm Jahre von Moreton bay schon 50000£

Moreton Bay

(35000 Thl) betrug, obwohl Moretonbay nur 2 Jahre un the Downs 4-5 Jahre beweidet werden. Man hat ungefähr 4000 Stück Rindvieh nd die Pferde vermehren sich gleichfalls schnell. — Ich ritt an 90 Meilen Nordwestl. über die downs. Ausserordentl. weite Ebenen (25 Meilen breit, 50 Meilen (Engl.) lang) breiten sich am nördlichen Ufer des Condamine aus. Dieser Fluß scheint dem großen Flußsystem an zu gehören, welches sich nach Südwest nach South Australia in den See Alexandrina mündet.[b] — In den Ufern einiger Bäche fand ich die fossilen Knochen riesenhafter Thiere, welche nach dem Plane der Kangooroos gebildet scheinen. Die untere Kinnbacke ist besonders häufig; sie enthält zwei sehr lange horizontal liegende Vorderzähne nd vier sehr große Backzähne, jeder mit zwei Querleisten und mit einer Art Absatz. Keines der großen Thiere der bekannten Erdtheile stimmt mit dieser Zahnbildung besonders der Vorderzähne überein; doch viele der Einheimischen Thiere haben eine ähnliche Bildung.[c]— Ich fand die Zähne von 3 andern wahren Kangooroos und diese Knochen waren von Muscheln begleitet, welche noch jetzt in den Wasserhöhlen leben. Die lebensbedingungen dieser Thiere können also wenig von den gegenwärtig bestehenden verschieden sein. Doch solche riesenhafte Graßfresser mußten viel Wasser erfordern nd wahrscheinlich war das Austrocknen von Seen un Sümpfen die Ursache ihres Verschwindens. Wahrschl. leben ähnliche Thiere noch jetzt in dem wasserreichern Innern zwischen den Wendekreisen Australiens. Alle fossilen Knochen welche ich fand waren nach dem Australischen Thiertypus gebildet nd ich zweifle ob die Elephanten Knochen, welche man vor giebt in Australien gefunden zu haben, wirklich diesem Erdtheile angehören.

Von den Downs ritt ich südlich nach Neu England, welches noch höher liegt, als die downs. Man nimmt an, daß der höchste Theil von Neu England 3000' über dem Meere erhoben ist. Auf diesem Hochlande entspringen die Wasser, welche gegen Westen zum Darling un zum See Alexandrina fließen, während die Östlichen Bäche sich fast alle in hohen

Fällen zu dem Küstenlande gegen Osten hinabstürzen. Ich besuchte die Fälle des Apsley, welche zwischen 3-500' hoch sind. Einige der Flüsse, haben einen langen Kanal in das Schiefergebirge eingefressen, indem das eigentliche Küstenland noch an fast 20 Meilen von dem Hauptfalle entfernt ist. Das ganze New England ist mit Schaaf u Viehheerden bedeckt. 500000 Schaafe weiden auf diesem Hochlande und doch ist es vielleicht kaum 10 Jahre alt, d. h. vor 10 Jahren war es noch eine unbekannte Wildniß, durch deren Wälder nur der Schwärze schweifte. Die Stämme der Schwarzen sind jetzt fast ganz geschmolzen, wenigstens ist ihre Unabhängigkeits gefühl gebrochen nd sie begnügen sich mit den Brocken welche vom [*sic*] dem Tische des Weißen Mannes fallen. Und so wird es überall sein, wo Europaeische Civilisation auf einmal un ohne Vorbereitung mit dem Wilden in Berührung kommt — nd so war es überall! Oft habe ich, als ich mich unter mächtigen Stämmen befand, mit Schmerzen der nicht zu fernen Zeit gedacht, wo viele dieser kräftigen Gestalten von der Kugel des Weißen durchbohrt sein werden, viele sich unter bösartigen Krankheiten zum zeitigen Grabe schleppen und endlich der Rest, siech nd verkümmert um die Wohnungen der Weissen herum bettelt, oder vor den öffentlichen Schenkhäusern auf springender Städte nach berauschenden Getränken giert. — Ist es nicht möglich diese Schwarzen Naturkinder zu civilisiren oder auch nur mit der Civilisation zu befreunden, so bin ich immer zu großer Freund meiner eigenen Raçe, ein wohlbevölkertes wohlregirtes Land von Weissen einem zwecklos hinlebenden Haufen Schwarzer vor zu ziehn und wir müssen in der Oberhand der Caucasier dasselbe Naturgesetz anerkennen, vermöge dessen die Hinde dem stärksten Hirsche folgt.

Die Bewohner von New England leben schon recht behaglich, während die Ansiedler der Downs nd um Moreton Bay District noch oft mit der einfachsten Buschhütte zufrieden sind. Viele der jungen Männer beginnen sich zu verheirathen und das Weib bringt das Gefühl des Anstandes, des Streben nach häuslichen Behaglichkeit, friedliche Gesselligkeit der Nach-

Moreton Bay

barn, ruhiges Denken in die Köpfe vieler dieser Jünglinge, welche wenn sie aus dem Busche zu den Küstenstädten kommen, sich der Ausschweifung und dem Tränke ergeben, wie der Matrose wenn er nach einer langen Seereise in den Hafen kommt. Eine fast durchaus männliche Bevölkerung in einer Ausdehnung von vielleicht 600 Meilen ist eine höchst interessante, obwohl nicht sehr befriedigende Erscheinung. Man findet nirgends mehr, warum Gott Even aus Adams Rippe schnitzte. Die Arbeiter, Schäfer Viehaufseher sind größtentheils noch Leute, welche wegen Vergehungen un Verbrechen hierher gesandt wurden, doch jetzt ihre Freiheit erhalten haben. Wenige von ihnen sind verheirathet. Sie haben kein Weib zu ernähren, keine Kinder zu versorgen, keine Anverwandten zu bedenken. Sie leben nur sich selbst! und da sie weder ernsterer Gedanken über die Zukunft, oder höherer Gedanken über die Gegenwart fähig sind, geht ihr ganzes Streben auf unmittelbarer Genuß — ⌜wenn sie nur immer genießen können⌝. So bald sie zu einer Schenke kommen, wird ihr Herr, ihr Dienst, ja die Strafe welche der Pflicht‑ verletzung folgt vergessen; der Schenkwirth ist ihr Vater, die Wirthin ihre Mutter, sie sind die Einzigen, welche sie freund‑ lich nd mit offenen Armen empfangen. Hier bleiben sie denn, bis der letzte Heller ihres Lohnes vertrunken ist, was glück‑ licher Weise nicht eben lange dauert und dann kehren sie zu ihrem Dienste zurück, um wiederum ein Jahr zu arbeiten nd dann das alte Spiel zu erneuern.

Von Neu England ging ich nach Port Stephens, welcher der Agricultural Companie gehört, einer Gesellschaft welche unter der Bedingung den Ackerbau nd die Vieh un Schaafzucht zu befördern an fast 2 Millionen [Morgen] Land zum Besitz erhielt. Man wählte zuerst das Land um Pt Stephens um einen Hafen zu haben; doch da es sich wenig für Schaafzucht eignete, gewährte die Regierung der Gesellschaft 1 Million Morgen Landes im Innern am Peel mit vortrefflicher Waide. Pt Stephens würde sich indessen sehr wohl zum Weinbau eignen, wenn nur die Gesellschaft diesem Kulturzweige

Aufmerksamkeit schenken würde. — Es ist sehr gut, daß Ihr mich nicht leiblich saht, als ich nach solangem Buschleben in Newcastle einritt. Meine Hosen waren so zerissen, daß ein rothes Wollenhemde, was ich als Rock anhatte, kaum die Schwächen der Hosen bedeckte. Man hielt mich für einen Schäfer, für einen Bushranger (Rauber) ud so w. — Ihr könnt Euch vorstellen, wie behaglich ich mich fühlte, als ich wieder reine Kleider anzog. ⌈Ich muß nun schließen;⌉ ich bin gesund nd hoffe daß auch Ihr es seid. Lebt wohl nd vergeßt nicht

Euren herzlich Euch liebender Schwager u Bruder
[*No signature.*]

⌈P.S.S. [*bottom of last page:*] Grüßt Mütterchen u Alle die Andern.

[*top of first page:*] Ich erhielt heut deinen Brief vom 23–24 July 1843. Herzlichen Dank für die angenehmen Nachrichten!! Ich erhielt .en Brief von Hilgenfelds vom 9tn July. Ich werde antworten sobald ich nach Sydney komme!⌉

Newcastle on Hunter's river, 14th May, 1844.
My dearest brother-in-law,

My faithful mare has covered a distance of more than 600 English miles bringing me back again, and on the way I have visited some of the most remarkable parts of the colony. From Brisbane town in the Moreton Bay district I went to see some interesting localities where there are outcrops of coal. Sandstone covers the coal, and it seems that coal deposits underlie the whole district. Through both the coal and the sandstone, however, [a rock has] erupted [to form] mountains and chains of mountains of quite a remarkable kind, so I visited them too. They are composed of the same rock as I found at the Glasshouses. German geologists have called it 'domite' because it forms dome-shaped mountains. The skirts of some of these chains are covered with dense brush, but it differs from the mountain brushes that I investigated in the Bunya Bunya district. They call it 'Rosewood scrub', because an acacia whose wood smells like roses or violets grows so well in it. On the plains over in the western parts of the country there's another acacia which the settlers call 'myall' (A. pendula). It has hanging branches like the weeping willow, and its wood too gives off a delightful smell of violets which is often quite noticeable when you're approaching myall scrub. It was in the rosewood scrub that I found another tree, a most remarkable oddity. It attains a height of about 45', is 15" thick at the base but swells out to 18 or 20" a yard higher up, whence it shoots up into a long spike with an unattractive crown of thin foliage. The trunk is so soft and spongy that you can cut easily and deeply into it with an axe. The young wood is very juicy and resembles the inside of cabbage stalks. It's edible and provides a complete meal for a hungry stomach. From my description you can see that it's rather like a French [wine] bottle in shape, although the neck of the bottle is extraordinarily drawn out. The settlers, of course, have called it the bottle tree. The leaves are pale green, long and lanceolate. I've not been able to find flowers or fruit although I felled a tree 5' in circumference with my own hands. From the basin of the Brisbane I ascended through a high pass to the plateau of the Darling Downs, which lies 1600' above sea level. This pass is called Cunningham's Gap after its discoverer, the botanist Cunningham.—The view from this pass, looking eastward, was extraordinarily varied and picturesque. I've seen nothing like it in Europe; for there before me, as the sun was setting behind the main range, lay isolated mountains of most remarkable form, in a haze of a most beautiful purple tint, whilst the chains beyond them faded off into paler and fainter blues, and the slopes of the main range fell steeply down in deep shadow like a stupendous amphitheatre. And there was a waterfall plunging down for 300' into a gorge.

During the night I was unlucky enough to lose the mare[a] that had the foal. I had been using her as my pack horse, loading her with my paper and the rock specimens I had collected. I spent eight whole days searching for her, but in vain. There was nothing to be done but load everything on to my other mare and to proceed on foot.—I went to see one of our countrymen, a Mr Friedrich Bracker from Mecklenburg, the superintendent of a sheep station belonging to a Mr ⌈Bolton⌉, and made Bracker's place the centre of my movements. The Darling Downs are wide, open, grassy plains, relieved by chains of low hills many of which are lightly wooded. The hills and the plains and the climate suit both *sheep and cattle extraordinarily well, and Bracker's flocks are a real example [of what sheep should be]. He runs Saxon Merinos, though most of the English [squatters] cross them with an English breed, the Leicester[shire] sheep. He maintains that the crossing spoils the wool, though the sheep become bigger and heavier in body, the Leicesters being a very big breed. To give you an idea of the speed of the spread of sheep rearing on the Downs and in the Moreton Bay district just let me quote from the statements of trustworthy people. They say that the exports from Moreton Bay for this year already amount to £50,000's worth (35,000 Rtl.), although sheep and cattle have been grazing in the Moreton Bay district for no more than 2 years and on the Downs only for 4 or 5. There are about 4000 head of cattle, and the number of horses is fast increasing. I rode on, over the Downs, for about 90 miles to the North-west. Extraordinarily extensive plains (25 miles wide and 50 miles (Engl.) long) spread out from the northern bank of the Condamine. This river seems to belong to the great river system which drains to the sea through Lake Alexandrina in South Australia.*[b] *In the banks of some of the creeks I found the fossil bones of gigantic animals which seem to have resembled kangaroos in build. The lower jaw is often to be found. It carries two very long, horizontally-disposed front teeth and four very big molars, each of which has two transverse ledges and a kind of heel. None of the larger animals of [other] parts of the world exhibit corresponding dentition, especially as regards the front teeth; but many animals native to this country do.*[c]—*I found the teeth of 3 other true kangaroos and they were associated with [fossil] mussels of kinds still to be found in the waterholes. This means that the conditions under which these animals lived could not have differed much from those prevailing now. Yet such huge herbivores must have needed a lot of water, and the drying up of lakes and swamps is probably what led to their extinction. There may be similar animals still living in better watered parts of Australia within the tropics. All the fossil bones I found belonged to kinds of animals peculiar to Australia, and I doubt that the elephant bones that are alleged to have been found here can really have come from this part of the world.*

From the Downs I rode southward to New England, which lies even higher than the Downs. The highest part of New England is assumed to be more than 3000' above sea level. In these highlands waters which flow westward to the Darling and Lake Alexandrina have their source; and brooks arise which flow to the East, nearly all of which plunge over high falls to reach the lowlands along the East coast. I went to see the falls of the Apsley, which are 300 to 500' high. Some of the streams have eaten their way into the slaty rocks of the highlands to form long gorges, for they still have more than 20 miles to flow from the highest fall to reach the true coastal lowland. The whole of New England is covered with flocks of sheep and herds of cattle. There are 500,000 sheep grazing on this highland country, yet I'd say that it's hardly 10 years old. I mean that 10 years ago it was still an unknown wilderness with nobody but the roaming blacks in its forests. And already the little bands of blacks have almost completely faded away. At the very least their spirit of independence has been broken, and they accept the crumbs that fall from the white man's table. And that will happen wherever European civilisation makes sudden contact with savages unprepared for it. Everywhere it has been the same. Often, when I've been with vigorous tribes [of blacks], I've thought sadly of the day that will not be long in coming, when many of these robust bodies will be pierced by the white man's bullet, when others, stricken by virulent diseases, will drag themselves to an early grave, and when those who survive, sickly and languishing, will finally come to begging at the white man's door or to craving for strong drink at public houses in the rising towns.—If it is not possible to civilise these black children of Nature, or even to reconcile them to civilisation, I am far too firm a believer in the race to which I belong ever to prefer a swarm of wayward, aimless blacks to a populous, orderly white country. In the predominance of the Caucasians we must acknowledge [the working of] the same law of Nature which ordains that the hind shall follow the strongest hart.

The inhabitants of New England are already leading quite comfortable lives, whilst the settlers on the Darling Downs and in the Moreton Bay district are still content with the simplest of bush huts. Many of the young men are now getting married, and the presence of women is arousing a sense of decorum, a desire for household comfort, for friendly accord between neighbours and for calmness of mind in many of these high-spirited young men. On getting down to the coastal towns from the bush they've been giving way to dissipation and drink like

Leichhardt's Letters

sailors who've just reached port after a long voyage. An almost exclusively male population strung out over nearly 600 miles is a very interesting manifestation but it does not express a very satisfactory state of affairs. Nowhere else could you find better reason why God carved Eve out of Adam's rib. The workmen, shepherds and stockmen are mostly people who were transported here on account of their crimes and misdemeanours and have served their time here. Few of them are married. They have no wives to support, no children to provide for, no relations to consider. They live for themselves, and, as they're capable neither of thinking seriously about the future nor better about the present, all they care for is immediate enjoyment—whenever they can get it. As soon as they come to a public house their employer, their duties, even the penalty for breach of obligations, are all forgotten. The publican and his wife, the only people who really welcome them with open arms, are father and mother to them, and there they remain, until the last farthing of their wages has gone on drink. Fortunately, this doesn't take long, and back they go, to put in another year's work, just for the sake of playing the same old game once again.

From New England I went [down] to Port Stephens. This place belongs to the Agricultural Company, an organisation which was granted nearly 2 million acres of land for undertaking to promote its agricultural and pastoral use. They first selected country on Port Stephens, so as to have a port; but, because the land proved to be poorly suited for sheep-rearing, the Government granted the company 1 million acres farther inland on the Peel, where the pasture is excellent. Port Stephens, however, has good country for wine-growing, if only the company would show some interest in this branch of agriculture.—It's just as well that you didn't see me in the flesh when I rode into Newcastle after spending such a long time in the bush. My trousers were so badly torn that a red woollen shirt that I was wearing as a coat hardly concealed the damage. People thought that I was a shepherd, a bushranger, and so on.—You can just imagine how I enjoyed the comfort of changing into clean clothes again. ⌜*But I must end this letter.*⌝ *I'm well, and hope that you are too. Good-bye, and don't forget*

Your affectionate brother-in-law and brother,
[No signature.]

⌜*P.S.S.* [bottom of last page:] *Love to mother and all the others.*
[top of first page:] *I received your letter of 23-24 July 1843 to-day.
Many thanks for very acceptable news! And there was a letter from the Hilgenfelds dated the 9th of July. I shall reply as soon as I get to Sydney.*⌝

105 [TO AUGUSTE L. HILGENFELD.]
Newcastle, den 15ten Mai 1843 [*for* 1844].ª

Meine vielgeliebte Schwester!

Das Dampfschiff brachte mir heute alle Briefe, welche während 3 Monaten in Sydney für mich angekommen [waren]. Und o der Freude! ein Brief von dir⌜, von Hilgenfeld u Schmalfuß befand sich unter ihnen⌝. Eine so liebliche freund- liche Frauenstimme hat nicht zu mir geredet, seitdem ich mein Vaterland verlassen habe. Sieh', wie mich das Schicksal umhertreibt, und wie der ganze Erdball mir heimisch wird. Selbst als ich noch ein Knabe war, schien es mir immer, als sollte ich einst Flügel der Morgenröthe nehmen, um zu den äußersten Meeren zu fliegen. Und fragst mich nun, meine gute Schwester, wie ich mich selbst fühle, ob sich denn das un- gestüme Streben auch mit Frieden in den eigenen Brust paare — so antworte ich, daß, obwohl die Leidenschaften in dem Busen des Mannes heftiger arbeiten und schwieriger zu be- siegen sind, dennoch das Herz sich noch rein und frei fühlt, und das Auge ohne Schmerzen in die Vergangenheit und mit Ruhe in die Zukunft blickt. Ich habe meine Lieben daheim gelassen, und dies verursacht mir Trauer; doch ich weiß, daß ich einen warmen Platz in ihren Herzen habe, und daß sie sich meiner nur als eines abwesenden Gutes erinnern. Was soll ich dir über meine Beschäftigungen sagen? Du efreust Dich der schönen Blumen und ihres Duftes, Du erfreust Dich des grünenden Baumes und seines Schattens, Du blickst über Wald und Flur der Erde zum gestirnten Himmel, und Du fühlst Dich von höhern Gefühlen bewegt, indem so viele Stimmen ⌜Dich⌝ von einem unendlichen Wesen, Dir un- bewusst, sprechen. Wenn Dich die Natur so freundlich bewegt, wie viel mehr muss sie es mir thun, der ich es mir zur Aufgabe mache, in ihre tiefsten Geheimnisse zu dringen und die ewigen Gesetzen zu entdecken, nach welchen sie so herr- lich, so großartig wirkt. Wäre es nicht Sünde, wenn ich nicht die Antwort gäbe, welche unser Erlöser seiner besorgten Mutter gab, als sie ihn im Tempel fand? Lasst mich! Ich bin in Diensten meines Vaters!^b — Schwerlich werde ich mir Schätze erwerben; doch meine Wissenschaft ist anspruchslos und führt mich bei den bessern Menschen leicht ein, welche mir dann freundlich helfen, so weit ich es eben erfordere. Ich komme so eben von einer 18 montal. Wanderung zurück, welche ich zu den wenigen bekannten Theilen der Colonie unternahm. Überall finden sich Schäfer und Hirten und oft wohlgebildete Männer. Oft lebte ich indessen einsam im Walde, nur von meinem Pferde und meinem Hunde be- gleitet — und niemals war ich weniger einsam. Meine Aus- gaben sind sehr gering, so lange ich im Busche zubringe — und in Sydney habe ich gute Freunde, welche mir ihr Haus zur Wohnung anboten. Dennoch zehre ich von dem Kapitale, welches Freund Nicholson zu meiner Disposition stellte, und da ich nicht Willens bin, ihn weiter zu belästigen, werde ich bald darauf denken müssen, mein Brot zu verdienen und mir die Mittel weiterer Wanderungen zu verschaffen. Wer weiß indessen, was der nächste Augenblick bringt; denn ist es für Euch schwierig selbst die nächste Zukunft zu wissen, wie viel schwieriger muß es für mich sein, der dem Schicksale und dem Zufalle weit mehr ausgesetzt ist, als Ihr. Ich danke Hilgenfeld für seinen brüderlichen Vorschlag, ich freue mich, daß er mich so hoch schätzt, und ich verspreche Euch zu schreiben, wenn Euch das Postgeld nicht zu theuer ist. Erinnert Euch wohl, daß der ferne Bruder mit seinen Gedanken Euch nahe ist, daß er mit seinen Briefen in Euere Gegenwart eintritt, während der Nahelebende, wenn er schweigt, eben so wohl bei den Wilden Australiens leben könnte. ⌜Vergesst auch nicht, daß ich, sollte ich nach Hause kommen, große Unannehm- lichkeiten von meiner Regierung zu fürchten habe. Ich bin nicht Willens, mich als Deserteur behandeln zu lassen; und doch könnte es so sein. Die Regierung hat außerordentlich thöricht gehandelt, und anstatt sich des fernen Landeskindes zu bedienen, ihre Museen und Sammlungen zu bereichern und sich über ferne Länder zu belehren, hat sie mich, wie ein elendes Geizhals, von sich gestoßen, als ich um Freiheit in meiner Militairpflicht bat. Dies ist eine außerordentlich wunde Stelle für mich, und ich schäme mich, im Auslande eine solche behandlung von meiner *mütterlichen Regierung* gestehen zu müssen. In London versagte man mir einen Preußischen Pass nach Paris; ich war gezwungen, einen Pass vom französischen Consul als Engländer zu nehmen.ᶜ In Paris wünschte ich wiederum mit der Regierung an zu knüpfen. Ich ging zum Gesandten und forderte einen Pass nach Italien; man wollte mir nur einen Pass nach Preußen geben. Was sollte ich machen? Sollte ich meine Verhältnisse mit meinem brüderlichen Freunde u Gönner Nicholson aufgeben, welche mir Frankreich, die Schweiz, Italien, England u fernen Welttheile versprach, um der unsinnigen Anforderung zu genügen, ein Jahr in Berlin zu dienen? Ich schrieb an die Regierung, setzte meine Grunde, meine Aussichten auseinander, und was war die Antwort? 'Es ist unmöglich für die Regierung meinen Urlaub weiter als bis 1840 zu verlängern'. Nicholson konnte sich eines 'Goddam it' nicht enthalten, und ich hoffte, daß diese kluge Regierung welche wahrscheinlich Haufen von Reisenden im Auslande hat, einst bedauern wurde, mich so zur Desertion gezwungen zu haben. Wäre ich reichen Leute Sohn gewesen, wahrscheinlich würde man auf das seltene Glück Rücksicht gewonnen haben, welches einem der Natur forschenden Landeskinde zu Theil wurde; doch da ich der Sohn eines armen Landmanns war, der keine Fürsprecher hatte, glaubte die weise Regierung, einen einjährigen Militairdienst nicht verlieren zu dürfen, um einen Reisenden in Neu-Holland dafür zu gewinnen. O Scham, Scham, Scham!! Vergebt mir, daß ich so lange darüber spreche; ich ärgere mich jedesmal, wenn ich daran denke. — So war ich dann gezwungen, meine wissenschaftlichen Verbindungen anderwärts zu wählen, und während [sic] meine Sammlungen nach Berlin zu schicken, gehen sie jetzt nach England u Paris.⌉ Ihr werdet aus meinen Briefen an Schmalfuss gesehen haben, daß dieses Land, obwohl jung, dennoch schon mit vielen Städten bedeckt ist, welche sich allmählig entwickeln. Was Europa an Lebensgenüssen bietet, findet sich auch hier. Eine Menge von Schiffen kommen und gehen nach allen Weltgegenden, Dampfschiffe laufen an den Küsten und einige Flüsse hinauf und verbinden fernliegenden Städte. — Der Busch war eine unbekannte Wildniss für mich, als ich meine Reise begann; ich weiß gegenwärtig, daß der geschäftige unternehmende Ansiedler ihn mit seinen Lastwagen nach allen Richtungen durchfurcht, daß seine Schaf- u Rinderheerden ihn überall hindurchziehen und durchweiden. Die Stationen liegen ungefähr 3–5 deutsche Meilen von einander entfernt, und überall wird der Wanderer gastfreundlich empfangen. Die Wohnungen sind in einigen Gegenden behaglich, in andern sind es Bretterhütten mit 2 od. 3 Stuben, durch deren Wände der Wind weht. Sie sind mit Rinde gedackt, indem diese sich leicht von mehrere Bäumen abschält. Der Reisende führt seine Wollendecke mit sich, da er oft im Freien schlafen muss. Doch die Luft ist so mild, das Klima so gesund, der Boden so trocken, daß man im Freien ebenso ruhig u angenehm schläft, wie in der Wohnung. Unzählige Male habe ich mich, so in meine Wollendecke gewickelt am freundlichen Feuer niedergestreckt und den Lauf der Sternbilder bewacht, die glänzend und herrlich über mir hinglitten. Die Sternbilder dieser Halbkugel sind verschieden von der Erde [sic], und nur wenige Sterne haben wir gemeinschaftlich. Doch derselbe Mond u dieselbe Sonne leuchten auch uns, obwohl nicht zu derselben Zeit. Dieses Land würde ich [sic] vielleicht eines der fruchtbarsten sein, ⌈welches es giebt,⌉ wenn die Luft mehr Feuchtigkeit enthielte und der Regen häufiger wäre. Doch von Zeit zu Zeit treten lange Dürren ein, während welchen die heiße Sonne alle Pflanzen versengt und weite Flächen fast in

Moreton Bay

nackte Wüsten verwandelt. Dann ein gutes Regenschauer — und der Boden bedeckt sich schnell wieder mit dem lieblichsten Grün. Weiter gegen Norden sind die Regen häufiger, und folglich der Boden fruchtbarer. Dichte Gebüsche bedecken die Gebirge und eine dichte Grasdecke den Boden. Das Vieh gedeiht vortrefflich auf solche Weide, obwohl die Schafe in zu großer Feuchtigkeit bedeutend leiden. In jenen Gebüschen habe ich mich lange Zeit aufgehalten und die verschiedenen Baumarten untersucht, welche dort wachsen. Riesenbäume von 200 Fuß Höhe ragen wie die Säulen des Himmels über eine Schaar niederer Bäumen 80–100′ Höhe hervor. Rebenartige Gewächse schlingen sich an ihnen hinauf und binden sie nach allen Richtungen hin zusammen, und oft durchwebt ein reicher Blumenflor ihre Kronen. — Wenige dieser Bäume haben essbare Früchte. Doch viele haben schön geäderte Holzen, welche einer hohen Politur fähig sind.

Erwecke und ernähre in deinen Kindern die Liebe zum Guten, und sei unbesorgt um ihre Zahl. Sieh' unsere Familie! Wir waren unser acht, und einige von Euch hatten hart zu kämpfen, ehe sie gute Wurzel fassten. ⌜Ich fürchte Barth hat nicht wohl daran gethan, die Schifffahrt zu verlassen, obwohl seine Anfälle von Gicht während der Schifffahrt höchst unangenehm waren. Daß Adolph ein tüchtiger Wirth geworden, freut mich sehr. Ich hoffe, er wird Alles aufbieten, Mütterchens Leben so leicht wie möglich zu machen. Kannst Du mir sagen, was aus den Kindern des Predigers Roedelius aus Bocks geworden ist. Es ist auffallend, wie die Eindrücke im Knabenalter sich gegen alle Eindrücke des spätern Lebens erhalten. Als kleiner Knabe hatte ich große Zuneigung zu Bertha, und das ältere Mädchen lachte über den kleinen Jungen; im 16n 17n 18n Jahre war Lottchen Bock meine Flamme. Gegenwärtig, nach so langen Zeit, nach einem so vielfach bewegten Leben, nach einen solchen Schaar anregender Eindrücke, erscheint Bertha immer wieder in meinen Träumen, und Lottchen nicht! Ist dies nicht recht sonderbar? — Und doch liebt ich Bertha als ein kleiner Junge u Lottchen als junger Mann. Weißt Du kluge Frau, mir es zu erklären? Was ist aus Ludwig Riemann geworden? —⌐

Nun lebt wohl, ⌐küsst Eure Kinder in meinem Namen, grüßt Gottfried, Raimund, Herrmann, solltet Ihr mit ihnen correspondiren, und schreibt mir bald einen andern Brief. Ich werde Euch in einiger Zeit, selbst ehe ich Antwort auf diesen Brief erhalte, von mir Nachricht geben.⌐

Eure
Euch herzlich liebender Br. u Schwager
Ludwig Leichhardt.

Newcastle, 15th May 1843 [for 1844][a]

My dear Sister,

To-day's steamer brought me all the letters that have been held for me in Sydney over the last 3 months. ⌐Amongst them there was one from Hilgenfeld, one from Schmalfuss⌐ and how excited I was to find one from you yourself! No other woman has spoken to me with such warm affection since I turned my back on our fatherland. But just see how far Fortune has carried me, and how much at home I am in any part of the world! Even when I was a boy I always had the idea that some day the wings of the morning would transport me to the uttermost parts of the sea. And now, my dear girl, if you want to know how I feel about it myself, and whether I manage to reconcile strenuous effort with a steady pulse, I can tell you that, although the passions of manhood are stronger and harder to control, one's heart can remain pure and aware of its freedom, and one's eye can contemplate the past without regret and the future without alarm. It saddens me to think that I left you dear folks all behind me over there; but I can feel the warmth of your remembrance, and I know that you count me as one of your far-off blessings. Yet what am I to tell you of the life I lead here? You say that you love the beauty and smell of the flowers; that you rejoice to see the trees coming into leaf and casting shade; that when you gaze across the woods and the fields, or look up from the ground to the starry sky, you are deeply moved, because you are receiving so many intimations of a hidden but infinite Being. If Nature stirs you to such pleasure, just

Moreton Bay

think how she must stir me, in my chosen task of penetrating her secrets and discovering the laws that govern the everlasting might and splendour of her workings! Would it not be sin in me to give you any other answer but that of our Redeemer to his anxious Mother when she found him in the temple? 'Wist ye not that I must be about my Father's business?'ᵇ I'm hardly likely to make much money; but my scientific activities are unassuming, and they easily gain me admittance to the best circles, where people prove willing to give me as much help as I ask for. I've just returned after wandering for 18 months through lesser known parts of the colony that I wanted to visit. Wherever you go you find shepherds and stockmen, and you often meet with well-educated men. Nevertheless I was often alone in the bush, with no other company but my horse and my dog—yet never was I less lonely. I spend very little so long as I stay in the bush—and in Sydney I've good friends who've invited me to be their guest. However, as I'm living on the money that my friend Nicholson placed at my disposal and so do not want to be a burden to him any longer, I shall soon have to think about earning my daily bread, and about finding the means that will provide for further journeys. Yet who knows what's around the corner? If it's hard for you two to foresee your immediate future, it's much harder for me, as I'm so much more at the mercy of chance than you are. Thank Hilgenfeld for his brotherly proposal—I'm delighted to learn that he thinks so well of me; and I promise you that I'll write, provided that you can really afford to pay the postage. Don't forget that your far-off brother is close to you in thought—that his letters bring him into your presence; and that one near at hand who has nothing to say might just as well be living with the Australian blacks. ⌜And never forget that, were I to return home, I would have to run the risk of getting into serious trouble with the government. I will not submit to being treated as a deserter, but that's something that could happen. The government dealt with me in an extraordinarily foolish way. Instead of making use of one of its sons abroad in the interests of its museums and [scientific] collections, and as an informant on remote countries, it cast me off, like a miserable skinflint, when I applied for exemption from military service. This puts me in an extremely difficult position, and I'm ashamed to have to admit to such treatment by my old woman of a government abroad. In London

I was refused a Prussian passport to Paris. I was obliged to get one from the French consul, [on which I was described] as an Englishman.^c *In Paris I wanted to re-open discussion with the government, so I went to the ambassador and asked for a passport to Italy. All they would give me was one for Prussia. What was I to do? Was I to give up my association with Nicholson, my brotherly benefactor—an association that held France, Switzerland, Italy, England and distant parts of the world in promise—just to comply with the unreasonable demand that I do a year's service in Berlin? I wrote to the government explaining my motives and my prospects, and what was the answer? 'It is impossible for the government to extend [your] leave beyond 1840'. Nicholson couldn't restrain a 'God damn it!', and I could only hope that this prudent government, which probably had thousands of subjects travelling abroad, would not [come to] regret having driven me so [hard] towards desertion. Had I been the son of rich parents, the golden opportunity that lay open to a German naturalist would probably have won consideration; but, as my father was a poor farmer who had nobody to sponsor me, the judicious government preferred to risk losing a traveller in New Holland rather than a year's military service. Shame, shame, shame on them! Forgive me for dwelling so long on the matter; but I feel angry whenever I think about it. After that, I was obliged to seek my scientific connections elsewhere, and [instead of] sending my collections to Berlin they're now going to London and Paris.*⁷ *From my letters to Schmalfuss you'll have seen that this country, though young, already has quite a number of towns that are gradually developing. What Europe offers for the pleasures of life can be obtained here too. Numerous vessels arrive and leave for all parts of the world. Steamers ply along the coast and go up some of the rivers, connecting towns that lie far apart.—The bush, to me, was the untrodden wilderness when I began my journey; but I know now that the drays of hard-working, enterprising settlers are leaving their tracks in all directions, and that their flocks of sheep and herds of cattle are moving about and browsing everywhere. The stations are from 3 to 5 German miles apart, and travellers are made welcome everywhere. In some parts of the country the living quarters are comfortable, but in others they are nothing but bark huts with 2 or 3 rooms that let the draught in through the walls. They're roofed with*

Moreton Bay

bark, which can easily be stripped from several [kinds of] trees. The traveller has to bring his blankets with him as he has often to sleep out in the open. The air, however, is so balmy and the climate so healthy that one sleeps as soundly and sweetly as he would indoors. Often and often I've rolled myself in my blanket, stretched out beside a cheerful fire, and watched the brilliant constellations go wheeling in splendour overhead. The constellations in this hemisphere are not the same as in [yours], and of the stars we have only a few in common. But the same Sun and the same Moon shine on us both, though not at the same time. This country would perhaps be one of the most fertile ⌜in the world⌝ were its atmosphere moister and its rains more frequent. From time to time, however, long drought sets in and the hot sunshine then scorches the vegetation and transforms considerable areas into almost naked desert. Then there's heavy rain—and in no time the ground is again covered with the most tender green. Farther North rain falls oftener, so the soil is more fertile. There's dense brush on the mountains and rich grass on the [open] ground. Cattle do exceedingly well on such pasture but sheep suffer badly where the ground is too wet. I have spent a long time in the mountain brushes, investigating the various kinds of trees that grow there. Gigantic trees of 200 feet high tower up like the supports of Heaven above a host of lower trees 80 to 100' high. Vine-like plants climb up them and spread [in festoons] in all directions, binding the trees together and interlacing their crowns with a display of flowers. Few of these trees bear edible fruit but many of them have beautifully grained wood that would take a high polish.

    Arouse and foster your children's interest in that which is good, and don't mind how many you have. Look at our own family! There were the eight of us, and some of you had to struggle hard before you took root firmly. ⌜I'm afraid that Barth has not done quite the best thing by giving up navigation, though his attacks of gout were most distressing when he was at sea. I'm delighted to hear that Adolph has become a capable farmer. I hope that he will do his utmost to make mother's life as easy as possible. Can you tell me what became of Pastor Roedelius's children, and of the Bocks? It's very striking how impressions made during boyhood can last, notwithstanding those of later years. When I was a little boy I was strongly attracted to Bertha, but she was a big girl and she only laughed at me. During my 16th, 17th and 18th years Lottchen Bock was the attraction. And now, after so long, after experience that has stirred me in so many ways, after such a host of exciting impressions, I see Bertha in my dreams again and again, but never Lottchen. Don't you think it's very strange?—For I was in love with Bertha as a little boy; with Lottchen as a young man. What have you, with your woman's wisdom, to say about it? What became of Ludwig Riemann?—⌐

Well, good-bye. ⌐Kiss your children for me, give my regards to Gottfried, Raimund and Herrmann if you still correspond with them, and write to me again soon. I'll let you have news of me some time, perhaps even before I get your answer to this letter.⌐

Your affectionate brother and brother-in-law,
Ludwig Leichhardt.

106 — ISAACS ESQ. / Darling Downs / Moretonbay.

Sydney the 3th June 1844

My dear Sir

I had not the pleasure of meeting you when I was visiting the Downs with the intention of studying the localities, where fossil bones have been found. I had however the satisfaction of looking over the collection of fossil bones, which you have made and the lower jaw of a gigantic animal which Richard Owen calls Dinotherium [austra]le would probably excite the desire of possessing it in the [most i]n different geologiste. You are a squatter and I am geologiste. Permit me therefor, to make you the following proposal: I promise you to give you a filly, the daughter of your [en]tire horse out of a Top mare which strayed from me when I left the district, but which is probably heard of at [p]resent or should she be lost a good saddle and bridle of English workmanship—under the condition that you let me have your collection of fossil bones. I heard you had the [in]tention of sending your collection home—I shall make you another proposal which perhaps pleases you better: I am ready to send you a collection of 300 well dried plants of the neigh-bourhood of Sydney and of Moretonbay under your name to any place in England, you wish to direct them to. Plants interest a far greater number of persons and there are so many beautiful ones in this colony, that such a collection would be equally gratifying to the scientific person and to the mere curious beholder.

Should you choose one of these proposals, you would oblige me infinitely and I should allways feel a great pleasure in serving you according to my abilities.

Believe me ever to be
[My dear] Sir
Most truly yours
L. Leichhardt

107 [TO HELENUS SCOTT, ESQ., Glendon, N.S.W.]

Sydney the 20th June 1844

My dear Sir

All the boxes have safely arrived and I am busy to arrang their contents. I never anticipated to find so many things waiting for me, the arrangement of which alone would require at least 6 months severe studies. But now the talk of an expedition to Port Essington comes into the bargain and turns my head almost upside down. I have not yet coloured the map[a] for you, but as I am so far advanced in an account of the geological structure of the colony, as to have done with my map, except as a map of reference, I shall send it back to you and whatever geologist wishes to verify, what I have described, may go to Glendon, look on the map and walk from Taingerring to Tyroman and from Poot yonny gun to Meranni, as I have done and then he may return to the map and add those improvements, which second observers will allways make. I have on receiving your letter, immediately communicated with Mr Robertson of the botanical garden and he has promised, to send you whatever he has of vines in his garden of the desired varieties. But the poor man seeems to lie on his deathbed; he suffers of rheumatism and probably of dropsy.—I do not yet see any chance of my going to Cambden, which probably would take me a week. If I go to Paramatta, it will be to practise observations with the sextant to find latitude and longitude, which will enable me to steer through the interior. Do you know any person on the Hunter, who has mules, well broken in and in good condition? Should I start without Sir Thomas, and many wish it for my sake, I should take 6 mules to carry provisions and 12 horses, for every person 2.—

I have seen Mr Robert Scott and I am sorry to say, that his health seems very much impaired. The change which I observed during the 2 years of my absence is very great and I fear that if he is going on in this restless life, he will not stand it long.

Many compliments to Mrs Scott, to Misses Rusden and to my dear little friends Saranna and Nene to whom I shall soon write a long long letter and

Believe me ever to be my dear Sir
Yours most sincerely
L. Leichhardt

P.S. Next Monday I shall pay 2£ on your account into the commercial bank. The orders on you amount to 1£ 2sh.— Though I dont think that 18sh. will pay what I have taken from your stores, I shall expect your bill and then pay the rest.

108 [TO PROF. RICHARD OWEN, Royal College of Surgeons, London.]

Sydney, 10th July, 1844.

My dear Sir,

You have very probably forgotten the German student to whom you were so kind as to give a letter of introduction to Sir Thomas Mitchell in Sydney. I am desirous of riveting my name more deeply in your memory, and in order to do so, I take the liberty of sending you one or two specimens of the collection of fossil bones I made in Darling Downs. It is the young animal of the gigantic Pachyderm, which once lived near and in the swamps and lagoons, which must have covered these rich plains. The imperfect state of the specimen which came under your observation made you believe, that it was an animal with the dentition of the Dinotherium; but this specimen as well as the two teeth, tied together and belonging to the lower jaw of the old animal, will show you that it was an animal of different incisors and really more allied to the dentition of the kangaroo than to any other animal. Is it not curious that a great number of Australian animals, very different in the formation of the body, have a certain resemblance in the lower jaw, and particularly in the two horizontal incisors? I would call it the Australian type, and should say that the large fossil jaw was formed on the same plan on which that of the Kangaroo, of the Opossum, of the Flying squirrel of the Colonists, and of the Koala, is formed without pretending that the animal to which it belonged was either a gigantic Kangaroo or a gigantic Opossum. The old animal showed at the end of the incisors a used surface, which leads to the supposition, that two or four upper incisors acted on it. The old animal has only four molar teeth, every one with two transversal ridges without any longitudinal connection. The young animal has the false molar, which it seems to cast as it grows older. The two branches of the old jaw are intimately united. You will be curious to know the locality in which these bones are found. The Darling Downs are extensive plains formed by broad shallow valleys, without trees, covered only with grass and herbage, which grows luxuriantly on the rich black soil in which concretions of carbonate of lime are frequently found. Ranges of low hills forming long simple lines with sudden slopes and flat topped cones accompanying these valleys and bear an open forest formed of various species of rather stunted eucalyptus. All these hills are formed by basaltic rock, containing frequently crystals of peridot, and being often cellular, sometimes real scoriae. The base of the rock is, however, feldspathic, and as the peridot is frequently absent, the rock becomes a uniform grey rock which forms a white globule before the blow-pipe, and which I should class amongst the trachytes or phonolites.

The plains are filled with an alluvium of considerable depth, as wells dug 50 or 60 feet deep have been still within it. The plains and creeks in which fossil bones have been found are Hodgson's Creek, Campbell's Creek, Isaac's Creek, and Oaky Creek. They pass all into or through immense plains on the west side of the Condamine into which they fall. The bones are either found in the bed of the Creek, particularly in the mud of the dried up water holes, or in the banks of the creeks in a red loamy breccia, or amongst beds of pebbles, containing many trachytic pebbles probably of the Coast Range from the west side of which these creeks descend. In the banks of the creeks you find, at first, the rich black soil of the plain about three feet thick, then layers of clay and of loam here and there, particularly at Isaac's Creek, with marly concretions of strange irregular forms, often of considerable thickness, though not extending far horizontally; the loam containing little broken pieces of ironstone (breccia) is equally local. Below these lie the beds of pebbles. The bones are found in the breccia, generally near the concretions, but not with them, or they are amongst the pebbles.

A very interesting fact is the presence of univalve and bivalve shells which live still in the neighbouring waterholes, in the same beds in which the bones are found. They are either intimately united with the bones by a marly cement, or they occur independent. The greatest depth in which bones were found is 12 feet; at Oaky Creek we found them at the surface. Besides the bones of the gigantic animal, there are lower jaws and different parts of the body of four other kangaroos, many of them little different from the living, and probably identic with those of Wellington Valley. It seems to me, that the conditions of life can have very little changed, as the same shells live still in similar waterholes. The want of food can scarcely be the cause of their disappearing, as flocks of sheep and cattle depasture over their fossil remains. But as such a herbivore must have required a large body of water for his sustenance, the drainage of these plains or the failing of these springs, the calcareous waters of which formed the concretions in the banks of

Moreton Bay

the creeks, has been, probably, the cause of their retiring to more favourable localities, and, I should not be surprised, if I found them in the tropical interior, through which I am going to find my way to Port Essington. I have put besides a caudal vertebra into the little box, more in order to fill it than as valuable to you, as Sir Thomas Mitchell told me that he had sent you a fine collection of almost every part of the body.

I have sent an account of the geological features of the Colony north of Hunter River to my friend Dr William Nicholson in Newcastle-upon-Tyne, and I hope he will translate and publish it in one of the geological journals. I recommend it to your favour; and should it be found deficient, as every first attempt must be, I have to add, that I am a man without any means, having been enabled to come to this Colony by my generous friend Dr W. Nicholson, and living here as the bird lives, who flies from tree to tree, living on the kindness of a friend fond of my science, or on the hospitality of the settler and the squatter. With a little mare I travelled more than 2,500 miles in zigzag from Newcastle to Wide [Bay], being often groom and cook and washerwoman, geologist and botanist, at the same time, and I delighted in this kind of life; but I feel too deeply that ampler means would enable me to do more, and to do it better.

When you next hear of me, it will be either that I am lost or dead, or that I have succeeded to penetrate through the interior to Port Essington.—Believe me ever to be,

My dear Sir,
Yours most truly,
Ludwig Leichhardt.

109 [TO GAETANO DURANDO, 4 Rue Copeau, Paris.]

Sydney the 12th July, 1844.

Mon cher Durando

Pourquoi n'est-il pas possible de vous tirer tout de suite de ce petit mansarde No 4 Rue Copeau, de vous mettre sur un cheval et d'aller ensemble directement à l'Interieur de la nouvelle Hollande — pourquoi n'est il pas possible, de vous avoir ici, que vous m'aidiez dans mes études et sur mes voyages. Ah mon ami — si j'avais les moyens, et si vous aviez de désires également vifs, qu'auparavant, nous serions bientôt ensemble. J'ai arrangé une partie de 6 hommes (4 blancs, 2 noirs). Nous sommes montés sur 6 chevaux, et 6 mulets portent nos provisions. — Nous avons un chemin de 2000 miles anglaises devant nous et Dieu le sait, si nous serons capable de les parcourir. J'ai envoyé une collection au Musée du Jardin des Plantes, et j'éspère, qu'on en sera content. Il ne faut pas oublier, que cette collection est faite dans un pays, ou il y avait de risque pour ma vie, et que j'avais besoin de tout mon energie, de souffrir les fatigues, le faim et la soif, et d'être tout ensemble: géologue, botaniste, tailleur des bois et des arbres, cuisinier, blanchisseuse, palefrenier et désecheur de mes plantes; que je vivais souvent 8–14 jours dans la solitude des bois, moi seul avec mon cheval et ma chienne. — Si je n'avais pas été supporté par des amis excellens, par des hommes hospitaliers, il ne m'aurait été jamais possible de faire un tel voyage ou de ramasser tant de choses. Vous les verrez; je les ai recuilliés sur un étendue de 600 milles de longueur et de 300 milles de largeur. —

J'ai envoyé mes 'Contributions à la géologie de la Nouvelle Hollande'[a] à Dr. Nicholson 2 Eldon Square Newcastle. Au moment, ou l'ouvrage est publié (dans Jamisons Journal?),[b] il vous en enverra un exemplaire. Ayez la bonté de le montrer a Brogniart, à Constant Prevost, à Cordier. Ma collection de rockes [sic], que j'ai envoyé au Musée trouve une explication dans cet ouvrage. J'y donne beaucoup de coupes géologiques. Monsieur Reinauld se chargera peut être de l'étude des roches, dont quelques unes sont bien intéressantes. Ainsi la roche que j'ai appellée 'Hornblende porphyry', Hornblende rock de la station de Monsieur Archer — la relation chimique entre les Domites, Phonolithes Trachyte et Basaltes. — les porphyres silicieuses de l'Hunter pp. — Un catalogue des rocks d'après la nomenclature de Cordier me serait très agréable — et alors les fossiles — Mons Reynauld aura beaucoup à faire. Qu'il publisse mon ouvrage avec additions d'après ses recherches sur les roches — voila un bon ouvrage! — Et vous? — Vous pouvez étudier les plantes, qui me paraissent nouvelles et dans l'enveloppe desquelles j'ai écrit 'pour Lindley et pour Hooker'. J'ai pensé qu'il serait mieux, de laisser tout la collection ensemble au lieu de la diviser. J'ai un herbier très large et il y en a beaucoup de plantes que je ne connais pas. Quand je reviens de Port Essington je les étudierai avec plus de loisir. Mon herbier est très riche en Legumineuses. Les Proteacées ne sont pas frequentes à Moreton bay — il y en a 4-5 arbres appartenants à cette famille. — J'ai mis un prix de 50£ Sterling sur toute la collection ensemble.

Après vous avoir écrit ma première lettre j'écrivais une autre à monsieur Hooker en lui envoyant les descins d'un fungus qui avait été trouvé par mon très cher ami Mons. Lynd. J'ai attendu une reponse, mais en vain. Il était donc très simple, que je n'avais beaucoup d'envie de lui écrire de nouveau ou de lui envoyer une collection de plantes. —

On me blamera peutêtre, que je n'ai pas envoyé un bon specimen de Castanospermum Australe ou d'Erythrina. J'en ai beaucoup et je les enverrai, si le Musée n'en a pas. Mais en arrangeant mes plantes, je mis un grand nombre de specimens appartenans aux bûches selon les familles; ainsi les Myrtacées et les legumineuses se trouvent entre 2-3-400 specimens et mon temps étant si court, il m'était impossible de les chercher. Et pourquoi est il si court, vous me demandez. Et bien — les pluies tropicales commencent au milieu de Janvier. Le temps necessaire pour finir mon expedition . . . 5 mois; il faut donc être à Port Essington avant que les pluies commencent et il faut donc que je me mette en route au commencement de l'août.

Il est bien possible que je resterai dans cette colonie pour toujours — il est possible que mes os blanchirons sur les plaines de l'interieur. Pourtant mes sentimens me disent, que je réussirai dans mon entreprize — et mon experience me dit le même. Vous trouverez dans mon ouvrage que j'ai trouvai de beaux ossemens fossiles sur les Darling Downs. J'ai envoyé un échantillon à Mons Richard Owen, parcequ'il était le seul, qui me donna une lettre d'introduction à Sir Thomas Mitchell dans cette colonie. Je ne suis pas sans mes adversaire[s]. Ce ne sont pas des ennemis personels, mais ce sont des enne[mis] contre les étrangers. Les amis que j'ai, sont des hommes excellens. Vous me demandez encore: avez vous vu Mons Vereau? [Je] vous répond! Non! je ne l'ai pas vu. Il vit hors de la ville, très l[oin] et moi je travaille de 7 heures du matin jusqu à 12 heures de la nuit. Il savait que j'étais revenu. J'aurais été heureux de le vo[ir] et de lui communiquer des plantes. Mais il ne venait pas et moi je n'avais pas du temps de le chercher. Il s'est marié il y a quelques jours. Peutêtre nous nous rencontrerons un jour. Mais je n'ai que peu d'amis; et il me paraît qu'il en a be[au]⟨coup; peut être je lui serais de peu de service.

⟨Vous avez parlé de plantes pour Mons. [*heavily deleted*]. Vous parliez de[s] plantes de Sydney; mais pour celles même 5 fr. par centaine est un prix tout à fait hors de question. Il ne paye pas le papier qu'il faut avoir pour les désècher. — 1500 plantes formeraient peut êt[e] toute la flore de Sydney, et vous n'attendez pas d'avoi[r] une collection parfaite pour 75 fr.?⟩ Je n'ai pas de temps de faire des collections des plantes de Port Jackson; mais je suis convaincu, que vous ne pourriez jamais l'avoir pour moi[ns] de 1000 fr. — Petites collections de 3–400 specimens, com[me] nous en avons envoyée une à Mons. Arnott chirurgien du Rhin — est facilement faite — mais alors les difficultés commencent et il requiert plus de temps de trouver la dernière plante de la collection, que toutes les plantes précédentes.

Adieu mon cher ami et n'oubliez pas

Votre ami
L. Leichhardt.

P.S. Faites mes complimens a Mons. et Mad. Witherill *et à Reynauld.*

Moreton Bay

My dear Durando,   Sydney the 12th July 1844

Why can't we drag you out of that little attic at No. 4, Rue Copeau at once, mount you on horseback, and set straight off for the interior of New Holland? Why can't I have you here, so that you could help me in my studies and on my journeys? I'm sure, my friend, that if I had the means, and you were still as eager as you used to be, we'd soon be together again. I've organised a party of 6 people (4 whites and 2 blacks). We have 6 horses to ride and 6 mules to carry our provisions.—We've 2000 ahead of us miles to cover, and God alone knows if we'll manage to do it. I've sent a collection to the Museum of the Jardin des Plantes, and I hope that they'll be satisfied with it. Don't forget that it was made in country where my life was at some risk, and that I needed all my strength to endure fatigue, hunger and thirst, and to be my own geologist, botanist, woodcutter, tree feller, cook, washerwoman, groom and plant-drier; that I often had to spend a week or a fortnight alone in the bush with no company but that of my horse and my dog. Had I not had the support of excellent friends and hospitable people I would never have been able to undertake such a journey or to make such big collections. Just wait until you see them! They come from a tract of country 600 miles long and 300 miles broad.—

I've sent my 'Contributions to the Geology of New Holland'[a] to Dr Nicholson at 2 Eldon Square, Newcastle[-upon-Tyne]. As soon as it is published (in Jamison's Journal?)[b] he'll send you a copy. Would you be so kind as to show it to Brongniart, Constant Prévost, and Cordier? My collection of rocks, which I've sent to the Museum, is explained in this memoir, which also contains a number of geological sections. Perhaps M. Reynauld would undertake to study the rocks. Some of them are very interesting—there's one I've called 'hornblende porphyry', the hornblende rock from Mr Archer's station; there's the chemical relationship between the domites, phonolites, trachytes and basalts; there are the acid porphyries of the Hunter, etc. A list of the rocks according to Cordier's nomenclature is something I'd be glad to have.—Then there are the fossils—M. Reynauld will find a lot to do. It would be wonderful if he would publish my memoir with his own supplementary remarks on the rocks.—And what about you? You can work on the plants which I believe to be new. You'll find them in an envelope in which I've put a note saying 'For Lindley and for Hooker'. But it's occurred to me that it would be better to keep the collection intact rather than to disperse it. I've a very big herbarium, and there are numerous plants in it that I don't know. I shall have more time to give to it when I get back from Port Essington. It's very rich in Leguminosae. The Proteaceae are not common at Moreton Bay—there are 4 or 5 trees belonging to the family there.—I'm asking £50 sterling for the entire collection.

After the first time I wrote to you I wrote to Mr Hooker, enclosing drawings of a fungus found by my dear friend Mr Lynd. I hoped that he would reply, but in vain. So I simply didn't want to write to him again or to send him a collection of plants.

I suppose that they'll complain that I haven't sent them a good specimen of Castanospermum australe or of Erythrina. I've plenty of them and will send some if the Museum needs them. But when sorting my plants I arranged a whole lot of specimens corresponding to the samples of wood, according to families—the Leguminosae and Myrtaceae, for example, are now in batches of 2, 3 or 400 specimens, and as my time's now so short I've not been able to look for particular genera. And why, you may ask, have I so little time? Well, the tropical rains set in by the middle of January. The time required to bring my expedition to its destination . . . 5 months. Therefore, as I must get to Port Essington before the rains set in, I shall have to set out at the beginning of August.

It's quite likely that I shall stay in this colony for good—I may even leave my bones to lie whitening on the plains far inland. But I feel no doubt about the success of my undertaking, and my experience also gives me confidence. You will see from my memoir that I found some beautiful fossil bones on the Darling Downs. I have sent a specimen to Mr Richard Owen, because he alone gave me a letter of introduction to someone in the colony, one to Sir Thomas Mitchell. I am not without antagonists. They are not personal enemies but just people who object to foreigners. My friends are all excellent people.

Again you ask if I've seen M. Verreaux yet. The answer is no, I have not. He lives a long way out of town, and as for me, I'm at work from 7 in the morning until midnight. He knows that I've come back.

Moreton Bay

I would be glad to see him and to let him have some plants, but he has not come, and I've not had the time to go to see him. He got married a few days ago. I may meet him some time. However, I've only a few friends and he seems to have quite a number. Perhaps I'd be of little use to him.

⟨You spoke of plants for Mons. [heavily deleted]. You mentioned Sydney plants, but even for these an offer of 5 francs a hundred is quite unacceptable. It would not cover the cost of the paper I'd need for drying them.—1500 plants would very likely embrace the whole flora of Sydney, but you're not expecting a perfect collection for 75 fcs.?⟩ I've not the time for assembling collections of the plants of Port Jackson, but I'm positive that you'll never get [what you want] for less than 1000 fcs. Small collections of 3 to 400 specimens, such as we've sent to Mons. Arnott, a surgeon on the Rhine, are easily made, but beyond that the trouble begins, and it may take more time to find the last plant needed than it took to find all the rest.

Good-bye, my dear friend, and don't forget

Your friend,
L. Leichhardt.

P.S. My compliments to Mr and Mrs Witherell and to Reynauld.

110 MR PAMPLIN / botanical bookseller / 9 Queen street / Soho square / London

Sydney the 15th July 1844

My dear Sir

I addressed you a letter under the 12th July 44, announcing you the sending of 8 boxes by the Ganges. Of these boxes 5 were addressed to the Museum of the garden of plants in Paris, one to the botanical garden of Schoenberg, Berlin University and two to my friend Dr. Will. A. Nicholson 2 Eldon Square Newcastle upon Tyne. I hope you have an agent in Hamburg for the box intended for the Berlin University; but should that not be the case the Prussian ambassador or consul will readily tell you, how to address it.—I send this letter, in case the former should be lost. The boxes contain specimens of wood, specimens of plants, fossil shells and bones and impressions of fern⁄leaves.

I remain my dear Sir Your most obedient
humble servant
Ludwig Leichhardt

III SIR THOMAS MITCHELL M.C. / &c. &c. /Lind⁄say House [Sydney, N.S.W.]

Sydney the 24th of July 1844

My dear Sir

I called on you twice, without finding you at the office and only my increasing and pressing work prevented me from calling again. I have sent the jaw bone of the young animal and the used forteeth of the old one to Mr Richard Owen and I hope that my explication and description of the locality will satisfy him. My plants and geological specimens are arranged, my woods are sent home, my geological account of the visited country is written and I myself, loath of the confined life of Sydney shall repair to my old haunts and see how far I am able, with my limited means, to explore the country in the direction of the North West. My intention is to follow the West Range, of which I told you, which probably forms the Northern brim of the System of the Darling. Should the consent of Home government to your intended expedition come—and my starting shows you my doubts and my impatience—may we meet in the Interior, which I consider my home, as I have no other one. What you can bear down by the force of your equipment, I can only by patience and by perseverance—I shall not do it so brilliantly, but I shall show by my unremitting efforts, how great my desire is, to investigate the nature of this continent, to which I have consecrated my life.

Believe me ever to be
My dear Sir
Most respectfully Yours
L. Leichhardt

112 [TO A FRIEND.]

[Sydney, early August, 1844.]

... All the arrangements are made, and by the liberal contributions of several gentlemen, my outfit has become more complete than my limited means allowed me to anticipate. I have kept an exact account of these contributions, and I shall add a list of them to the account of my journey, should it be successful enough to bestow credit as well on those who accompany me, as on those, who were liberal enough to support me. The names of my companions are Mr James Calvert, Mr Roeper [sic], John Murphy, (a lad of about sixteen years) Philips, a government man, who hopes to obtain a pardon by his good behaviour, and a black fellow. My provisions and ammunition are calculated to last five months, which time I hope will be sufficient to reach either Port Essington, or to bring us back to Darling Downs, should any insurmountable obstacle prevent our farther progress. I take thirteen horses, six for riding, and seven to carry our provisions. I bought a strong but light spring cart, which will serve to carry part of our provisions to the out-stations, and perhaps beyond, should a country similar to the Downs, extend farther to North and North West. The saddles and pack-saddles, which form a considerable item in my equipment, are excellently done, and I wish to express my obligations to Mr John Knox, of Upper George-street, who has paid the most scrupulous attention to their making. My progress will be slow. I shall always try to avoid difficulties by longer roads, than to grapple with them at once. As the neighbourhood of ranges is generally better provided with water and game, I shall invariably keep to them. I anticipate the greatest difficulties from scrubs, from swamps, and from the high temperature as we approach the tropics. Captain King gave me a description of that country, and he believes that there will be greater difficulties in crossing the swampy sea coast of Port Essington, than on our whole journey. Every one of us will partake equally in the necessary work. I shall avoid every intercourse with black fellows, and instead of inviting them to come near me, I shall endeavour to increase their natural fear of every thing unknown to them by legitimate means. These are the outlines of my equipment, and of the principles on which I start. I am well aware of the difficulties which await me, but during the last two years I have been in an excellent school, and have learnt to overcome them by patience and perseverance...

113 [TO S. H. A. MARSH, ESQ., Sydney.]

Sydney the 12th of August [1844.]
My dear friend

I expected to have a day to spare and to come to Ammington to see you; but my time has been so incessantly occupied, at first with the arrangement of my collection and afterwards with the preparations for my journey, that I now find me at the eve of starting for Moreton bay, without having been able to see Mrs Marsh. I hope to find you still here in New South Wales, when I return, which will be in a years time and I shall then make good, what I was not able to do during the present short stay in Sydney.

I bid you farewell for the next 12 months and with my best regards to Mrs Marsh I remain

My dear friend
most sincerely yours
L. Leichhardt

APPENDIXES

CONTRACTIONS AND ABBREVIATIONS USED IN APPENDIXES

ADB	*Allgemeine Deutsche Biographie.*
advt.	advertisement; advertised.
ALS	Autograph letter, signed.
Arch. Off. NSW	Archives Office of New South Wales.
Br.	*Dr Ludwig Leichhardt's Briefe an seine Angehörigen* (1881).
ck.	Australian tributary watercourse.
Col. Sec.	Colonial Secretary.
coll.	Collection(s).
dft.	Draft.
dim.	Diminutive.
DL	Dixson Library (Sydney).
DM	Deutsches Museum von Meisterwerken der Wissenschaft und Technik (München).
Engl.	In English.
facs.	Facsimile.
Fr.	In French.
fragt.	Fragment.
Froriep	Froriep's *Fortschritte der Geographie und Naturgeschichte* (Weimar).
Gg	German in Gothic handwriting.
Gi	German in Italic handwriting.
GSA	Geographical Society of Australasia (later Royal GSA).
I	Partly in Italian.
JRGS	*Journal of the Royal Geographical Society* (London).
jnl.	Journal.
JPRAHS	*Journal and Proceedings, Royal Australian Historical Society.*
JPRSNSW	*Journal and Proceedings, Royal Society of New South Wales.*
Kew	Hooker correspondence, Australian letters, Herbarium of the Royal Botanic Gardens, Kew, Surrey.
L.	Friedrich Wilhelm Ludwig Leichhardt.
MBC	*Moreton Bay Courier.*

Appendixes

misc.	Miscellaneous.
MM	*Maitland Mercury.*
n, nn	Note notes.
NSW or N.S.W.	New South Wales.
PLSNSW	*Proceedings, Linnean Society of New South Wales.*
Pol.	*Dr Ludwig Leichhardt's Letters from Australia,* translated by Louis L. Politzer (1944).
PP	Parliamentary Papers.
PRGS	*Proceedings, Royal Geographical Society.*
RGSA	Royal Geographical Society of Australasia.
RHS Queensland	Royal Historical Society of Queensland.
SMH	*Sydney Morning Herald.*
stn.	Australian sheep or cattle station.
Tgb.	*Tagebuch, Tagebücher.*
trsl.	Translation, translated.
Warrnambool	Public Library and Museum, Warrnambool, Victoria.

Reference by name of author only, is to Section I(*b*) of the Bibliography unless otherwise stated.

Call numbers are given for holdings in the Deutsches Museum (München), at Kew (Surrey), and in the Mitchell and Dixson Libraries (Sydney).

APPENDIX I

TABLE OF EVENTS
BEARING ON THE LIFE AND REPUTATION
OF F. W. LUDWIG LEICHHARDT

(Abstracted from Leichhardt's letters, his journals, the journals of his associates, contemporary printed works and newspapers.)

October 1841–July 1844

III. SCIENTIFIC RECONNAISSANCE IN NEW SOUTH WALES

(a) Around Sydney

1841 Oct.	26	Leichhardt sails from Cork in the *Sir Edward Paget*.
— Nov.	?	He begins to study navigation and to keep the ship's meteorological log; he assists in the treatment of sunstroke and scarlet fever.
— Dec.	7?	*Sir Edward Paget* speak barque *Planter*; Leichhardt boards the latter and meets Dieffenbach, naturalist to the New Zealand Company.
— —	?	He begins to give lectures to idle passengers.
1842 Feb.	12	Drought breaks in parts of New South Wales.
— —	14	*Sir Edward Paget* enters Port Jackson, N.S.W.
— —	?	Leichhardt takes lodgings in Marsh's house, Bligh Street, Sydney; he meets W. Kirchner, Dr Charles Nicholson and Rev. W. B. Clarke; presents his letter of introduction to Mitchell.
— Mar.	?	He presents his collection of minerals to the Australian Museum. He meets Dr Stuart (Military Hospital).
— —	28	He makes a botanical excursion towards Botany Bay.
— Apr.	2	He dines with Sir Thos. Mitchell.
— —	3	Makes excursion with Lt. R. Lynd to the North Shore and the waterfall.
— —	early	He informs people that he has taken no university degree; falls in love with Marianne Marlow; takes several pupils.
— —	10	Complains of eye-strain.
— —	11	H.M.S. *Fly* sails from Falmouth for Australian waters.
— —	18	Leichhardt visits Rev. W. B. Clarke at Parramatta.
— —	25	He applies for the vacant position of Superintendent of the Botanic Gardens, Sydney.

Appendixes

1842	May	7	Excursion with Lynd to the Botany marshes.
—	—	13	Is informed that his application for the position at the Botanic Gardens has been rejected.
—	—	14	Excursion with Lynd to 'the other side of Darling Harbour'.
—	—	23	Attends a meeting of the Debating Society.
—	—	24	Queen's Birthday. Leichhardt attends a ball at Government House.
—	—	26	He meets the Rev. G. K. Rusden.
—	—	27	Lynd invites Leichhardt to share his quarters at the Military Barracks. Leichhardt sends copy of meteorological observations taken at South Head to Prof. Dove (Berlin).
—	June	2	Leichhardt moves to the Military Barracks.
—	—	4	Excursion with Lynd to the Botany sandhills.
—	—	19	Excursion to Bondi.
—	—	23	Leichhardt sends drawings of fossil plants from Nobby's Island to Durando and offers to collect for the Muséum d'Histoire Naturelle.
—	—	25	Excursion to South Head.
—	July	1	Leichhardt delivers the first of a course of weekly lectures on botany at the Sydney Mechanics' School of Arts.
—	—	2	Excursion over the Surry Hills to the Botany marshes.
—	—	?	Leichhardt meets Robert and Walker Scott of Newcastle, N.S.W. and their sister (wife of Dr James Mitchell).
—	Aug.	5	Last of his lectures on botany.
—	—	late	Botanical excursions with senior boys of Sydney and Australian Colleges. Leichhardt is offered assistant-mastership at former, and a tutorship in the family of Dr Mitchell.
—	—	—	Excursion to 'the Cumberland plain'.
—	—	29	Leichhardt lends £50 to James Murphy.
—	Sept.	1	Meets John Carne Bidwill.
—	—	5	Sends description and drawings of a fungus to Hooker.
—	—	6	Assigns his rights in a projected book to his family.
—	—	11	Excursion to the North Shore with Dr Mitchell's family.
—	—	18	Long excursion in the bush alone; botanical and geological observations.
--	—	19	Leaves Sydney by sea for Newcastle with Walker Scott.

Appendixes

(b) *Around Newcastle on foot*

1842 Sept.	20	They arrive at Newcastle. Leichhardt visits Stockton.
— —	21	Excursion to the South-west.
— —	23	Pleasure trip to Ash Island.
— —	25	Leichhardt meets Rev. C. P. N. Wilton. Studies coastal geological section at Nobby's Island and southward.
— —	27	To the Valley of Palms with Walker Scott.
— Oct.	1	Studies coastal section at the South head of Newcastle; begins sea-bathing.
— —	2	Fishing from the rocks with Walker Scott.
— —	4	To Telligerry with Major Crummer.
— —	5	Returns to Newcastle.
— —	8	At Nobby's
— —	15	To W. Scott's cattle station at Minmi.
— —	16	To the foot of the Sugarloaf.
— —	18	Lynd arrives at Newcastle.
— —	21	To Ash Island with Lynd, W. Scott and — Bolton.
— —	26	Meets J. S. Calvert (passenger in *Sir Edward Paget*).
— —	28	5.45 a.m., earthquake at Newcastle.
— Nov.	1	First municipal elections held in Sydney. Calvert brings Leichhardt birds for dissection.
— —	4	Leichhardt determines strike of volcanic dykes near Newcastle.
— —	5	To Calvert's holding.
— —	6	To Red Head and 'the lagoon'.
— —	7	Returns to Newcastle.
— —	11	Mentions a business proposal made by Walker Scott.
— —	14	Sets out for Brisbane Water with — Flood (Postmaster); they reach Brook's holding, near Boolaroo.
— —	15	At Threlkeld's estate, Ebenezer, on Lake Macquarie.
— —	16	At Holden's estate, Newport, L. Macquarie. They spend the night in Carter's hut.
— —	17	At Sidebottom's hut, just North of Wyong.
— —	18	Through F. A. Hely's estate to Gosford on Brisbane Water.
— —	20	Through Forster's station near entrance to Tuggerah Lakes to a point on the coast opposite Bird Island.
— —	21	At the entrance to L. Macquarie (the constable's hut).
— —	22	They return to Newcastle, Leichhardt receives news of death of his sister Mathilde and learns that Strzelecki is about to leave Newcastle.

Appendixes

1842 Nov.	24		Lynd touches at Newcastle on way to Port Macquarie.
—	—	28	Leichhardt studies the supply of artesian water at Ash Island.
—	—	29	He sets out for Point Stephens.
—	—	30	At Crummer's hut, Telligerry.
—	Dec.	1	Bushfires in swamps. Leichhardt goes along the sea-shore to Point Stephens.
—	—	2	He returns along Stockton Beach to Newcastle.
—	—	5	On a wallaby hunt.
—	—	9	He leaves Newcastle on foot for Calvert's holding. Buys a horse from Calvert.

(c) In the Hunter–Goulburn Valley on Horseback

—	—	9	Leichhardt rides from Calvert's holding to Scott's station at Minmi.
—	—	10	He ascends the Sugarloaf with — Callaghan.
—	—	12	Returns to Minmi.
—	—	13	To Maitland with — Rorke
—	—	14	At Harper's Hill.
—	—	16	Leaves Harper's Hill. Thunderstorms. Through Wyndham's to Dawson's at Belford.
—	—	17	Excursion with — Porter to Black Creek and Blind Creek.
—	—	?	With Porter to Kelman's vineyard.
—	—	20	Arrives at Glendon.
—	—	21	Excursion to Jump up Creek.
—	—	22	Excursion downstream to Bell's.
—	—	24	Excursion up Glendon Brook to the coal.
—	—	26	Excursion towards Singleton.
—	—	28	Leichhardt sets out on a geological tour of the Scott estate.
—	—	31	On the way to Mt Tangorin.
1843 Jan.	2		At Merannie.
—	—	3	Tolka Tolka and Stanhope.
—	—	6	At Mt Tyroman.
—	—	7	At Bukembelong (Blackguard Hill).
—	—	8	At Jack Shea.
—	—	9	At Wargungunnie.
—	—	10	Leaves Poot yung gun.
—	—	?	Goes on, to the Paterson and Allyn Rivers.
—	—	15	Returns to Glendon.
—	—	22	At Boorah station.

Appendixes

1843 Jan.	23	At Merannie. Is attacked by a bullock.
— —	24	Sets out for Mt Royal.
— —	25	Climbs Dyrring from St Clair (Capt. Mayne's station).
— —	26	Makes camp in a hollow tree on flank of Piri.
— —	29	At Merannie.
— —	30	Returns to Mayne's station. Rain sets in.
— Feb.	2	Returns to Piri.
— —	14	Returns to Glendon.
— —	18	Begins to press wine at Glendon.
— Mar.	4	Leaves Glendon. First notices the great comet of 1843.
— —	5	At Ravensworth (Bowman's station). Proceeds up Bowman's Creek to Scrumlo.
— —	6	Sandy Creek.
— —	7	Rouchel Brook.
— —	9	Returns to Ravensworth.
— —	11	To Bengalla near Muswellbrook. Meets — Luther.
— —	?	To Segenhoe.
— —	?	To Page's River.
— —	?	To Scone.
— —	?	To Mt Wingen.
— —	?	To Dart Brook (Hamilton's).
— —	?	To Myall Hill.
— —	?	Returns to Bengalla.
— —	17	Leaves Bengalla.
— —	18?	Through Bettington's to Wybong Creek.
— —	19?	Gammon Plains. Spends the night in the bush.
— —	20?	Krui River. Spends another night in the bush.
— —	21	Through Cassilis to Four Mile Creek (Barclay's). Arrives at Dalkeith?
— —	?	Excursion from Dalkeith to Munmora River.
— —	?	Excursion to Talbragar River.
— —	24	Leaves Dalkeith with — Lawson for Rotherwood (Steele's station).
— —	25	Across Norfolk Island Creek and Colly Blue Creek to Lawson's Koolah station.
— —	26	They leave the Koolah station and go (*via* Bowen Creek?) to Lawson's new station on top of the Liverpool Range.
— —	31	At Blaxland's hill. Rain sets in.
— Apr.	1	At Lawson's (new station?).

Appendixes

IV. TO THE MORETON BAY DISTRICT

1843 Apr.	5	Leichhardt leaves Lawson's, goes to Breeza (Andrew Lang's station) and thence down the Mooki (Conadilly).
— —	6	At Carroll Ford (Howe's stn.) on the Namoi.
— —	7	Rain. Proceeds down the Namoi and spends night with — Davidson's team.
— —	8	Through Keepit (W. S. Bell's stn.) to Borah (Campbell's stn.) on Huskisson's Creek.
— —	9	Loses his way to Ogilvie's. Returns to Huskisson's on Manilla Ck. (upper Manilla River?).
— —	10	Through Collet's to Ogilvie's on 'the Gwydir' (Horton's River).
— —	11	Reaches F. T. Rusden's (on upper Horton below Gwydir Falls); finds Hentig in charge. Sees last of great comet.
— —	13	Climbs mountains near the station.
— —	14	To 'the western chains'.
— —	19	Rusden comes from 'Stockyard plain'.
— —	26	Rides past Bundock's, Bell's and Pagan's ('now Pringles') to Ottley's.
— —	28	Crosses the 'Gwydir' (Horton's River)—on way back to Pringle's?
— May	7	At Rocky Creek (Pringle's stn.) Murray-Prior in charge.
— —	10	Writes to W. and M. Nicholson and to Lynd.
— —	13	Rides with — Stoney towards Ogilvie's.
— —	18	Rides with Murray-Prior through 'Buluru' (Paleroo?) Stockyard Nobby (Eulowrie).
— —	?	In the 'Rocky Mountains'.
— —	22	Leaves Rocky Creek, passes the 'Gwydir' and goes to Ottley's on the Big River (Gwydir proper).
— —	24	Proceeds through Beatty's and Morris's up Myall Ck. to Dangar's 'Hunger station'.
— —	25?	Crosses Dangar's Plains and Byron Plains.
— —	26	Through Wyndham's (Bukulla) on 'the MacIntire' to Blaxland's (1) on Frazer's Ck.
— —	27	Crosses 'the Beardy' (for which he mistakes the Severn) and passes by Wilks's stn. to Blaxland's (2).
— —	28	Passes Mayne's stn. 'on the McIntire' (for which he mistakes Rocky Ck., a tributary of the Severn), to Hetherington's stn. at Bonshaw.

Appendixes

1843 May	31	'On the banks of the Severn' (Dumaresq of the present day) at Cox's stn. 'Buluru' (Borella), '103 miles from the Gwydir' (Borella is slightly over 100 miles from the junction of Horton's River with the Gwydir).
— June	3	Rides to McDougal's (and back to Cox's).
— —	5	Murray-Prior overtakes him at Cox's.
— —	7	They proceed together, past Brown's to Hargrave's stn.
— —	8	They 'make 25 to 30 miles more' (Mosquito Ck.).
— —	9	They reach Pitt's stn. on Canal Ck.
— —	10	They cross the Condamine between Russell's and Leslie's stns.
— —	?	The pass between Mt Rubieslaw and Broxburn Sugarloaf on Hodgson's Ck.
— —	13	They ride 'down the new road over the east range' from Alford's house of accommodation to Mocatta's stn. on Lockyer Ck.
— —	15?	At Smith's house of accommodation.
— —	16	At Bell's stn., Laidley's Plains. Leichhardt meets one of the Hodgson brothers.
— —	17	At Owen's stn.
— —	18	At Chambers's stn. 25 miles from Owen's.
— —	19	Through Neal's stn. 'Ipswich' to Limestone (now Ipswich). At a house of accommodation. Leichhardt meets J. Kent.
— —	20	At Brisbane. Leichhardt buys a chestnut mare from Murray-Prior.
— —	22	Leichhardt at the German Mission, Zion Hill, near Eagle Farm.
— July	5	At Brisbane? Kent shows Leichhardt Dixon's map. Leichhardt meets John Campbell.
— —	13	Leichhardt mentions Petrie (Andrew?).
— —	18	Leaves Brisbane. Meets Thomas Archer at the Mission Stn.
— —	21	He sets out with Archer for 'the Bunya country'.
— —	22	Arrives at Durundur.
— —	25	Sets out for Wide Bay.
— —	26	Leaves Mackenzie's (Kilcoy).
— —	27	Crosses the Conondale Range ('the Bunya Bunya Mtns.').
— —	30?	Reaches the 'first station' 24 miles from the head stn. on the Wide Bay (Mary) River.

Appendixes

1843 July	31	Reaches Eales's head stn., Tiaro.
— Aug.	5	Rides to Congitin or Scabby Sheep Stn.
— —	9	Spends a night in the bush.
— —	11	At 'Korura'.
— —	13	In Obi Obi (Ubi Ubi) country.
— —	14	Is back on Mackenzie's run.
— —	16	Returns to Durundur.
— —	17	Receives letters offering him positions at Sydney College and at Ash Island, Newcastle.
— —	31	Sets out, with David Archer and two blacks, to visit the Glasshouse Mtns. They ascend Beerwah.
— Sept.	2	They return to Durundur.
— —	6	To Neurum Neurum (Mt Neurum?). Rain sets in and lasts for 6 weeks.
— —	7	Goes with T. Archer to 'Tschentschillim' (Chin-chillim), the lambing stn. for Durundur.
— —	10	Leichhardt 'has a plan for Swan River' (Archer diary).
— —	14	Sets out, with D. Archer and 3 blacks, for the Bunya country (Blackall Range?).
— —	24	Sets out again, with D. and J. Archer, for 'Durval' (Toorbul) and Brieves (Bribie) Island.
— Oct.	3	They return to Durundur.
— —	8	An excursion with a black abandoned on account of rain.
— —	11	To Mackenzie's run.
— —	13	An excursion to the North.
— —	14	To Bigge's Midday Ck. Stn.
— —	24	Leaves Bigge's Reedy Ck. Stn., passes by Mt Esk and Jones's stn. (Manbouri).
— —	25?	At Wingate's stn. (Tarampa).
— —	26	Past Sandy Ck. and Lockyer's Ck. to Murray-Prior's stn. (Rosewood). Report of Select Committee on proposed expedition to Port Essington published in Sydney.
— —	28	Returning from Murray-Prior's finds, when 16 or 18 miles from Archer's, that his horses have strayed. Returns to Durundur on foot.
— —	30	At Bigge's stn. again. Goodwin's stn. attacked by blacks.
— Nov.	5	To Scott's stn. near Mt Esk (at 'junction of Stanley and Archer's Creeks', i.e. junction or upper Brisbane River and Stanley Ck., respectively).

Appendixes

1843 Nov.	11	To M'Connel's (Cressbrook). 'McConnel has made a map of Stanley Creek.'
— —	15	To McConnel's Sugarloaf (Brisbane Range).
— —	16	To Balfour's, the highest station on Stanley Ck., 8 miles North of Cressbrook.
— —	17	Emu Ck.
— —	18	Crosses the Brisbane Range eastward to Mackenzie's.
— —	19	Returns to Durundur.
— —	23	Leichhardt has already heard of the proposals for the Port Essington expedition.
— Dec.	3	Rides down Wararba Ck. through Capt. Griffin's stn. to Brisbane with (D.?) Archer.
— —	6	Meets Petrie again.
— —	9	Returns to Durundur.
— —	10	Receives delayed letter from Sir Thos. Mitchell about the exped.
— —	12?	Operates on a child (David) for hydrocoele.
— —	16	Rides to Mackenzie's.
— —	18	Rides on to Bigge's. Finds his horses 4 miles below Scott's.
— —	27	Goes with J. Archer, Mr Waterton and 3 blacks to the bunya feast in the Baroon Pocket (Blackall Range).
— —	30	Returns to Durundur.
1844 Jan.	7	Leichhardt 'has immense collections' (Archer diary).
— —	10	Goes to Brisbane with J. Archer and Mr Waterton.
— —	19	Tropical rains set in. Leichhardt now aware of Gipps's delay of the Port Essington expedition.
— —	24?	He treats the child David for malarial fever.
— Feb.	6	Thomas Archer and others have tertian fever. Heavy rains.
— Mar.	3?	Leichhardt leaves Durundur.
— —	4	Goes from Brisbane through Cooper's Plains to Canoe Ck.
— —	5	To coal-workings 3 miles from (Woogooroo?, Dr Simpson's) with — Wiseman.
— —	6	At Limestone. Sees J. Kent again.
— —	9	To Wilson's stn., then to Flinders's Peak. Camps out.
— —	10	Through Neale's stn. to Cameron's. Climbs Mt Fraser.
— —	11	At foot of Mt Edward with Cameron.
— —	12	To Coulson's (Kolson's).
— —	13	At Mt Greville (Gravel). His chestnut mare and foal stray. He searches for them for a week in vain.

Appendixes

1844 Mar. 20	Passes through Cunningham's Gap.
— — 23	Past Macdonald's stn. to Bracker's (Rosenthal).
— — 30	Sets out with Fairholme in search of fossil bones. They visit Campbell's Ck. (Glengallen), King Plains, Hodgson's Ck. (Eton Vale), Isaac's Ck. (Gowrie), Oakey Ck. (Hugh Ross's stn.) and Coxen's stn., where they find bones.
— Apr. 2	To Russell's stn. (Cecil Plains) where Leichhardt records that he has found people willing to accompany him on an expedition.
— — 7?	They return to Bracker's.
— — 16	Leichhardt goes to Maryland (Marsh's stn.).
— — 17	To Gardner's (N.N.E. of Tenterfield).
— — 18	Past Mackenzie's and Wiseman's to a camp in the bush.
— — 19	Past 'a bluff mountain between Mackenzie's and Windeyer's' to another camp in the bush.
— — 20	Past Turner's, near the site of Glen Innes, to Boyd's (Stonehenge) on 'Beardy Plains'.
— — 21	Past Masters' near Ben Lomond (to another camp in the bush?)
— — 22	To Dumaresq's.
— — 23	Past Macdonald's to Armidale (the Commissioner's house).
— — 25	Past Dangar's to Cruickshank's.
— — 26	Past Jenkins's and Thompson's to T. G. Rusden's (Europambela).
— — 30	Excursion to Apsley Falls.
— May 1	From Rusden's to McIvor's (near Denne's Sugarloaf?).
— — 2	Crosses the divide between the Hastings and the Gloucester–Manning drainage to Cox's. Observes serpentine.
— — 3	To Ashall's, where he meets — Turnbull.
— — 4	Through Gloucester to Turnbull's.
— — 5	To Stroud, headquarters of the Australian Agricultural Company. Meets Capt. P. P. King.
— — 8?	To Raymond Terrace (Bolton's).
— — 9	Reaches Newcastle.
— — 15	John Archer visits him.
— — 17?	Leaves Newcastle for Glendon.
— — 26?	Returns to Newcastle.
— — 29?	Arrives back in Sydney.
— June 3	Writes to F. N. Isaac.

796

Appendixes

1844 June	20	Reports that his boxes have arrived from Glendon, and promises to send his coloured geological map to Helenus Scott.
— July	9	Sends fossil bones to Owen by *Ocean Queen*.
— —	10	Reports that he has sent his geological memoir to W. Nicholson.
— —	12	Sends 8 boxes of specimens to Pamplin, London, by *Ganges*.
— —	17	*The Australian* (Sydney) prints short account of his journey to Moreton Bay.

APPENDIX II

ANNOTATED CALENDAR OF LETTERS

March 1842 – August 1844

50   1842, March 23. ALS, Gg, DM/II, 1; Br. 30, Pol. 1.
- [a] The editors of 1881 changed Leichhardt's clearly written 'Neuseeländern' (Maoris) to 'Neuholländern (Australian blacks); but Mrs Charles (Louisa Anne) Meredith reported, in her *Notes and Sketches of New South Wales... from 1839 to 1844* (London, 1844, p. 36), that the harbour master's boat at Port Jackson was manned by Maoris; and Peck reported the same of the pilot boat in 1848 (Peck, 1850, p. 33). Emmeline De Falbe (née Macarthur) remembered the Port Jackson Maori boatmen.
- [b] Stephen Hale Alonzo Marsh (c. 1805–88), a harpist. He set up as a professor of harmony and importer of pianos, spending ten years at Sydney. He went to Melbourne in 1852, and to San Francisco in 1872. He toured the Presidencies of India with the violinist Ravac in 1846.
- [c] Major Sir Thomas Livingstone Mitchell (1792–1855) served in the Peninsula War. He went to New South Wales in 1827, and led official expeditions into the interior in 1831, 1835, 1836 and 1845. Eyre, Sturt, Mitchell and Leichhardt were the most important explorers by land in Australia during their time. Mitchell, however, regarded Leichhardt not only as his rival but also as a foreign upstart. Mitchell was a gifted man—artist, linguist, botanist, palaeontologist, surveyor and inventor. He and his son Roderick were in advance of the times in their dealings with the Australian blacks. The father, however, behaved in so extraordinary a manner after his expedition of 1845–6 as to leave doubts about the state of his mind at the time. His biographers have not yet appreciated the complexity of his character or the variety of his performance. See Foster, Moore, Cumpston.
- [d] Dr (afterwards Sir) Charles Nicholson (1808–1903), M.D. Edinburgh, went to Australia in 1834. He was deeply interested in natural science and education, and became President of the School of Arts (Sydney) and Chancellor of the University of Sydney. He was a member of the first Legislative Council of New South Wales, and was the first Speaker of the Queensland Legislature. See Dallen.
- [e] Wilhelm Kirchner was Consul for Hamburg at Sydney, and a member of the committee for the German Mission at Moreton Bay. He wrote *Australien und seine Vortheile für Auswanderer* (1850).
- [f] For the German Mission see Lang (1847), ch. XI, and W. N. Gunson, 'The Nundah missionaries', *JRHS Queensland*, VI, 3 (1960–1), 511–39.

Appendixes

51 1842, March 25. ALS, Gi, ML/A. 1383/1; typed trsl. in ML/A. l. 69; copy, Gg, Tgb. 1842, ML/C. 145. The suppressed passages were omitted not from the letter but from the copy. Comparison of the two is revealing and gives an indication of the kind of information that Leichhardt recorded in his journals. They contain nothing about Nicholson's graduation at Berlin, or about his own discussion of the question of his military service. Like his meeting with Dieffenbach on the high seas, these were things that he would never forget.

a The ship *Sir Edward Paget* met the barque *Planter* on 7 Dec. 1841 in Lat. 12° 23' S., Long. 32° 32' W. A party from the ship, which evidently included Leichhardt, boarded the barque at noon (*Monatsber. Ges. Erdk. Berlin,* N.F. 3, 1845-6 (1846), p. 66). For Ernst Dieffenbach, see Johannes Anderson, *The Lure of New Zealand Book Collecting* (Auckland, etc., 1936).

b Augustin-Pyramus de Candolle (1778-1841) was the most distinguished member of a family devoted to botany. He was professor at Geneva (where a street commemorates his name). His son, Alphonse de Candolle, published the *Prodromus systematis naturalis regni vegetabilis,* 21 parts in 24 vols. (Paris, 1824-74).

c The short-lived *Tasmanian Journal of Natural Science,* in format and layout resembling the *Philosophical Magazine.* It is of great interest in the history of science in Australia.

d The hypothesis does not fully hold. The dunes of the Sydney district, now mostly built over, were accumulations of blown sand, but they lie in places over sandy or loamy estuarine deposits of Recent age (see Etheridge, David and Grimshaw, *JPRSNSW,* 30 (1897), 30, 158.

e James Stuart (1802-42). See Musgrave, Iredale and Whitley, *Australian Zoologist,* XII, 2 (July 1955), 120-31.

f The Tank Stream, now confined to a concealed drain.

g In allusion to Luke ix. 33: '. . .let us make three tabernacles; one for thee, one for Moses, and one for Elias . . .'

h Sir Richard Owen (1804-98), first Hunterian Professor at the Royal College of Surgeons. He attended Georges Cuvier's lectures in Paris in 1831 and became the leading British authority on fossil avian and mammalian skeletons. Appointed Superintendent of the Natural History Departments of the British Museum in 1856.

52 1842, April 10. Dft. in *Tgb.* 1842, ML/C. 145.

a Sir Edward Deas-Thomson (1800-79), born at Edinburgh and went to Sydney at the end of 1828 as Colonial Secretary and Registrar of Records. Member of the Legislative Council of New South Wales 1837. Benefactor and Chancellor of the University of Sydney. See Philip Mennell, *The Dictionary of Australian Biography . . .1855-1892* (London, 1892). See Dallen.

b Col. George Barney (1792-1862). Went to New South Wales in 1835

Appendixes

and was appointed Colonial Engineer. He was Lieutenant-Governor of the experimental Colony of North Australia in 1846, Chief Commissioner for Crown Lands in 1849 and succeeded Sir Thomas Mitchell as Surveyor General, New South Wales, in 1855. (*Australian Encyclop.*, see also Cumbrae-Stewart, 1919).

c James Anderson, see Froggatt, *JPRAHS*, 18 (1932), 122–3 and J. H. Maiden, *ibid.* 14 (1928), 1–42.

d Lt. Robert Lynd (1800–51) went to Sydney with the 63rd Regiment and was appointed Barrack Master. He was a somewhat hypochondriac bachelor with a serious interest in literature and botany. With a house at his disposal he was able to show great hospitality to chosen friends. Leichhardt copied several of Lynd's poems into his own journal for 1842. As one of them was a rather stiff translation of Goethe's *Der Fischer*, and as Lynd corresponded with relatives of Leichhardt in Hamburg (*Monatsber. Ges. Erdk. Berlin*, N.F. 5, 1847–8 (1848), 130–1) it seems likely that he spoke German. He was transferred to New Zealand in 1847 and died at Auckland in 1851. His papers, which probably included a sheaf of important letters from Leichhardt, have not yet been traced. Part of his botanical collection was sold to the British Museum (Natural History) by Capt. Drury, R.N., in 1858.

e Marianne Marlow.

f Probably the widespread Newer Basalts of Victoria, which contain beautiful deposits of white minerals (zeolites) in veins and cavities.

53 1842, May 17. Dft., Gg, *Tgb.* 1842, ML/C. 145.

a Genesis xxix. 26–9.

b Allan Cunningham (1791–1839), botanist and explorer in whose footsteps Leichhardt was to follow on his first journey to Moreton Bay. See Ida Lee, *Early Explorers in Australia* (London, 1925), chs. VI–XI, XV, XVII and XIX; also Geo. H. Mitchell, 'Allan Cunningham', *JP Parramatta Hist. S.* 4 (1935), 5–19.

c Alexander Macleay (1767–1848), a Scot who became Honorary Secretary of the Linnean Society. He was Colonial Secretary to New South Wales from 1825 to 1837 and became first Speaker of the first representative Legislative Council of New South Wales in 1843. He did much to establish the Public Library of New South Wales, and was a member of the committee for the Botanic Gardens, Sydney. The question of a curator for the botanic gardens came up again in 1844, when Leichhardt's name was again put forward. A letter from William Macarthur of Camden to Hooker, dated 5 Aug. 1844, and one from Macleay dated 6 Feb. 1848 (Kew, lxxiii, 240 and 248 respectively) show that Macleay objected to Leichhardt because the latter was a foreigner, and, that Macleay did not hesitate to override the committee.

d See Lemprière.

e Who lost his life during the great eruption of Vesuvius in A.D. 79.

800

54   1842, May 27. Dft., Gg, *Tgb.* 1842, ML/C. 145. This letter was delivered in Germany only after an unexplained delay of more than four years. It is one of the most important of Leichhardt's letters, for it brought him to the notice of German geographers, and it established him as the first of the Australian explorers to take the field with a sound hypothesis concerning the character of the interior of the continent. He was convinced that it was arid. Mitchell took boats with him in 1845; and Sturt never renounced his belief in the existence of an inland sea (see M.A., review of Cumpston's *Charles Sturt* in the *Geogr. J.* 119 (1953), 113). The draft possibly represents two letters posted about the same time (see *Monatsber. Ges. Erdk. Berlin,* N.F. 1847–8 (1848), p. 6; but if it does, I am unable to distinguish between them. [M.A.]

a Heinrich Wilhelm Dove (1803–79), Professor of Natural Philosophy at Berlin, was one of the founders of modern meteorology. In 1828 he recognised the senses of rotation of cyclones and anticyclones in the two hemispheres. His books *Das Gesetz der Stürme* and *Verbreitung der Wärme* are classics in the history of meteorology. (See Sir William Napier Shaw, *Manual of Meteorology*, vol. 1, Cambridge, 1926, portrait.) On 7 November 1846 Dove communicated the contents of Leichhardt's letter to the Gesellschaft für Erdkunde in Berlin (*Monatsber. Ges. Erdk. Berlin,* N.F. 4 (1846–7 (1847)), 215). It was considered important enough for publication, and an abstract of it appeared in *Frorieps Fortschritte der Geogr. u. Naturges.* 2, no. 4 (Feb. 1847), cols. 113–14.

b Rev. William Branwhite Clarke (1798–1878), an Englishman of classical education and wide interests who went to Australia in 1839. He was Principal of the King's School at Parramatta, N.S.W., when Leichhardt met him. Clarke's activities were on the heroic scale. He travelled widely and worked out the broad relations in the stratigraphical geology of eastern Australia. Through his general observations in natural history he developed a voluminous correspondence with scholars overseas, and through his journalism he became the critic of geographical advances in Australia. 'Perhaps no one in Australia', said the *Sydney Morning Herald* (27 Jan. 1881), 'has devoted more attention to the subject of exploration than the late Rev. W. B. Clarke...' He correlated Leichhardt's result with those of other explorers of the time by land and sea, and demonstrated again and again the care and accuracy of Leichhardt's work. When jealousy of the German was acute Clarke wrote in his private notebook (ML/Uncat. MSS, set 141) 'In Australian Newspaper appeared an article maliciously false against Leichhardt headed The Herald The Atlas & Sir Thomas Mitchell—30 Jan. 1847—when in Herald appeared an article on *Australian & the Surveyor General* vindicating inter alia Leichhardt 4 Feb. 1847. W.B.C.' As Clarke's journalistic work has never been collected it lies forgotten, and his unflagging defence of Leichhardt is now hardly known. There is no study of Clarke on record which does full justice to his industry

Appendixes

and his contribution to our knowledge of New South Wales. They call for a biographer of both classical and scientific training. See Jervis, 1944-5. It is only fair to say that Clarke could be no less amusing on Leichhardt than Leichhardt on Clarke. When Leichhardt left Sydney on his last expedition he left some of his possessions, including most of his Australian journals, in Clarke's care. Clarke's descendants deposited them in the Mitchell Library, Sydney, in 1948.

c Capt. (afterwards Rear-Admiral) Phillip Parker King (1791-1856) was born at Norfolk Island. He entered the Navy in 1807. In 1817 he was sent to survey the intertropical coasts of Australia. In 1825 he was appointed to the *Adventure*, and served under Capt. Robert FitzRoy, R.N., during the voyage of the *Beagle* and *Adventure*. He returned to Australia in 1832, becoming Resident Commissioner for the Australian Agricultural Company in 1839. He lived at Tahlee House, near Carrington, on Port Stephens, N.S.W. There is no satisfactory account of his career on record. It is evident, from the *Athenaeum* and from the records of the Royal Geographical Society that he was in London at the end of 1848, and the Stanley correspondence shows that he was still there in June 1849 but arrived back in Sydney at the end of April 1850. Leichhardt was familiar with King's *Narrative of a Survey of the intertropical and western coasts of Australia* (2 vols., London, 1827). King, and his son Phillip Gidley King, remained loyal friends of Leichhardt.

55 1842, May 27. Dft., *Tgb.* 1842, ML/C. 145.

56 1842, June 16. Dft., *Tgb.* 1842, ML/C. 145.
 a For the position of Curator of the Botanic Gardens, Sydney.
 b Maj. T. L. Mitchell, *Three Expeditions into the interior of Eastern Australia*, 2nd ed. (London, 1839), vol. 2, pl. 48, p. 368.
 c Baron Andreas von Baumgartner, editor of the *Zeitschr. f. Physik und Mathematik* (Vienna) from 1826 and author of a standard work, *Die Naturlehre...etc.*, 7th ed. (Vienna, 1842).

57 1842, June 23. ALS, ML/C. 161. Dft., Fr., in *Tgb.* 1842, ML/C. 145; partial copy in Durando's handwriting, Kew, lxxiii, 239; faulty trsl. *London J. Bot.* 4 (1845), 279-80.
 a Robert Brown: *Prodromus floræ Noviæ Hollandiæ et insulæ Van Diemen...* 2nd ed. (Norimbergæ, 1827). Brown was the botanist who accompanied Flinders on his voyage to Terra Australis, 1801-03.
 b John Carne Bidwill (1815-53): an English botanist who was collecting in Australia and New Zealand. See Maiden, 1908; Froggatt, *JPRAHS*, 18 (1932), Lennon, McKinnon 1940.
 c Almost certainly *Glossopteris*, first described by Adolphe Brongniart (*Histoire des végétaux fossiles*, 2 vols. Paris, 1828).

Appendixes

58 1842, c. July 17, cont'd 25th. Dft., Gg, *Tgb*. 1842, ML/C. 145.

[a] See Fowles. The School is still functioning (1967), though mainly as a lending library.

[b] See Lemprière. Democritus travelled over a great part of the Old World in search of knowledge and returned home in the greatest poverty.

[c] See Lemprière, according to whom it was not Midas but one of his servants who told the secret of Midas's ears to the Earth.

[d] Thomas Barker established the second steam flour mill in the colony, near Darling Harbour, Sydney, in 1823. See Norman Selfe, *JPRSNSW*, 34 (1900), pp. xxi–xxii, and Joseph Harding, *Barker's Mill*, n.p., n.d. (? Stanmore, N.S.W., 1945).

[d] The brothers Scott were wealthy landowners in the Hunter River valley. Robert Scott had a house in Sydney. There is not much on record about this important family.

[f] Dr James Mitchell (1792–1869) lived in Sydney but had estates in the Hunter River valley. He married Miss A. M. Scott and became the father of David Scott Mitchell (1836–1907), who founded the Mitchell Library, Sydney.

[g] Most likely the well known 'Prospect laccolite', which consists of essexite (see *JPRSNSW*, 45 (1912), 445).

[h] Capt. Alexander Machonochie, R.N. (1787–1860), who held advanced views on the treatment of convicts. See J. V. Barry, *Alexander Machonochie of Norfolk Island* (Melbourne, 1958), and *Geogr. J.* 126 (1960), 459–68.

[i] *Corypha* is an old synonym for *Livistonia*. *L. australis* was felled for timber around Port Jackson in the earliest days of the colony (see John Cobley, *Sydney Cove 1789–1790*, Sydney, 1963, pp. 96, 98) and was later exploited for the manufacture of 'cabbage-tree' hats. The palm was known as the 'cabbage-tree palm' on account of its edible growing point. Not to be confused with the 'cabbage palm' (*Cordyline* sp.) of New Zealand.

59 1842, September 5. ALS, Kew, lxxiii, 228.

[a] Hooker identified this fungus as *Aseroe rubra*, and published Leichhardt's description of it (*London J. Bot.* 3 (1844), 191–2, Tab. v). A plant appeared on a shady bank on the weather side of my own house at Balgowlah, N.S.W., during the wet Spring of 1963. It was identified as *Aseroe rubra* at the herbarium of the Royal Botanic Gardens, Sydney, on 14 November 1963, though the lobes were more rounded and the colours more varied than in Leichhardt's description. Growing nearly flat on the ground, about the diameter of a penny, it resembled the small starfish to be found in the rock-pools along the coast near Sydney. The upper surface had a scaly, enamelled appearance, mottled in black, red, cream and yellow, like the colours of the American king snakes and coral snakes. [M.A.]

[b] See note 57(a).

[c] De Candolle's *Prodromus* (see note 51(b)).

[d] Adrien de Jussieu, *De Euphorbiacearum generibus...etc.* (Paris, 1824).

Appendixes

^e John Lindley (1799–1865), was Professor of Botany at London. The work in question was probably his book *A Natural System of Botany* (London, 1836).

^f Achille Richard (1794–1852) was botanist in the *Astrolabe*, 1826–9. He is best known for his work on the flora of New Zealand.

60 1842, September 6. ALS, Gg, DM/II, 2; *Br.* 31, Pol. 2.

^a Friedrich Ludwig Georg von Raumer (1781–1873), a Prussian civil servant who travelled widely in Europe and America. He published numerous historical works and books of travel.

^b This person was most likely the Henri or Henry (? Heinrich) Böcking who accompanied Leichhardt to the Peak Range in 1846–7.

61 1842, September 26. Dft., Engl., *Tgb.* 1842, ML/C. 145.

^a Karl Alexander Anselm Freiherr von Hügel (1795–1870), son of an Austrian nobleman and a Scottish mother. He became a very distinguished diplomat and botanist. From 1831 to 1836 he travelled through the eastern Mediterranean countries to India, thence to Ceylon and the East Indies, then to Australia, New Zealand and Oceania, and arrived home with immense collections. He established gardens in Vienna where he cultivated Australian flowering plants, of which some were still being propagated in the Imperial gardens at Schönbrunn in 1905. He was in Western Australia in 1833, and visited Illawarra, N.S.W. (see Wells, p. 211). His Western Australian plants were described by Endlicher in collaboration with Bentham, Fenzl and Schott in 1837–8. See *Charles von Hügel / April 25, 1795–June 2, 1870*, privately printed, 2nd issue (Cambridge, 1905), and J. H. Maiden, *J. West Australian Nat. Hist. Soc.* VI (Feb. 1909).

62 1842, October 2. Dft., Engl., *Tgb.* 1842, ML/C. 145.

^a A small valley that terminates at a small lagoon 6 miles along the coast South of Newcastle. The cabbage-tree palm had evidently been cut out here also.

^b Major James Henry Crummer, an Irishman who served in the 28th Regiment of Foot and who saw many years of active service before going to New South Wales in 1835. He was Police Magistrate and Superintendent of Convicts at Newcastle from 1837 to 1849, Police Magistrate at Maitland 1849–58, and at Port Macquarie 1858–64, where he died in 1867. See Heaton; see Backhouse (1838).

63 1842, October 12. Dft., Gg, *Tgb.* 1842, ML/C. 145; printed in part in *Berlinchen Nachr. von Staats- und Gelehrten Sachen*, Beil., 27 Jan. 1856, cols. 110–12, and in Zuchold (1856).

^a Wilhelm Kirchner was Consul for Hamburg at Sydney, and a member of the committee for the German Mission at Moreton Bay.

^b Leichhardt was mistaken, but the age of the beds was not known in his

Appendixes

time. The Newcastle coal-measures are now considered to be Permian in age. Some of the impressions of ferns from the Australian Permian and Carboniferous rocks do, nevertheless, resemble ferns of the present day.

64   1842, October, 26, cont'd before 28th. Dft., Gg, in *Tgb.* 1842, ML/C. 145.

a   Cf. note 63(b).
b   There is little on accessible record about the Dawsons of the Hunter River. They were connected, at first, with the Australian Agricultural Company.
c   The Rev. Charles Pleydell Neale Wilton: one of the earliest of the many geologists who have worked in the Hunter River district. His paper 'A sketch of the geology of six miles of the South-east line of the coast of Newcastle in Australia' (*Phil. Mag.,* N.S. 1, July–Dec. 1832, pp. 92–5, fig.) is of interest here. One of the seams of coal was on fire when he made his observations, but Leichhardt made no mention of the fire in 1842.

65   1842, begun October 31, cont'd November 1, 5, 7, 8 and 9. Dft., Gg, in *Tgb.* 1842, ML/C. 145.

a   The waterfall was probably Harnett's Falls, formerly one of the attractions of Mosman's Bay, Sydney, but now deprived of all scenic value by suburban development. See MSS, Carroll. There is a painting of the falls as they appeared in 1874, by S. T. Gill, in the Dixson Library, Sydney.
b   Leichhardt seems to have thought that there were two systems of jointing to be seen in the rock platform just South of the mouth of the Hunter; one, a criss-cross network of regional extent, the other a localised, radial system due, in his opinion, to sharp, local earthquake shock. The miniature ironstone walls that give a honeycombed appearance to the platform are due to the leaching of iron from the mass of the rock and its deposition in the fissures, rather than to the action of volcanic gases.
c   James Snowden Calvert (1825–84) and his brother William were steerage passengers in the *Sir Edward Paget* when Leichhardt went to Australia in 1841. Of all Leichhardt's companions on his expeditions James Calvert was the only one who fully appreciated what Leichhardt was attempting to do, and who became his scientific disciple. James Calvet eventually contributed to the knowledge of Australian botany. He remained unswervingly loyal to Leichhardt and his few letters to the press are of more importance as testimony than the whole flood of journalistic writings except those of Clarke. It is evident, from this and the preceding letter, that the Calvert and Dawson huts, if not one and the same, were not far apart. Another student of Leichhardt has informed me that there was a foster relationship between one generation of Dawsons and the Calvert boys. For Calvert see Heaton.
d   A dangerous whirlpool near the Sicilian side of the Straits of Messina. See Lemprière.
e   M'Gill, properly Biraban, was an intelligent black who assisted the Rev. L. Threlkeld of Lake Macquarie in the analysis of the Awabakal dialect.

Appendixes

See Backhouse (1843, pp. 379-81), *JPRAHS*, 25 (1939), portrait, p. 376; and *Australian Dict. Biogr.*

f Ankaeos or Anchaeus. See Lemprière. The story is the origin of the proverb 'There's many a slip twixt the cup and the lip'.

g For an early account of wine-growing in New South Wales see Anon. (1838), pp. 168-72; and for an account contemporary with Leichhardt see Hodgkinson, pp. 179-95. The activities of Mons de Ligny were reported in the *Australian* on 19 and 22 March 1842.

h In his will Leichhardt bequeathed to his mother 'Two Hundred and Fifty Pounds which I lent to Mr Frederick Luther of Regentville Cumberland New South Wales'. In his note-book opened in 1842 (ML/C. 154) he listed, amongst books he had lent, the loan of vols. 18-25 of Goethe's *Sämtliche Werke* to a Mr Priddle and all the rest to Luther. The set (last edition corrected by the author) has recently come into the market, and on 19 Dec. 1960 I identified Leichhardt's signature on the title-page of each volume for Messrs Berkelouw, antiquarian booksellers, Sydney.

i The Macarthurs of Camden were the most influential family in New South Wales in its earliest days. See Heaton; see Onslow. William Macarthur brought six German wine dressers to Camden in 1839.

j Launcelot Edward Threlkeld (1788-1859), a missionary who gained his first experience in the Society Islands. When stationed at Lake Macquarie he prepared a grammar of the local (Awabakal) dialect (*An Australian Grammar*, Sydney, 1834) and translated part of the Scriptures. He was a member of the committee for the German Mission at Moreton Bay. See Backhouse (1834, pp. 381-3); see Champion.

k The Helys were amongst the earliest settlers in the Gosford district, N.S.W. See Swancott. F. A. Hely had been Superintendent of Convicts. His son, Hovenden Hely, was to become closely associated with Leichhardt.

l See Lemprière. The theme of frustration has already appeared in this letter (Ankaeos), but it seems likely that Leichhardt here meant Sisyphus, not Tantalus.

66 1842, November 10. ALS, Gg, DM/II, 3; *Br.* 32, Pol. 3.

a The same time-table was still being observed more than 70 years later.

67 1842, November 11. Dft., Engl., *Tgb.* 1842, ML/C. 145.

a James Murphy was a Welshman who emigrated to Australia with his family in the *Sir Edward Paget* in 1841. Before leaving Sydney in 1842 Leichhardt had given the Murphys medical treatment and had already lent James Murphy £50 and Lynd had upbraided him for doing so (*Tgb.* 1842). In his will he bequeathed to his mother 'Two Hundred and Fifty Pounds which I lent to Mr James Murphy plasterer of Surry Hills Sydney'. Murphy must have been a man of some ability because he became Mayor of Sydney in 1860. When Leichhardt left Sydney on his last journey he left many of his possessions, including personal papers, in Murphy's care. Murphy deposited them in the Australian Museum in 1853, where Krefft eventually discovered them.

Appendixes

The son, John Murphy (1829–70), had some talent for drawing and became closely associated with Leichhardt. A grand-niece of his, now living in Sydney, has assured me that the family has never held anything against Leichhardt. She has lent me the MS journal kept by John Murphy on the Port Essington expedition, but withholds permission to copy it or to quote from it. The journal is ably written, the handwriting mature but more beautiful than legible. Some of the misspellings appear to have been derived from Gilbert, and here and there the very words of Murphy are the words of Gilbert. The drawings are simple but effective, and that of a fish, drawn on 19 Jan. 1845, is the earliest figure of a barramundi (*Scleropagus Leichhardti*, Gunther, determined as *Osteoglossum* sp. by Leichhardt). It was not from Leichhardt that we learned that Murphy was deformed, but from Pemberton Hodgson (pp. 243, 286) who alluded merely to deformity; and to the Rev. J. Dunmore Lang, who described him as 'a young man from Ireland, having a hunch-back...' (Lang, 1847, p. 307). The name of John Murphy raises the question of the authorship of the illustrations in Leichhardt's published journal of the Port Essington expedition, which, except for the portraits of the two blacks by Rodius (see Moore) and the view of Victoria, Port Essington by H. S. Melville, is unacknowledged. Except for two views on the Mackenzie and two on the South Alligator River, where the skylines are bold, there are no other topographical illustrations in the book beyond the bold skylines of the Peak Range. Murphy's journal was not continued after 13 April 1845, and his hill profiles are all of the bold skylines of the Peak Range. All the plates were engraved by H. Melville, and S. H. Melville was the artist on *H M S Fly* (for the Melvilles see Moore). I suggest that S. H. Melville, from his knowledge of Australia, worked Murphy's profiles up into finished drawings. [M.A.]

b Now a suburb of Newcastle, N.S.W.
c The wife of Dr James Mitchell and mother of David Scott Mitchell who endowed the Mitchell Library, Sydney.
d Thomas Henry Braim, Principal, Sydney College, from 1842, and author of *A History of New South Wales*, 2 vols. (London, 1846). For Sydney College see Lang (1837, vol. 2, ch. VIII) and W. W. Burton, *The State of Education and Religion in New South Wales* (London, 1840), pp. 131–8.

68 1842, November 12. ALS, Gi, ML/A. 1383/1. Typed trsl. in ML/A. l. 69; copied in part into *Tgb*. 1842, ML/C. 145, Gg.

a A Presbyterian college founded by the Rev. J. Dunmore Lang; see Lang (1837, vol. 2, ch. VIII, and Burton, *op. cit.* note 67(d), pp. 139–42).

69 1842, November 24. Dft., *Tgb*. 1842, ML/C. 145.

a The Rev. Edward Rogers, Anglican pastor at Gosford, N.S.W.

70 1842, November 24. Dft., Gg, *Tgb*. 1842, ML/C. 145.

a Leichhardt's sister Mathilde.
b The Newcastle formations are Permian, those of Sydney Triassic in age.

Appendixes

^c There seems to be confusion between William Macleay, the nephew, and William Sharp Macleay, the eldest son, of Alexander Macleay.

71 1842, December 3. ALS, Engl., ML/Leichhardt papers, purchase of 1957. This letter was thought to be addressed to John Roper, but he had not reached Australia at that time.

^a John Skinner Prout (1806–76), the well-known artist and lithographer. who was then in Sydney. See Moore.

72 1842, December 4. Printed, *Berlinchen Nachrichten von Staats- und Gelehrten Sachen*, 1856, no. 26, cols. 112–15; Zuchold (1856).

^a Newport came to nothing.
^b Now Ourimbah Creek.
^c This is the earliest reference to artesian water in Australia of which I am aware.
^d The 'porphyries' are now recognized as andesites, and are of Carboniferous age.
^e Oxley had the same kind of providential experience on 18 Oct. 1818, when he found a small boat in good condition buried in the sand near The Three Brothers on the coast of New South Wales.
^f The Rev. William Grant Broughton (1788–1853), Archdeacon of New South Wales and Van Diemen's Land 1829, Bishop of Australia 1836.

73 1843, January 16. ALS, DM/II, 4; *Br.* 33, Pol. 4.

^a The Scott estate at Glendon was one of the most extensive in the Hunter River district. See Maps, Dixon 1837. This map also shows some of the other properties mentioned in this and succeeding letters, and in Leichhardt's journals.

74 1843, January 16. Dft., *Tgb.* 1842–3, ML/MSS. 683/1/1.

^a This is a puzzling statement. In Leichhardt's time a distinction was still being drawn between trap and basalt. Many hills in eastern New South Wales are capped with basalt, but not the Minmi sugarloaf. In his *Tagebuch* for 1842 (ML/C. 145, 18 Oct.) Leichhardt wrote 'Sugar Loaf... welche auf den Karte Bucklands als aus Trapp bestehend angegeben ist' (The Sugar Loaf... which is shown as consisting of trap on Buckland's map). The Rev. William Buckland (1784–1856) was a diluvialist and had been one of Lyell's teachers at Oxford. It is not known that he was ever in Australia, though he worked on Australian material, particularly on specimens collected by Oxley. Geological maps of part of the Newcastle district either by Buckland or Dr Charles Nicholson are unknown, but it is possible that the latter had plotted geological information on Mitchell's *Map of the Colony of New South Wales* (1834).

^b W. Harper was a surveyor employed by the Department of Lands, N.S.W. in 1821. He produced a map of Sydney in 1823. (*Fide* Librarian, Dept. of

Appendixes

Lands, Sydney.) Mrs Harper was making the most of things, as 'the late Surveyor General', Mitchell's predecessor, was John Oxley (1781–1828). The name of Harper's Hill has been confused with that of the Harpurs, another family well known in the Hunter Valley in Leichhardt's time.

c Leichhardt was here in the heart of what was to become wine-producing country. The names of Kelman and Wyndham were to become well-known amongst those of Hunter River wine-growers (see Max Lake, *Hunter Wine*, Brisbane, 1964, p. 25).

d For the limestone concretions of the district see H. G. Raggatt, *P. Linnean S., New South Wales*, 54 (1929), 149–61.

e Of the aboriginal names here recorded by Leichhardt only Tangorin, Merannie, Tyroman and Dyrring survive on modern maps. Even Jack Shea is absent from the 1-inch sheet. But Leichhardt's record sounds quite convincing.

f The Rev. George Keylock Rusden (1786–1859), a school master at Leith Hill, Surrey, went to Australia in 1834 as Colonial Chaplain and was stationed at Maitland. He had a large family and three of his sons became distinguished men. Two of them, Francis Townsend and Thomas George, both squatters, were friends of Leichhardt and became members of the first parliament elected under responsible government in New South Wales (1856). The third son, George William, became the historian of Australia and New Zealand. A fourth, Henry Keylock, made the most of his recollections of Leichhardt 49 years after the latter disappeared and when the former was 71 years of age. The Rusdens claimed to have considerable knowledge of Leichhardt, but, except for his short visit to F. T. Rusden's station in April 1843, he spent no more than a few days with any of the others at any time. Nor is Galbraith's supposed recollection of Leichhardt being confined to a darkened room for 10 days with sandy blight at T. G. Rusden's station to be trusted. Leichhardt was never at Europambela for more than several days. The Rev. George Keylock Rusden shared common ground with Leichhardt as a student of Oriental languages. The biographical record of the Rusden family would amply repay study. For the father see Heaton.

75 1843, February 15. Dft., *Tgb.* 1842–3, ML/MSS. 683. 1/1.

a Probably Charles Bolton, Landing Waiter in the Customs Department at Newcastle; but there was a Rev. R. T. Bolton, chaplain at Hexham near Newcastle in 1839.

b A Mons. and Mme de Vonge and child sailed from Sydney for the South Seas in the *Cintra* on 19 Oct. 1842 (*Australian*, 21 Oct. 1842).

c For the pseudomorphs see David, Taylor, Woolnough and Foxall, *Rec. Geol. Surv. New South Wales*, 8 (1905–8), 161–79. James Dwight Dana, who visited New South Wales with the United States Exploring Expedition in 1839–40 was the first to notice their occurrence in Australia. Leichhardt appears to have been the second.

d Leichhardt was working along the strike of the Carboniferous and

Appendixes

Permian deposits of the region. For a general account see Osborne, *P. Linnean S. New South Wales*, 54 (1929), 436–62.

e *Manuel d'Actinologie ou de Zoophytologie...avec un atlas de 100 planches*, by Henri-Marie Ducrotoy de Blainville, 2 vols. (Paris, 1834).

76 1843, February 19. Dft., *Tgb.* 1842–3, ML/MSS. 683. 1/1.

Leichhardt's movements as reported in this letter can be closely followed on the Singleton, Camberwell and Woolooma sheets of *Australia* 1:63,360.

a Capt. William Colburn Mayne (1808–1902) arrived in New South Wales in 1839. Not much is known about him. His station on the flanks of Dyrring was near the settlement of St Clair on Glennie's Creek.

b The wild cattle of St Clair were still known thirty years later. See W. S. Campbell (1920, p. 265).

c On an expedition to Barrington Tops (just north of Mount Royal), which was organised by the Department of Zoology of the University of Sydney in 1924, I recognised a rock which we called granodiorite at the time. I handed my report on the geology and topography of the district to Professor Lancelot Harrison just before my departure for Europe in 1925, but Professor Harrison died soon afterwards and the records of the expedition have been lost, except for a form-line map which I prepared, which has been superseded by official survey. [M.A.]

d Leichhardt's 'Piri' is called 'Pieries Point' on the 1-inch sheet (Woolooma). Blake has confused it with Mount Carrow on his fig. 11.

e In full: Forsan et haec olim meminisse juvabit (This suffering will yield us yet / A pleasant tale to tell), Virgil, *Aeneid*, I, 203, trsl. Conington.

f Most likely the Hentig who accompanied Leichhardt on his last journey. Very little is known about Arthur M. Hentig. According to Glenville Pike (1949, p. 272) he was a German Russian, which could imply that he came from one of the Baltic provinces of the Russian Empire. According to the *Sydney Morning Herald* (24 Jan. 1881) he worked for Busby at Cassilis, N.S.W. There are two letters from him to Leichhardt in ML/MSS. 683. 2, written in English.

77 1843, February 21. Printed, *Berlinchen Nachrichten von Staats- und Gelehrten Sachen*, Beil. 6 Feb. 1856; Zuchold, 1856.

78 1843, March 2. Dft., *Tgb.* 1842–3, ML/MSS. 683. 1/1.

a Farewell, flourish, and forget me not. (Perhaps an echo from Horace, *Epistolae*, I. 6. 67.)

79 1843, May 24. ALS, ML/C. 161.

a Bengalla, not far south of Muswellbrook, N.S.W., according to letter 65 (begun on 31 Oct. 1842), would be the estate of 'Herr Captain Scott'.

b Wm. A. G. Bloomfield gives information about Wilkinsons in the Hunter River district in his *Cessnock 1826–1954* (Cessnock, 1954), pp. 34–6,

Appendixes

but there is no mention of a Captain Wilkinson. Possibly a connection of the Rev. Samuel Wilkinson (see Heaton).

c Leichhardt was not then in a region that has produced lead in appreciable quantity; but there is a railway station named Leadville about 25 miles west of Dalkeith, and native lead has been found in the Peel valley to the north (W. B. Clarke, *Papers Relative to Geological Surveys*, Sydney, 7 May 1853).

80 1843, c. late March or early April. Autograph fragment, ML/A. 1383. 1. Wrongly identified by Politzer as being addressed to Capt. P. P. King.

81 1843, May 10. Printed, Lang, *An Historical and Statistical Account of New South Wales*, 4th ed. (London, 1875), vol. 2, pp. 234–5. This, and later surviving fragments can be placed, thanks to notes, recording letters despatched, on fly-leaves of journals for 1843–4. From the time when he left Dalkeith until he reached the Moreton Bay district Leichhardt was beyond the reach of official mail services and was travelling along the frontier of settlement. The full record of this part of the journey is in his journals.

a Leichhardt heard, or remembered, the gentleman's name as Mawson, but I can find no record of squatters named Mawson in the district. Dixon (Maps, 1837) shows grants of land to W. Lawson and to N. Lawson along the 'Coolabwarragundy River' but his topography is hard to reconcile with that of modern maps. Proeschel's atlas shows the position of 'Lawson's model station' near where Coolah now stands and Leichhardt certainly visited 'Mawson's Koolah station'. It is well known that William Lawson was a pioneer in the Mudgee district.

82 1843, June 23. Printed. Lang (1847, pp. 81–2, 471).

83 1843, June 27th. ALS, DM/II, 5; Br. 34, Pol. 5. In the *Briefe* (p. 127) this letter is assigned to January — 'd. 27 Januar'. The editors have not only changed Leichhardt's style, for he wrote clearly 'den 27ten Juny', but they have also misread his Juny as Jany, an impossible contraction for Januar, and a strange mistake as Neumayer spoke English. And nobody, translator or critic, has noticed that he could not have ridden from Glendon, whence he had last written home, to Moreton Bay between the 16th and the 27th of January.

84 1843, July 13. Dft. in *Tgb.* 1842–3, ML/MSS. 683. 1/1.

a Thomas Lodge Murray-Prior (1819–92): Was in charge of Pringle's station on Rocky Creek, Liverpool Plains, when Leichhardt stayed there from 7 to 22 May 1843. He and Leichhardt travelled from Cox's station on the 'Severn' (Dumaresq) to Brisbane in June 1843. Leichhardt then

Appendixes

bought a chestnut mare from Murray-Prior. The latter became a squatter and a prominent figure in public life in Queensland. Leichhardt lent him £50 in 1846, which was refunded to Lynd (Murray-Prior to Lynd, 10 Sept. 1847, in ML/MSS. 682/2). His sister Rosa (Mrs Campbell Praed) is responsible for the story of Leichhardt's supposed passionate attachment to Lucy Nicholson. For Murray-Prior see Hannah, Roderick.

85 1843, July 24. ALS, ML/C. 161; dft. in *Tgb.* 1842–3, ML/MSS. 683. 1/1.

a John Kent, Deputy Assistant Commissary General, and Superintendent of the Government Stock Establishment, Moreton Bay (Lang, 1848).

b For some of the squatters mentioned in this and later letters, see Gregor, Lennon, McKinnon.

c Parts of the Conondale Range and country to the north of it, where the so-called 'bunya' pine (*Araucaria Bidwilli*) grows. See Petrie.

d See John Campbell.

e The Archers were a large Scottish–Norwegian family, four of whom emigrated to Australia in 1833. They settled first at Wallerawang, but went north in 1841, to Durundur, where Leichhardt found them. They moved again to a station on Emu Creek (near where Cooyar now stands) in 1845, then to the Burnett district (1848) and finally to Gracemere near Rockhampton (1855), where Alister, a son of James, resided until he died in 1964. The four brothers of 1843 are, in order of age, Charles, John, David and Thomas. The fullest particulars of the family are given in the ML copy (A. 3876, no. 7 of 9 existing copies) of their collected letters. See MSS, Archer, the brothers; see also Thomas Archer; William Clark. Thomas Archer (1823–1905) went to England in 1878 and became Agent General for Queensland in 1881. He was the father of William Archer the dramatic critic who translated the works of Ibsen. Thomas left a valuable record of Leichhardt at Durundur in his book *Recollections of a Rambling Life*.

f Ottley's station was near the junction of Myall Creek with the Gwydir. Leichhardt was there on 26 April and again on 22 May 1843.

g Mackenzie, later Sir Evan Mackenzie, had a run to the West of Durundur.

h F. T. Rusden's station 'Lindesay' was near the Gwydir Falls on what was then regarded as the head of the Gwydir but is now called Horton's River. Leichhardt was there from 11 April 1843 until some time after the 19th (*Tgb.*), when Hentig was in charge. Leichhardt made it a base for a geological reconnaissance of the 'western chains' (Nandewar Mountains, as distinct from the Nandewar Range). He continued his investigation from Pringle's station into the 'Rocky Mountains' between the 7th and the 22nd of May 1843. His Stockyard Nobby appears to be Pound Mountain. Leichhardt here anticipated Jensen (*PLSNSW*, 32 (1907–8), 842–914, maps).

86 1843, August 7. Printed fragments Lang (1847, pp. 82–3).

Appendixes

87   1843, August 27. ALS, DM/II, 6; *Br.* 36, Pol. 7.
 a   Altered to hüten in the *Briefe* (p. 133).
 b   This statement is emphasised in the *Briefe* (p. 134). On his return to Sydney from his first journey to the Moreton Bay district Leichhardt was regarded as an excellent bushman and a man of experience in dealing with the blacks (see *Australian*, 17 July 1844, and Flanagan, vol. 2, p. 95). Other men of experience, like the Archer brothers, Thomas Petrie and F. N. Isaac, who had been with him in the bush, accepted him as one of themselves; and people in touch with his contemporaries, like Samuel Sidney (*PRGS*, 1, 1855–7 (1857), pp. 322–3), William Howitt (Howitt, vol. 2, ch. 1), and Steele Rudd (cf. note 90(b)), confirmed their judgment. A later generation, however, has spread the belief, now widely held in Australia, that he was a very bad bushman indeed. The charge was developed mainly by Ernest Favenc (1846–1908), who, like R. Logan Jack, was a small child when Leichhardt disappeared. Favenc was a minor explorer, journalist, novelist and historian, and his word carried great weight in Australia at the turn of the last century. He based his charge mainly on a very interesting and scientifically objective statement on the supposed instinctive ability of animals to find water (Leichhardt, *Journal of an Overland Expedition* etc., pp. 181–2); and, in a passage preposterous both in its misunderstanding of Leichhardt's language and in Favenc's elaborate conditions of test (Favenc, 1888, p. 167), he convinced the undiscerning that Leichhardt made 'obvious errors in the very rudiments of bushcraft', though what the rudiments were, he did not state. Favenc was associated with John F. Mann on the Council of the defunct New South Wales Branch of the Royal Geographical Society of Australasia.
 c   The Rev. Carl Wilhelm E. Schmidt and the Rev. Christopher Eipper.

88   1843, September 4. Printed fragment, Lang (1847, p. 83).
 a   The Glasshouse Mountains, a group of volcanic plugs, are conspicuous for many miles around and are visible from the hills just North of Brisbane. They were named by Cook. The first to visit them was Flinders in 1799 (*A Voyage to Terra Australis* etc., London, 1814, vol. 1, pp. cxvi–cxix and chart ix). Beerwah, the highest plug, was first climbed by Andrew and Thomas Petrie (see Petrie, pt. 2, pp. 251–2), then by the Government surveyor Burnett, and after him by Thomas Archer, q.v. Here again, as in the Nandewar Mountains, Leichhardt's geological reconnaissance anticipated Jensen's work (*PLSNSW*, 31 (1906–7), 73–173, map).

89   1843, September, *c.* end. Printed fragment, Lang (1847, pp. 83–4). Quoted by Petrie.
 a   Leichhardt is here referring to the Bribie Island blacks, whom he called elsewhere the 'Nynga-Nynga blacks'. They had a bad reputation. According to Petrie *ningi-ningi* means oysters in the Maroochy dialect.

Appendixes

90 1843, October 19. Printed fragment, Lang (1847, pp. 84-5).

^a The Toorbal district, opposite Bribie Island.

^b A change in nomenclature has occurred with reference to the headwaters of the Brisbane River. Leichhardt's 'southern head of the Brisbane' is now the Stanley River, which he called Archer's Creek; and his 'Stanley Creek' is now the upper Brisbane River. Leichhardt was at Bigge's Mid-day Creek station on 14 Oct. 1843 and left the head station (Reedy Creek) on 24 Oct. He was back at Bigge's on 30 Oct., and was there again on 18 Dec. 1843, probably for some days on each occasion. It was one of Bigge's men named Johnstone whom Steele Rudd questioned, many years afterwards, about Leichhardt's qualifications as a bushman. 'Bad bush-man!' repeated Johnstone with a look of astonishment—'when he'd be away all day in the gorges and scrub by himself and come home to the station in the dark? He couldn't do that if he was a bad bushman, could he?' (See Rudd; and cf. Sidney, 1855-7, and Howitt, vol. 2, ch. 1.)

91 1843, November 8. Printed fragment, Lang (1847, pp. 85-6).

^a 'I am a great admirer of Leichhardt's most valuable pioneering work in the geological field and I, personally, have had the experience of finding that my interpretation of Mt Tarampa, an interesting volcanic neck, had been anticipated by him many years ago' (Professor W. H. Bryan to Aurousseau, 2 April 1958).

92 1843, November 24. Printed fragment, Lang (1847, pp. 87-8).

^a This seems clear proof that the upper Brisbane was then known as Stanley Creek. M'Connel's station 'Cressbrook' was located on it, and Leich-hardt reported that M'Connel had made a map of Stanley Creek (*Tgb.*). For M'Connel see A. J. M'Connel, Mary M'Connel; McKinnon; Pownall (inaccurate).

^b This delightful under-statement conceals the fact that his wanderings took him far to the West, beyond Emu Creek and nearly to where Cooyar now stands.

93 1843, December 11. ALS, ML/MSS. 1009; dft. in *Tgb.* 1843-4, ML/MSS, 683. 1/3. Printed in part in sale-catalogue of Messrs Sotheby & Co. London, 10 April 1962.

^a Commissioner for Crown Lands, Moreton Bay. He had a station named 'Woolston' near Woogaroo in the Ipswich district. See Lennon, McKinnon.

94 1844, January 6. ALS, ML/C. 161; dft. in *Tgb.* 1843-4, ML/MSS. 683, 1/3; transcript in Durando's handwriting (not in full), Kew, lxxiii/239; trsl. in *London J. Bot.* 4 (1845), 280-90.

^a The 'burning mountain' of Wingen was already well known (see Wilton, 1833; Anon. 1838, pp. 149-51; Mitchell, 1839, vol. 1, ch. 1 and fig. 5).

Appendixes

According to the late Sir Edgworth David the Greta seam of the Lower Marine coal measures has been on fire for centuries at Wingen (oral statement).

b Leichhardt was undoubtedly using Mitchell's map of 1838 (see Maps). From this point onward names of rivers may apply to streams other than those that bear them in our own time. For instance, Leichhardt's Severn is our Dumaresq (see R. C. Hamilton, p. 327).

c The Nandewar Mountains, not the Nandewar Range.

d The brigalow is a suckering acacia which used to be a formidable obstacle to travel. See Skerman. It grows on such good soil that it has come under discussion of recent years. The Royal Society of Queensland held a symposium 'The brigalow country, past, present and future' on 26 Nov. 1962.

e This was a reasonable hypothesis in 1843. Leichhardt himself was to invalidate it.

f For the Darling Downs see Jay, Hall, Morgan.

g Not to be confused with Lake Macquarie.

h Now the Mary River.

i See Maps, Dixon (1842).

j Leichhardt first saw the gigantic comet of 1843 as he was leaving Glendon, and saw the last of it from F. T. Rusden's station near Gwydir Falls. This comet was fully visible only in the Southern Hemisphere. It 'appears to belong to a kind of family group composed of the comets of 1668, 1843, 1880, 1882 and 1887, each distinct in itself but pursuing, at irregular intervals, almost the same path. The period of the comet of 1882 has been determined at approximately 700 years and much the same period has been assigned to that of 1843 (probably it is between 400 and 800 years)'— Secretary, Royal Observatory, Greenwich to Aurousseau, 5 Sept. 1934. There is a picture of the comet in the *Tasmanian J. Nat. Sci.* 2 (1846) facing p. 155. Charles Harpur (1813–68) wrote a moving ode to this blazing comet.

95 1844, January 9. Printed fragment, Lang (1847, pp. 89–90).

a Waterston, according to the *Durundur Diary* (see Archer, the brothers), for 10 Jan. 1844, but possibly Waterton, the brother of Charles Waterton of *Wanderings in South America*, who was in New South Wales at about this time (see Henderson, vol. 2, p. 203). *The Australian* recorded the arrival of Edward Waterton at Sydney from Swan River on 16 April 1830, and of a Mr Waterton on 12 Aug. 1833. A Robert Waterston subscribed to the Christ Church building fund, Sydney, on 2 Jan. 1841.

b Now called the Baroon Pocket (Blackall Range).

96 1844, January 19. Printed fragment, Lang (1847, p. 90).

97 1844, February 2. ALS, DM/II, 7; Br. 37, Pol. 8.

a I have not succeeded in tracing the author of this prophecy, but am much indebted to the Reference Librarian of the Public Library of New South Wales for assistance in my efforts. (*Later*: the prophet was Irving, M.A.)

Appendixes

^b The diagram, which looks like an unintentional oblique stroke across the page, adds little to the description. Footnotes in the *Briefe* state that the comet was seen from many places in Europe in full daylight on 28 Feb. 1843 (p. 136), and that only one tail was observed (also p. 136). Eyre (*Journals of Expeditions of Discovery into Central Australia and Overland*, London, 1845, vol. 2, pp. 358–9; and Stokes (q.v., vol. 2, pp. 512–13) have left vivid accounts of the comet, but neither they nor other observers, so far as I am aware, have mentioned a second tail.

^c 'The finest things here are the *potatoes*. You never saw anything like them at home, so large and white and good' (Charles Stanley to Catherine Stanley, Hobart Town, 16 Feb. 1847).

^d 'One of the most curious features of the landscape is the bush fires. Almost every day we see the blue smoke curling up from some of the hills . . . This dry weather these fires are almost continual' (Eliza Stanley to Mary Stanley, 18 Feb. 1847).

98 1844, February 6. Dft. in *Tgb.* 1843–4, ML/MSS. 683. 1/3.

^a The four Archers of Durundur were then unmarried, but the late Alister Archer (letter to Aurousseau, per secretary, 5 Oct. 1964) told me that a married couple named Fraser was employed at the station at that time, and thought that the child (which Leichhardt called David in his *Tagebuch*) must have been theirs.

^b This was a very early use of quinine for malaria in Australia. The curative property of cinchona bark was known to the Jesuits in Peru in 1630, but quinine was chemically isolated only in 1820, and cinchona trees were not cultivated on a commercial scale until 1865 (*Encyclop. Brit.*).

^c For a general account of 'boiling down' see Hodgkinson, pt. III.

^d Leichhardt wrote on medical matters only to his medical friends Little and Nicholson, but Thomas Archer showed how valuable his medical services were at Durundur, and Flanagan reported that he treated the blacks (1862, vol. 2, p. 95).

99 1844, March, c. 2nd. Dft. in *Tgb.* 1843–4, ML/MSS. 683. 1/3.

^a Leichhardt seems to touch here on one of the fundamental reasons for dissension amongst the Prussian Lutherans of his time. Some of them went to Australia less on account of religious persecution than because of inability to agree amongst themselves. Even in Australia schismatic movements occurred amongst them. See Anon. 1933.

100 1844, March 26. Dft. in *Tgb.* 1843–4, ML/MSS. 683. 1/3. Printed in *PRGSA* (New South Wales and Victorian Br.), 1884 (1885), 70–2.

^a Mitchell's letters to Leichhardt are preserved in ML/MSS. 683. 2.

101 March 27. Printed fragment, Lang (1847, pp. 91–2).

^a Cunningham's Gap.

Appendixes

102 1844, April 13. Printed fragment, Lang (1847, pp. 92-3).

ᵃ Mrs Cotton assumed that Leichhardt recruited his party for the Port Essington expedition on his way outward to the Moreton Bay district, which invalidates her chronology.
ᵇ William Nicholson.

103 1844, May 10. ALS, ML/A. 1383. Incomplete dft. in ML/C. 163, vol. 2.

ᵃ From this statement it is clear that there was a movement in the Moreton Bay district in support of a private expedition to Port Essington. This is confirmed by a report in the *Australian* for 16 Nov. 1843, that stock holders to the north were already raising a subscription for such an expedition, and by Howitt's statement on Leichhardt's offer to lead such an expedition (Howitt, vol. 2, p. 3).

104 1844, May 14. AL not signed, DM/II, 8; *Br.* 38, Pol. 9. Passages copied into ML/C. 163, vol. 2.

ᵃ Chisholm ascribed the loss to carelessness (Chisholm, 1941, p. 90). Where is the evidence?
ᵇ Cf. note 94(e). Leichhardt had already modified his opinion, possibly after discussions with squatters. A party from New England had reconnoitred as far as the Dawson River in 1842, if Eldershaw's report of 1854 be true (see F. Eldershaw, no. 9, esp. p. 162).
ᶜ This is an excellent illustration of the range of Leichhardt's knowledge and the retentiveness of his memory. With none of his works of reference at hand, he saw at once, from the dentition, that he was handling marsupial remains.

105 1844, May 15. MS copy in the handwriting of Auguste Hilgenfeld. DM/not numbered; *Br.* 35, Pol. 6. Auguste's letter, to which this was the reply, is preserved in ML/MSS. 683. 2.

ᵃ This letter belongs not to 1843 but to 1844. The statement 'I've just returned after wandering for 18 months through lesser known parts of the colony...' should have caused an alert editor or translator to question the date. Whether the mistake is due to Leichhardt's inadvertence or his sister's, is hard to say, but the editors of 1881 introduced the word *waren* (*Br.* p. 129, line 12 from top) to make the sentence grammatical; they printed 'der Eurigen' (p. 131, line 9 from top) where the copy has 'der Erde' the latter making little sense; and the copy has the word 'ich' between *würde* and *vielleicht* on p. 131 (line 12 from top), which they rejected.
ᵇ Luke ii. 49.
ᶜ The story of how it was done is told in a letter from William Nicholson to Little (13 July 1838, in ML/A. n. 6). The consul, 'after gently signifying

Appendixes

that we should have to draw on our purse strings for the honour he was about to do us', and 'far from suspecting aught in the crackjaw name' said, of his own accord 'mais...naturellement né à Bristol'.

106   1844, June 3. ALS. In ML/A. l. 69. 2/12 (Peppercorn purchase of 1958).

107   1844, June 20. ALS. ML/A. 1383/1.

   a Leichhardt's geological map of the Glendon district is not amongst the Scott papers in the ML, and is not known to the Geological Survey of New South Wales (Government Geologist, N.S.W. to Aurousseau, 18 Aug. 1965).

108   1844, July 10. Printed in 'Geological Surveys, Rept. x, App. 5', in *Votes and Proceedings of the Legislative Council* [N.S.W.], 1853, II. Reprinted in *Papers presented to both Houses of Parliament, Dec. 1854* (London, 1855). Quoted, by Sidney 1853, by Heising 1855, by Zuchold 1856, by Blair 1881. Reprinted in part, with some editing, by Owen 1877-8. Sidney, Heising, Zuchold, and Lang (1847) all made misleading statements about its publication. It was evidently read by Owen at the Cambridge meeting of the British Association for the Advancement of Science in 1845 but was not published in the report of the meeting. An abstract of Owen's remarks was printed in the *Annals and Mag. Nat. Hist.* 16 (1845). Leichhardt's material could not have reached Owen in time for the York meeting of the BAAS in September 1844, to which so many have ascribed its mention.

109   1844, July 12. ALS, ML/C. 161. Extracts, in Durando's handwriting, Kew, lxxiii, 239; garbled trsl. in *London J. Bot.* 4 (1845), 290-1.

   a His *Beiträge zur Geologie von Australien*, of which the draft is preserved in his [Note-book opened 1842], ML/C. 154, pp. 235-62 and 302-30. Nicholson sent the MS to Jameson in Edinburgh (William Nicholson to Leichhardt, 18 Nov. 1845), who rejected it because there were too many plates. He then submitted it to the Royal Geographical Society, but received 'a letter from Horner' hedged with discouraging conditions. The name of Horner does not appear in the published rolls of Fellows and Officers of the Royal Geographical Society. He then sent it to Professor Heinrich Friedrich Link (1767-1851) in Berlin, who accepted the collection of plants which Leichhardt had offered to the University of Berlin, and undertook to publish an extract from the memoir. This appeared in the first volume of the *Zeitschrift der deutschen geologischen Gesellschaft*. The full MS then came into the hands of Sir Robert Schomburgk (the explorer of Guiana, and brother of Otto and Richard Schomburgk) who passed it to Professor H. Girard 'during Leichhardt's

Appendixes

lifetime'. Girard eventually had it published by the Naturforschende Gesellschaft in Halle in Leichhardt's memory, in the year 1856, when Zuchold's biography of Leichhardt appeared. Clarke had the *Beiträge* translated by G. H. F. Ulrich of the Geological Survey of Victoria (active 1859–77) and published the translation in *The Australian Almanac* for 1867 and 1868, where it lay forgotten. Robert Logan Jack (1844–1921), not to be confused with R. Lockhart Jack, complained of the 'extraordinary *media* chosen by Clarke for the publication of his writings' (Jack, 1922, vol. 1, p. 198), but there were no serial publications for geological work in Australia in Clarke's earlier days and few in Europe. Dr F. W. Whitehouse (formerly of the University of Queensland) acquired a copy of the *Beiträge* in Germany in 1922, and he reported (typed copy of part of a letter, probably to Sir T. W. Edgworth David written before 1931, State Library of Victoria, Leichhardt papers 12/3) that it was 'a pioneering work, extraordinarily accurate in detail, showing Leichhardt as a geologist of considerable merit and a very accurate observer... A few little things that we are finding out now in S.E. Queensland that no one seemed to know of before are set down by Ludwig as a matter of course'. To myself he has written (Whitehouse to Aurousseau, 12 April 1958) 'Leichhardt remains to me one of our greatest explorers and one of our very few scientific explorers. In my own travels I have passed over most of the tracks that Leichhardt pioneered; and I have been vastly impressed by his insight and his almost invariably correct interpretation of the geological evidences he saw in those times, when there were little means of correlating them.' (Cf. Sonter, for the testimony of another critic competent to judge the matter.) Dr Whitehouse subsequently translated Leichhardt's remarks on the Brisbane Schists for the Government Geologist of Queensland (Denmead to Aurousseau, 9 July 1964). It is astonishing that Mrs Cotton, so sympathetic to Leichhardt and a geologist herself, could have written 'He gave them little lectures on the botany they could not altogether understand, and the geology that was really beyond his reach' (Cotton, 1938, p. 135).

b The *Edinburgh New Philosophical Journal*, conducted by Professor Robert Jameson.

110 1844, July 15. ALS, National Library of Australia, Canberra, A.C.T.

111 1844, July 24. ALS, ML/MSS. 1009; printed in part, sale-catalogue, Messrs Sotheby & Co., London, 10 April 1962.

112 1844, August, early. Printed, *Australian*, 14 Aug. 1844; *Port Phillip Patriot*, 5 Sept. 1844.

113 1844, August 12. ALS, State Library of Victoria (framed). Printed in facs., *Trans. Phil. Inst. Victoria*, 4, 1859 (1959–60), Proc., p. x; and *Queensland Geogr. J.* 60 (1960–1), p. 44; wrongly described by both as 'Leichhardt's last letter'. Addressee Marsh the harpist, but wrongly identified by the ML as Marsh the squatter.